Lecture Notes in Mathematics

Edited by J.-M. Morel, F. Takens and B. Teissier

Editorial Policy
for the publication of monographs

1. Lecture Notes aim to report new developments in all areas of mathematics and their applications – quickly, informally and at a high level. Mathematical texts analysing new developments in modelling and numerical simulation are welcome.

   Monograph manuscripts should be reasonably self-contained and rounded off. Thus they may, and often will, present not only results of the author but also related work by other people. They may be based on specialised lecture courses. Furthermore, the manuscripts should provide sufficient motivation, examples and applications. This clearly distinguishes Lecture Notes from journal articles or technical reports which normally are very concise. Articles intended for a journal but too long to be accepted by most journals, usually do not have this "lecture notes" character. For similar reasons it is unusual for doctoral theses to be accepted for the Lecture Notes series, though habilitation theses may be appropriate.

2. Manuscripts should be submitted (preferably in duplicate) either to Springer's mathematics editorial in Heidelberg, or to one of the series editors (with a copy to Springer). In general, manuscripts will be sent out to 2 external referees for evaluation. If a decision cannot yet be reached on the basis of the first 2 reports, further referees may be contacted: The author will be informed of this. A final decision to publish can be made only on the basis of the complete manuscript, however a refereeing process leading to a preliminary decision can be based on a pre-final or incomplete manuscript. The strict minimum amount of material that will be con-sidered should include a detailed outline describing the planned contents of each chapter, a bibliography and several sample chapters.

   Authors should be aware that incomplete or insufficiently close to final manuscripts almost always result in longer refereeing times and nevertheless unclear referees' recommendations, making further refereeing of a final draft necessary.

   Authors should also be aware that parallel submission of their manuscript to another publisher while under consideration for LNM will in general lead to immediate rejection.

3. Manuscripts should in general be submitted in English. Final manuscripts should contain at least 100 pages of mathematical text and should always include

   – a table of contents;
   – an informative introduction, with adequate motivation and perhaps some historical remarks: it should be accessible to a reader not intimately familiar with the topic treated;
   – a subject index: as a rule this is genuinely helpful for the reader.

   For evaluation purposes, manuscripts may be submitted in print or electronic form (print form is still preferred by most referees), in the latter case preferably as pdf- or zipped ps-files. Lecture Notes volumes are, as a rule, printed digitally from the authors' files. To ensure best results, authors are asked to use the LaTeX2e style files available from Springer's web-server at:

   ftp://ftp.springer.de/pub/tex/latex/svmonot1/ (for monographs) and

   ftp://ftp.springer.de/pub/tex/latex/svmultt1/ (for summer schools/tutorials).

   Additional technical instructions, if necessary, are available on request from lnm@springer.com.

Lecture Notes in Mathematics

1925

Marcus du Sautoy · Luke Woodward

Zeta Functions
of Groups and Rings

 Springer

Marcus du Sautoy
Luke Woodward
Mathematical Institute
University of Oxford
24-29 St Giles
Oxford OX1 3LB, UK
dusautoy@maths.ox.ac.uk
luke.woodward@talk21.com

ISBN 978-3-540-74701-7   e-ISBN 978-3-540-74776-5

DOI 10.1007/978-3-540-74776-5

Lecture Notes in Mathematics ISSN print edition: 0075-8434
                                 ISSN electronic edition: 1617-9692

Library of Congress Control Number: 2007936935

Mathematics Subject Classification (2000): 20E07, 11M41

Cover design: *design & production* GmbH, Heidelberg

Printed on acid-free paper

9 8 7 6 5 4 3 2 1

springer.com

To our families

Preface

The study of the subgroup growth of infinite groups is an area of mathematical research that has grown rapidly since its inception at the Groups St. Andrews conference in 1985. It has become a rich theory requiring tools from and having applications to many areas of group theory. Indeed, much of this progress is chronicled by Lubotzky and Segal within their book [42].

However, one area within this study has grown explosively in the last few years. This is the study of the zeta functions of groups with polynomial subgroup growth, in particular for torsion-free finitely-generated nilpotent groups. These zeta functions were introduced in [32], and other key papers in the development of this subject include [10, 17], with [19, 23, 15] as well as [42] presenting surveys of the area.

The purpose of this book is to bring into print significant and as yet unpublished work from three areas of the theory of zeta functions of groups.

First, there are now numerous calculations of zeta functions of groups by doctoral students of the first author which are yet to be made into printed form outside their theses. These explicit calculations provide evidence in favour of conjectures, or indeed can form inspiration and evidence for new conjectures. We record these zeta functions in Chap. 2. In particular, we document the functional equations frequently satisfied by the local factors. Explaining this phenomenon is, according to the first author and Segal [23], "one of the most intriguing open problems in the area".

A significant discovery made by the second author was a group where all but perhaps finitely many of the local zeta functions counting normal subgroups do not possess such a functional equation. Prior to this discovery, it was expected that all zeta functions of groups should satisfy a functional equations. Prompted by this counterexample, the second author has outlined a conjecture which offers a substantial demystification of this phenomenon. This conjecture and its ramifications are discussed in Chap. 4.

Finally, it was announced in [16] that the zeta functions of algebraic groups of types B_l, C_l and D_l all possessed a natural boundary, but this work is also yet to be made into print. In Chap. 5 we present a theory of natural boundaries of two-variable polynomials. This is followed by Chap. 6 where the aforementioned result on the zeta functions of classical groups is proved, and Chap. 7, where we consider the natural boundaries of the zeta functions attached to nilpotent groups listed in Chap. 2.

The first author thanks Zeev Rudnick who first informed him of Conjecture 1.11, Roger Heath-Brown who started the ball rolling and Fritz Grunewald for discussions which helped bring the ball to a stop. The first author also thanks the Max-Planck Institute in Bonn for hospitality during the preparation of this work and the Royal Society for support in the form of a University Research Fellowship. The second author thanks the EPSRC for a Research Studentship and a Postdoctoral Research Fellowship, and the first author for supervision during his doctoral studies.

Oxford,   *Marcus du Sautoy*
January 2007   *Luke Woodward*

Contents

1

Introduction

1.1 A Brief History of Zeta Functions

Zeta functions are analytic functions with remarkable properties. They have played a crucial role in the proof of many significant theorems in mathematics: Dirichlet's theorem on primes in arithmetic progressions, the Prime Number Theorem, and the proofs of the Weil conjectures and the Taniyama–Shimura conjecture to name just a few.

Many different types of zeta function have been defined. We summarise below some of the more significant ones.

1.1.1 Euler, Riemann

In the eighteenth century a number of mathematicians were interested in determining the precise value of the infinite series

$$1 + \frac{1}{4} + \frac{1}{9} + \frac{1}{16} + \cdots + \frac{1}{n^2} + \cdots , \tag{1.1}$$

the sum of the squares of the harmonic series. Daniel Bernoulli suggested 8/5 as an estimate for its value, but it was Leonhard Euler who first gave the precise value of this sum. To do this, Euler defined the *zeta function*

$$\zeta(s) = \sum_{n=1}^{\infty} n^{-s}$$

for $s \in \mathbb{R}$, $s > 1$. The infinite sum (1.1) is then the zeta function evaluated at $s = 2$. However Euler was able to do more than just give the value of $\zeta(2)$. He gave a formula for the zeta function at every even positive integer:

$$\zeta(2m) = \frac{2^{2m-1}\pi^{2m}|B_{2m}|}{(2m)!} .$$

As an acknowledgement of the support the Bernoulli family had given him, he was able to identify the rational constants B_{2m} as the Bernoulli numbers discovered by Daniel's uncle Jacob. Since $B_2 = 1/12$, it follows that $\zeta(2) = \pi^2/6$. To this day, nobody has been able to find a comparable expression for the zeta function at odd integers. It is not even known if $\zeta(3)$ is transcendental.

Euler also discovered the *Euler product identity*. If one sets

$$\zeta_p(s) = \sum_{n=0}^{\infty} p^{-ns} = \frac{1}{1 - p^{-s}} \, ,$$

then

$$\zeta(s) = \prod_p \zeta_p(s) \, ,$$

where the product is over all primes p. This identity is fundamental to the connection between the zeta function and the primes. As well as encapsulating the Fundamental Theorem of Arithmetic, it also offers a simple analytic proof of a classical result on primes: the fact that the harmonic series $1 + 1/2 + \cdots + 1/n + \cdots$ diverges means that there must be infinitely many primes.

The zeta function converges for $s > 1$ but diverges at $s = 1$. Later, Bernhard Riemann, inspired by Cauchy's work on functions of a complex variable, considered the zeta function as a function on $\mathbb{C}$. By doing so, he could analytically continue the zeta function around the pole at $s = 1$, and obtain a function meromorphic on the whole complex plane. The pole at $s = 1$ is simple and is the only singularity of the zeta function. Furthermore, Riemann showed that this zeta function satisfies a *functional equation*. If one sets $\xi(s) = \Gamma(s/2)\pi^{-s/2}\zeta(s)$, where $\Gamma(s)$ is the gamma function, then

$$\xi(s) = \xi(1 - s) \, . \tag{1.2}$$

This analytically-continued function is now known as the *Riemann zeta function* in honour of Riemann's achievements with it.

Since the zeta function is nonzero for $\Re(s) \geq 1$, the only zeros of the Riemann zeta function with $\Re(s) \leq 0$ are the trivial zeros at negative even integers. Hence the only other zeros are those within the *critical strip*, $0 < \Re(s) < 1$. Riemann famously hypothesised that all the zeros lie on the *critical line* $\Re(s) = \frac{1}{2}$. Hardy and Littlewood [33] have since proved the existence of infinitely many zeros on the critical line and Conrey [3] has proved that more than 40% of the zeros lie on the line. At the time of writing, the most recent computer calculation [27] seems to have confirmed that the first ten trillion (10^{13}) Riemann zeros are on the line. Despite all this evidence, it is still not known whether a zero lies off the line.

Such is the importance of this Hypothesis that there is a considerable body of mathematical work which depends on the truth of this Hypothesis. Its proof would simultaneously prove numerous other theorems for which its truth has had to be assumed. Furthermore, its status as one of the Clay Mathematics Institute Millennium Prize Problems would also earn its author a million-dollar prize.

Hadamard and de la Vallée Poussin were also able to utilise the power of the Riemann zeta function. By showing that the Riemann zeta function is nonzero on $\Re(s) = 1$, they independently proved the Prime Number Theorem, that

$$\lim_{n \to \infty} \frac{\pi(n) \log n}{n} = 1 \,,$$

where $\pi(n)$ is the number of primes no larger than n.

1.1.2 Dirichlet

In the meantime, Dirichlet was taking the concept of the zeta function in a new direction. His major innovation was to attach a coefficient a_n to each term n^{-s}. Recall that the Riemann zeta function is defined for $\Re(s) > 1$ by

$$\zeta(s) = \sum_{n=1}^{\infty} n^{-s} \,.$$

A Dirichlet character with period m is a function $\chi : \mathbb{N}_{>0} \to \mathbb{C}$ that has the following properties:

- χ is totally multiplicative, i.e. $\chi(1) = 1$ and $\chi(n_1)\chi(n_2) = \chi(n_1 n_2)$ for all $n_1, n_2 \in \mathbb{N}_{>0}$.
- $\chi(m + n) = \chi(n)$ for all $n \in \mathbb{N}_{>0}$.
- $\chi(n) = 0$ if $\gcd(n, m) > 1$.

The *Dirichlet L-function* of χ is defined by

$$L(s, \chi) = \sum_{n=1}^{\infty} \chi(n) n^{-s} \,.$$

Using these L-functions, Dirichlet proved that if $\gcd(r, N) = 1$, the arithmetic progression r, $r + N$, $r + 2N$, ... contains infinitely many primes. Furthermore, his proof yields the additional result that the primes are in some sense evenly distributed amongst the congruence classes of integers coprime to N. In honour of this achievement, any function of the form $f(s) = \sum_{n=1}^{\infty} a_n n^{-s}$ is called a Dirichlet series.

If $m = 1$ then χ is the trivial character, hence $L(s, \chi) = \zeta(s)$, the Riemann zeta function once again, which we know can be meromorphically continued to $\mathbb{C}$. If $m > 1$, $L(s, \chi)$ can be analytically continued to an entire function on $\mathbb{C}$. Indeed, the fact that $L(s, \chi)$ is nonzero at $s = 1$ for nontrivial characters χ plays a key part in Dirichlet's proof. A functional equation of $L(s, \chi)$ which takes a similar shape to (1.2) can also be given, however its statement is less succinct than that satisfied by the Riemann zeta function. We refer the interested reader to the section on Dirichlet L-functions in [37].

The multiplicativity of the characters χ leads easily to an Euler product for the Dirichlet L-function,

$$L(s,\chi) = \prod_p \frac{1}{1 - \chi(p)p^{-s}} \; .$$

Indeed, it is easy to see that any Dirichlet series where the sequence (a_n) grows at most polynomially in n and is totally multiplicative (i.e. $a_m a_n = a_{mn}$ for all $m, n \in \mathbb{N}$) satisfies such an Euler product.

1.1.3 Dedekind

The zeta functions described above have had predominantly number-theoretic applications. It was Dedekind who was perhaps the first to use zeta functions for an algebraic purpose. For K a finite extension of the rational numbers $\mathbb{Q}$, the *Dedekind zeta function* of the field K is defined by

$$\zeta_K(s) = \sum_{\mathfrak{a}} |\vartheta_K : \mathfrak{a}|^{-s} \; ,$$

where $|\vartheta_K : \mathfrak{a}|$ is the index of the ideal $\mathfrak{a}$ in the ring of integers ϑ_K and the sum is over all nonzero ideals $\mathfrak{a}$ in ϑ_K. Again, this zeta function extends to a meromorphic function on $\mathbb{C}$, with a simple pole at $s = 1$.

Perhaps one of the most remarkable properties of the Dedekind zeta function is the *class number formula*, which encodes the class number of the field in the residue of the pole of $\zeta_K(s)$ at $s = 1$. If $\Delta(K)$ is the discriminant of the field K, R_K the *regulator* of K, u the order of the group of roots of unity within the ring of integers ϑ_K, r_1 (resp. r_2) is the number of real (resp. the number of pairs of complex conjugate) embeddings of K and h_K the class-number of K, then

$$\mathrm{Res}_{s=1}(\zeta_K(s)) = \frac{2^{r_1}(2\pi)^{r_2} h_K R_K}{u\sqrt{|\Delta(K)|}} \; .$$

As with the Riemann zeta function and Dirichlet L-functions, the Dedekind zeta function satisfies a functional equation. Let $n = |K : \mathbb{Q}|$, the degree of the field extension, and put

$$\Xi_K(s) = \left(\frac{\sqrt{|\Delta(K)|}}{2^{r_2}\pi^{n/2}} \right)^s \Gamma\left(\frac{s}{2}\right)^{r_1} \Gamma(s)^{r_2} \zeta_K(s) \; .$$

Then $\Xi_K(s) = \Xi_K(1 - s)$.

1.1.4 Artin, Weil

Dedekind's zeta function considers finite extensions of the rational numbers $\mathbb{Q}$. E. Artin considered zeta functions connected to finite extensions of global fields of characteristic p. One particular example he considered was the field $K = \mathbb{F}_p(x)(\sqrt{x^3 - x})$, i.e. the field of rational functions with coefficients in $\mathbb{F}_p(x)$ extended by adjoining $\sqrt{x^3 - x}$. Let R be the integral closure of $\mathbb{F}_p[x]$ in K. Artin considered the zeta function

$$\zeta_R(s) = \sum_{\mathfrak{a} \trianglelefteq R} |R : \mathfrak{a}|^{-s} \ .$$

If one sets $y = \sqrt{x^3 - x}$, then quite clearly we have an elliptic curve $y^2 = x^3 - x$. Artin found that the zeta function $\zeta_R(s)$ was encoding the number of points on this elliptic curve. In particular,

$$\zeta_R(s) = (1 - p^{-s}) \exp\left(\sum_{m=1}^{\infty} \frac{N_{p^m} p^{-ms}}{m} \right) ,$$

where

$$N_{p^m} = |\{\, (a, b) \in \mathbb{F}_{p^m}^2 : b^2 = a^3 - a \,\}| + 1 \ .$$

The extra term is necessary to count the point at infinity in projective space. Furthermore, Artin could show, for this elliptic curve and about 40 others, that

$$\exp\left(\sum_{m=1}^{\infty} \frac{N_{p^m} p^{-ms}}{m} \right) = \frac{(1 + \pi_p p^{-s})(1 + \bar{\pi}_p p^{-s})}{(1 - p^{-s})(1 - p^{1-s})}$$

for a certain pair of complex conjugate numbers π_p and $\bar{\pi}_p$ which depend on the elliptic curve. Hasse later extended this result to all elliptic curves, and Weil to all smooth projective curves of arbitrary genus. Indeed, this property that the zeros of the zeta function satisfy $|\pi| = p^{1/2}$ is known as the *analogue of the Riemann Hypothesis* for the zeta function.

Weil was inspired by his work to consider the zeta function of an arbitrary smooth projective variety X defined over a finite field $\mathbb{F}_q$. This is defined analogously to Artin's zeta function, but omitting the factor $(1 - p^{-s})$, by

$$\zeta_X(s) = \exp\left(\sum_{m=1}^{\infty} \frac{N_{q^m} q^{-ms}}{m} \right) ,$$

where N_{q^m} is the number of points on X over the field $\mathbb{F}_{q^m}$. In particular, $\zeta_X(s)$ was conjectured to always be a rational function in q^{-s}, and to satisfy the functional equation $\zeta_X(n - s) = \pm q^{(\frac{1}{2}n - s)C} \zeta_X(s)$, for some constant C which can be given explicitly in terms of geometrical invariants of X. Weil was also able to formulate a strategy for proving these conjectures. He observed that if one has a suitable cohomology theory similar to that for varieties defined over $\mathbb{C}$, the conjectures follow from various standard properties of this cohomology theory. This observation motivated the development of various cohomology theories and eventually led to the development of the l-adic cohomology by Grothendieck and M. Artin, successfully employed by Deligne to confirm these conjectures.

1.1.5 Birch, Swinnerton-Dyer

If one has a polynomial equation over $\mathbb{Z}$, one can reduce it modulo p to give a variety defined over a finite field. So, given the zeta functions for the reductions mod p, what do we get when we multiply them all together? Does this 'global' zeta function tell us anything about the solutions of the original polynomial over $\mathbb{Q}$ or $\mathbb{Z}$?

In the case where X is an elliptic curve defined over $\mathbb{Q}$, such a global zeta function has been defined. If E is an elliptic curve over $\mathbb{Q}$, the L-function of E is defined by[1]

$$L(E, s) = \prod_{p \nmid 2\Delta} \frac{1}{1 - a_p p^{-s} + p^{1-2s}} \,,$$

where Δ is the discriminant of E, N_p is the number of points on E mod p and $a_p = p - N_p$. This Dirichlet series converges for $\Re(s) > \frac{3}{2}$ and thanks to the complete proof of the Taniyama–Shimura conjecture [1], it is known that $L(E, s)$ can be analytically continued to an entire function. A functional equation relating $L(E, s)$ and $L(E, 2-s)$ also follows from Taniyama–Shimura. It was conjectured by Birch and Swinnerton-Dyer that E has infinitely many rational points if and only if $L(E, s)$ is zero at $s = 1$, and furthermore the torsion-free rank of the Mordell–Weil group of points on E over $\mathbb{Q}$ is the order of the zero at $s = 1$. Coates and Wiles [2] have proved that if $L(E, 1) \neq 0$ then E has only finitely many rational points, and it has since been shown that the conjecture is true for $r \leq 1$ [5]. However the rest of the conjecture remains open. Like the Riemann Hypothesis, the Clay Foundation offers a million-dollar prize for the proof of this conjecture.

1.2 Zeta Functions of Groups

By no means is the above a complete list of zeta functions. We have omitted more than we have included, for we simply do not have the space to list them all. The final chapter of the Encyclopedic Dictionary of Mathematics [37] is

---

[1] There are factors associated to the primes $p \mid 2\Delta$ but for simplicity we ignore them.

a good place to start for those keen to know more about the panoply of zeta functions.

Furthermore, the Encyclopedic Dictionary also lists four basic properties a zeta function should ideally satisfy:

(ZF1) It should be meromorphic on the whole complex plane
(ZF2) It should have a Dirichlet series expansion
(ZF3) There should be some natural Euler product expansion
(ZF4) It should satisfy a functional equation

All the zeta functions we listed above satisfy all four of these properties. It may also be of interest to determine the residue of the zeta function at a pole, whenever such a singularity exists.

In this book, we consider these criteria for a relative newcomer to the family of zeta functions, zeta functions of groups and rings. We cannot expect that these zeta functions will reach the same lofty heights as the zeta functions presented above, but we do hope the reader agrees with our viewpoint that there is interesting mathematics concerning zeta functions of groups.

1.2.1 Zeta Functions of Algebraic Groups

The first example of a zeta function of a group is associated to a $\mathbb{Q}$-algebraic group $\mathfrak{G}$ with a choice of some $\mathbb{Q}$-rational representation $\rho : \mathfrak{G} \to \mathrm{GL}_n$. The zeta function $Z_{\mathfrak{G},\rho}(s)$ of $\mathfrak{G}$ has been defined as the Euler product over all primes p of the following local zeta functions defined by p-adic integrals with respect to the normalised Haar measure $\mu_{\mathfrak{G}}$ on $\mathfrak{G}(\mathbb{Z}_p)$:

$$Z_{\mathfrak{G},\rho,p}(s) = \int_{\mathfrak{G}_p^+} |\det(\rho(g))|_p^s \, \mathrm{d}\mu_{\mathfrak{G}}(g) \,,$$

where $\mathfrak{G}_p^+ = \rho^{-1}\left(\rho\left(\mathfrak{G}(\mathbb{Q}_p)\right) \cap \mathrm{M}_n(\mathbb{Z}_p)\right)$ and $|\cdot|_p$ denotes the p-adic norm.

The definition of the zeta function of an algebraic group goes back to the work of Hey [35] who recognised that the zeta function attached to the algebraic group GL_n could be used to encode the subalgebra structure of central simple algebras. In the 1960s, Tamagawa established in [56] the meromorphic continuation of the zeta functions of Hey attached to GL_n. Subsequently, Satake [50] and Macdonald [43] considered zeta functions of other reductive groups. But it is the work of Igusa [36] in the 1980s that established explicit expressions for the local factors of Chevalley groups which allow for some analysis of the analytic behaviour of the global zeta functions. In particular his work shows that the zeta function is built from Riemann zeta functions and functions of the form

$$Z(s) = \prod_{p \text{ prime}} W(p, p^{-s}) \,, \tag{1.3}$$

where $W(X,Y) \in \mathbb{Z}[X,Y]$, with $W(X,0) = 1$. Further development of Igusa's work was made by the first author and Lubotzky [21] and [9] to more general algebraic groups. The motivation for our work came from the observation in [32] that zeta functions of algebraic groups were in fact counting subgroups in nilpotent groups, thus extending Hey's original motivation for the investigation of these functions.

In [32] Grunewald, Segal and Smith proposed a definition of a 'zeta function of a group G':

$$\zeta_G^{\leq}(s) = \sum_{H \leq G} |G : H|^{-s} .$$

The function may be viewed as a non-commutative version of the Dedekind zeta function of a number field where we sum over subgroups instead of ideals. The superscript $\leq$ in the zeta function emphasises that we are counting all subgroups within G; we shall define variants of this zeta function later. If the group is finitely generated (either as an abstract group or profinite group) then the following invariant is finite for every natural number n:

$$a_n^{\leq}(G) = |\{\, H : H \leq G \text{ and } |G : H| = n \,\}| .$$

We can then write the zeta function as a Dirichlet series satisfying condition (ZF2):

$$\zeta_G^{\leq}(s) = \sum_{n=1}^{\infty} a_n^{\leq}(G) n^{-s} .$$

These zeta functions were first introduced in the 1980s by Grunewald, Segal and Smith in [32] and studied in the particular case that G is a torsion-free finitely generated nilpotent group (a $\mathfrak{T}$-group for short). The nilpotency of G lends itself to a natural Euler product, thus satisfying condition (ZF3):

$$\zeta_G^{\leq}(s) = \prod_{p \text{ prime}} \zeta_{G,p}^{\leq}(s) ,$$

where $\zeta_{G,p}(s) = \sum_{n=0}^{\infty} a_{p^n}^{\leq}(G) p^{-ns}$.

One can also consider variants of these zeta functions in which one only counts subgroups H with a particular property, for example normal subgroups, whose associated zeta functions we denote by $\zeta_G^{\triangleleft}(s)$ and $\zeta_{G,p}^{\triangleleft}(s)$. One type of subgroup deserves special mention, namely those H whose profinite completions are isomorphic to the profinite completion $\widehat{G}$ of G. When G is nilpotent the associated zeta function counting these subgroups, denoted by $\zeta_G^{\wedge}(s)$, is (up to finitely many local factors) the same as the first zeta function of the algebraic group $\mathfrak{G}$ of automorphisms of G (or its associated Lie algebra) with an appropriate representation.

1.2.2 Zeta Functions of Rings

As well as introducing zeta functions of groups, Grunewald, Segal and Smith defined the zeta function of a not-necessarily-associative ring L additively isomorphic to $\mathbb{Z}^d$ for some d, by

$$\zeta_L^{\leq}(s) = \sum_{H \leq L} |L : H|^{-s} .$$

Zeta functions only counting ideals in L, and the corresponding local zeta functions, can be defined in a similar way, with the obvious notation. We can also define analogues of the pro-isomorphic zeta functions. $\zeta_L^{\wedge}(s)$ counts all subrings $H \leq L$ such that $H \otimes \hat{\mathbb{Z}} \cong L \otimes \hat{\mathbb{Z}}$, where $\hat{\mathbb{Z}}$ is the profinite completion of $\mathbb{Z}$, with the corresponding local zeta functions $\zeta_{L,p}^{\wedge}(s)$ counting subrings H of p-power index such that $H \otimes \mathbb{Z}_p \cong L \otimes \mathbb{Z}_p$.

Since these zeta functions are defined in an analogous way to those counting in groups, it is clear that these zeta functions have Dirichlet series expansions. Moreover, these zeta functions satisfy the Euler product

$$\zeta_L^*(s) = \prod_{p \text{ prime}} \zeta_{L,p}^*(s)$$

for all $* \in \{\leq, \lhd, \wedge\}$, regardless of whether L is nilpotent (or even soluble).

The motivating reason for introducing zeta functions of rings is to provide an alternative way of calculating zeta functions of groups. In [51], the Mal'cev correspondence between a $\mathfrak{T}$-group G and a nilpotent Lie ring L is detailed. In particular it is noted that L is additively isomorphic to $\mathbb{Z}^h$, where $h = h(G)$ is the *Hirsch length* of G, i.e. the number of infinite factors in any composition series of G. In [32] this correspondence was extended to show that

$$\zeta_{G,p}^*(s) = \zeta_{L,p}^*(s) \tag{1.4}$$

for $* \in \{\leq, \lhd, \wedge\}$ and for all but finitely many primes p depending only on the Hirsch length of G. For every calculation of a zeta function $\zeta_L^*(s)$ for L a nilpotent Lie ring, we obtain a zeta function (up to finitely many local factors) of the zeta function of the corresponding $\mathfrak{T}$-group. The linearity of the rings makes it considerably less difficult to calculate $\zeta_{L,p}^*(s)$ than $\zeta_{G,p}^*(s)$, although it cannot be said that these calculations are in general easy.

In the case that G is nilpotent of class 2, then we can short-circuit the Mal'cev correspondence. We define a Lie ring on G by setting $L = G/Z(G) \oplus Z(G)$, where $Z(G)$ is the centre of G, with the Lie bracket on L induced by the commutator on G. It is not difficult to see in this case that $\zeta_{G,p}^*(s) = \zeta_{L,p}^*(s)$ for *all* primes p.

Since there is no requirement that the rings are nilpotent, we may consider non-nilpotent Lie rings. Indeed, the first author and Taylor calculated in [24] the zeta function of the Lie ring $\mathfrak{sl}_2(\mathbb{Z})$. Furthermore, Chap. 3 is devoted to a family of soluble Lie rings.

1.2.3 Local Functional Equations

Many examples of local zeta functions of $\mathfrak{T}$-groups and Lie rings satisfy a *local functional equation* of the form

$$\zeta^*_{G,p}(s)\big|_{p\to p^{-1}} = (-1)^r p^{b-as} \zeta^*_{G,p}(s) \tag{1.5}$$

for $* \in \{\leq, \lhd, \wedge\}$, $a, b, r \in \mathbb{N}$, and for at least all but finitely many primes p.

For $* = \wedge$ it is known that the local zeta functions satisfy a functional equation of the form (1.5). This was proved by the first author and Lubotzky in [21]. This functional equation has its origins in symmetries for the associated building of the algebraic group [21].

In [59], Voll proves that the zeta functions counting all subgroups also satisfy functional equations. Voll also proves that the local ideal zeta functions of $\mathfrak{T}$-groups of nilpotency class 2 also satisfy functional equations. However, this result is best possible, as the following result demonstrates.

Theorem 1.1. *Let the Lie ring $L_{(3,2)}$ be given by the presentation*

$$\langle z, w_1, w_2, x_1, x_2, y_1 : [z, w_1] = x_1, [z, w_2] = x_2, [z, x_1] = y_1 \rangle \ ,$$

where, up to antisymmetry, all unlisted Lie brackets of basis elements are zero. For all primes p, the local zeta function $\zeta^{\lhd}_{L_{(3,2)},p}(s)$ satisfies no functional equation of the form (1.5).

Via the Mal'cev correspondence, we obtain a $\mathfrak{T}$-group $G_{(3,2)}$ of nilpotency class 3. For all but finitely many primes p, $\zeta^{\lhd}_{G_{(3,2)},p}(s)$ satisfies no functional equation. The zeta function $\zeta^{\lhd}_{L_{(3,2)},p}(s)$ is given explicitly on p. 49.

Chapter 4 is concerned with a reciprocity conjecture for p-adic integrals, 'Conjecture 4.5'. This conjecture can be used to predict when local zeta functions should satisfy functional equations, and the shape of the functional equation satisfied. It agrees with the results of Voll mentioned above. However, we have been unable to formulate this conjecture rigorously. There are technical preconditions which need to be satisfied, but we do not know what these preconditions are. However, we do believe that these conditions are always satisfied by the p-adic integrals representing local zeta functions of nilpotent Lie rings.

Assuming this conjecture, we list below the most significant consequences of it:

Theorem 1.2. *Let L be a Lie ring additively isomorphic to $\mathbb{Z}^d$ for some $d \in \mathbb{N}$. Assume Conjecture 4.5.*

1. *Under no further assumptions on the Lie ring L,*

$$\zeta^{\leq}_{L,p}(s)\big|_{p\to p^{-1}} = (-1)^d p^{\binom{d}{2}-ds} \zeta^{\leq}_{L,p}(s)$$

for all but finitely many primes p.

2. *Suppose L is nilpotent of class c. Let $\sigma_i(L)$ denote the ith term of the upper-central series of L, and put*

$$N = \sum_{i=0}^{c} \operatorname{rank}(L/\sigma_i(L)) \ .$$

Then either
(i) For all but finitely many primes p,

$$\zeta_{L,p}^{\triangleleft}(s)\big|_{p \to p^{-1}} = (-1)^d p^{\binom{d}{2} - Ns} \zeta_{L,p}^{\triangleleft}(s)$$

or
(ii) For all but finitely many primes p, $\zeta_{L,p}^{\triangleleft}(s)$ satisfies no such functional equation.
In particular, alternative (ii) only occurs if L has nilpotency class ≥ 3.

Equation (1.4) yields corresponding results for the local zeta functions of $\mathfrak{T}$-groups.

We also define a subset of nilpotent Lie rings within which we can determine whether alternative (i) or (ii) holds. This subset contains all class-2 nilpotent Lie rings, $L_{(3,2)}$ mentioned above, and many of the examples presented in Chap. 2. It also contains the free nilpotent Lie rings:

Theorem 1.3. *For $c, d \geq 2$, let $F_{c,d}$ be the free class-c-nilpotent Lie ring on d generators. Assume Conjecture 4.5. Then $\zeta_{F_{c,d},p}^{\triangleleft}(s)$ satisfies a functional equation of the form (1.5) for all but finitely many primes p.*

In Chap. 2 we document experimental evidence concerning the existence of these local functional equations. All this evidence counts in favour of Conjecture 4.5.

We also present a partial proof of a significant special case of this conjecture. This proof is not intended to be rigorous, merely to give some reason why the conjecture may be true.

1.2.4 Uniformity

Many of the examples of zeta functions of nilpotent groups calculated in [32, 28, 57, 64] can be written in terms of Riemann zeta functions and zeta functions of type (1.3). The remaining examples had local factors that depended on some finite division of primes. Indeed speculation in [32] hinted that the following could plausibly have a positive answer:

Question 1.4. Let G be a finitely generated nilpotent group and $* \in \{\leq, \triangleleft\}$. Do there exist finitely many rational functions $W_1(X, Y), \ldots, W_r(X, Y) \in \mathbb{Q}(X, Y)$ such that for each prime p there is an i for which

$$\zeta_{G,p}^*(s) = W_i(p, p^{-s}) \ ?$$

Such zeta functions are called *finitely uniform*. If additionally $r = 1$, we say the zeta function is *uniform*.

In [13] and [14] the first author has shown that this question in fact has a negative answer as the following Proposition indicates:

Proposition 1.5. *For each elliptic curve $E = y^2 - x^3 + x$, define a class-2-nilpotent group G_E by the following presentation, where all unlisted commutators are trivial:*

$$G_E = \left\langle x_1, \ldots, x_6, y_1, y_2, y_3 : \begin{array}{l} [x_1, x_4] = y_3, [x_1, x_5] = y_1, [x_1, x_6] = y_2, \\ [x_2, x_4] = y_1, [x_2, x_5] = y_3, \\ [x_3, x_4] = y_2, [x_3, x_6] = y_1 \end{array} \right\rangle$$

Then there exist two non-zero rational functions $P_1(X, Y)$ and $P_2(X, Y) \in \mathbb{Q}(X, Y)$ such that for almost all primes p,

$$\zeta^{\triangleleft}_{G_E, p}(s) = P_1(p, p^{-s}) + |E(\mathbb{F}_p)| P_2(p, p^{-s}) , \tag{1.6}$$

where $|E(\mathbb{F}_p)|$ is the number of points on E mod p.

The non-uniform behaviour therefore arises from the term $|E(\mathbb{F}_p)|$. To see where the elliptic curve is hiding in the presentation, take the determinant of the matrix with entries $[x_i, x_{j+3}]$ and you'll get the projectivised version of E.

1.2.5 Analytic Properties

We have so far considered zeta functions of groups and rings purely as formal beasts. So what of the convergence of this series as a function in the complex variable s? Such a Dirichlet series converges on some right half of the complex plane if and only if the invariant $a_n(G)$ grows polynomially in n. We now have a characterisation of groups of so called polynomial subgroup growth or PSG groups. In the category of abstract finitely generated groups, these are the virtually soluble groups of finite rank [41]. For pro-p groups, they are the p-adic analytic groups [40]. For profinite groups the description is slightly more complicated but the groups are essentially extensions of pro-soluble groups of finite rank by products of simple groups of Lie type with bounds on the rank and field degrees of the Lie groups involved [52]. These are the classes of groups for which our function defines an analytic function on the right half complex plane $\{ s \in \mathbb{C} : \Re(s) > \alpha_G \}$ where α_G is the abscissa of convergence:

$$\alpha_G = \limsup_{n \to \infty} \frac{\log(a_1(G) + \cdots + a_n(G))}{\log n} .$$

It is clear that the zeta function of a ring L additively isomorphic to $\mathbb{Z}^d$ has polynomial subring growth. This follows from the fact that subrings of L are subgroups of $\mathbb{Z}^d$. We shall use the notation $\alpha^{\leq}_G$ and $\alpha^{\triangleleft}_G$ for the abscissae of convergence of $\zeta^{\leq}_G(s)$ and $\zeta^{\triangleleft}_G(s)$, and similarly for Lie rings.

In Chap. 5, we consider the situations where we can analytically continue these analytic functions beyond their radius of convergence to meromorphic functions on the whole complex plane, so satisfying (ZF1). In the category of pro-p groups or for the local zeta functions $\zeta_{G,p}(s)$ this is possible because in general these are rational functions in p^{-s}:

Proposition 1.6 ([10]). *Let G be a finitely generated PSG pro-p group (i.e. a p-adic analytic group). Then $\zeta_G(s)$ is rational in p^{-s} and can be continued to a meromorphic function on the whole complex plane.*

Proposition 1.7 ([10]). *Let G be a finitely generated PSG group (i.e. a virtually soluble group of finite rank). Then, for all primes p, $\zeta_{G,p}(s)$ is a rational function in p^{-s} and can be continued to a meromorphic function on the whole complex plane.*

Combining these results for the local factors of zeta functions of algebraic groups and nilpotent groups, the local zeta functions score reasonably well against the conditions (ZF1)–(ZF4) for a zeta function.

Let us now return to the global zeta functions which are Euler products of these rational functions. Using the explicit expression (1.7), the first author and Grunewald [17] show that zeta functions of nilpotent groups always admit some analytic continuation beyond the region of convergence. The key to their analysis is the proof of an explicit expression for local factors which depends on counting points mod p on a system of varieties, and the use of Artin L-functions. This work also establishes the useful result that the abscissa of convergence of these zeta functions is always a rational number.

This analytic continuation allows us to apply the following Tauberian Theorem (see for example the Corollary on p. 121 of [47]) to zeta functions of groups and rings. This allows us to deduce the precise rate of subgroup/subring growth:

Theorem 1.8. *Let the Dirichlet series $f(s) = \sum_{n=1}^{\infty} a_n n^{-s}$ with non-negative coefficients be convergent for $\Re(s) > \alpha > 0$. Assume in its domain of convergence, $f(s) = g(s)(s-\alpha)^{-w} + h(s)$ holds, where $g(s)$, $h(s)$ are holomorphic functions in the closed half-plane $\Re(s) \geq \alpha$, $g(\alpha) \neq 0$ and $w > 0$. Then for x tending to infinity, we have*

$$\sum_{n \leq x} a_n = \left(\frac{g(\alpha)}{\alpha \Gamma(w)} + o(1) \right) x^{\alpha} (\log x)^{w-1} .$$

In [14] the explicit expression of [17] together with the formalism of motivic zeta functions developed in [20] is used to establish a hierarchy in the class of nilpotent groups according to the complexity of the varieties that arise in the explicit expression. The analysis of the following chapter can then be seen to apply to nilpotent groups at the bottom of this hierarchy where the varieties involved are nothing more complicated that $\mathbb{Q}$-rational varieties and hence the zeta functions are of type (1.3). Specifically we see how the general theory developed here applies to the early examples of [32] and [28] that led to the speculation of Grunewald, Segal and Smith that all nilpotent groups were at the bottom of such a hierarchy.

1.3 p-Adic Integrals

p-adic integrals are an immensely powerful tool used in the study of zeta functions of groups and Lie rings. Indeed, we have already seen them used to define the zeta function of an algebraic group. There are many other applications that these important tools have.

We shall introduce these integrals below. Before we do this, we must introduce a notion of 'size' of subsets of $\mathbb{Z}_p$. Let μ be the additive Haar measure on subsets $\mathbb{Z}_p$ normalised so that $\mu(\mathbb{Z}_p) = 1$. The key properties of this measure are that:

1. μ is *additive*, in that if S_1 and S_2 are disjoint measurable sets, then $\mu(S_1 \cup S_2) = \mu(S_1) + \mu(S_2)$.
2. μ is *translation invariant*, in that if S is measurable and $a \in \mathbb{Z}_p$, $\mu(a+S) = \mu(S)$.

As a consequence of these two properties, $\mu(p^m \mathbb{Z}_p) = p^{-m}$ for any $m \in \mathbb{N}$. There are p^m pairwise disjoint additive cosets of $p^m \mathbb{Z}_p$, all of which have the same measure, and the sum of the measures of all p^m cosets must be 1. Furthermore, all open subsets of $\mathbb{Z}_p$ are measurable, since the additive cosets of the form $a + p^m \mathbb{Z}_p$ form a base of neighbourhoods for the topology of $\mathbb{Z}_p$. Finally, by abuse of notation, we can extend μ to a Haar measure on $\mathbb{Z}_p^n$ for $n \in \mathbb{N}_{>0}$.

With a Haar measure in hand, we can now define the p-adic integral of a constant function. Let $\mathbf{x} = (x_1, \ldots, x_n)$ be n commuting indeterminates. If $f(\mathbf{x})$ takes the constant value c on the measurable set $S \subseteq \mathbb{Z}_p^n$, then

$$\int_S |f(\mathbf{x})|_p^s \, d\mu = \mu(S)|c|_p^s .$$

In other words, we simply multiply the constant value by the measure of the set on which the function is constant. For a nonconstant function $f(\mathbf{x})$, we split the domain of integration into pieces on which $|f(\mathbf{x})|_p$ is constant, and then sum the measure of each piece. In other words, if $v(x)$ denotes the p-adic valuation of x and $V_f(k) = \{ \mathbf{x} \in \mathbb{Z}_p^n : v(f(\mathbf{x})) = k \}$, then

$$\int_{\mathbb{Z}_p^n} |f(\mathbf{x})|_p^s \, d\mu = \sum_{k=0}^{\infty} \mu(V_f(k)) p^{-ks} .$$

These integrals can easily be generalised to include a factor $|g(\mathbf{x})|_p$ independent of s in the integrand, or to integrate over a (measurable) subset of $\mathbb{Z}_p^n$.

One particular type of p-adic integrals, *cone integrals*, are especially important. Let $f_i(\mathbf{x})$, $g_i(\mathbf{x})$ be polynomials in $\mathbf{x}$ for $0 \le i \le l$. The cone integral with *cone data* $\mathcal{D} = \{f_0(\mathbf{x}), g_0(\mathbf{x}), \dots, f_l(\mathbf{x}), g_l(\mathbf{x})\}$ is defined to be

$$Z_{\mathcal{D}}(s, p) = \int_{\Psi(\mathcal{D})} |f_0(\mathbf{x})|_p^s |g_0(\mathbf{x})|_p \, \mathrm{d}\mu \, ,$$

where

$$\Psi(\mathcal{D}) = \{\, \mathbf{x} \in \mathbb{Z}_p^m : v(f_i(\mathbf{x})) \le v(g_i(\mathbf{x})) \text{ for } i = 1, \dots, l \,\} \, .$$

The first application of p-adic integrals came with the proof that the local zeta functions $\zeta_{G,p}^*(s)$ and $\zeta_{L,p}^*(s)$ for $* \in \{\le, \lhd \wedge\}$ are rational functions in p^{-s} for all primes p. To prove this, Grunewald, Segal and Smith then showed that these local zeta functions can be expressed as 'definable' p-adic integrals. A deep theorem due to Denef [6] yields the required rationality.

Definable integrals were also employed by the first author in [11] to prove two significant results on enumerating p-groups:

- Firstly, let $f(n, p, c, d)$ be the number of finite groups of order p^n of nilpotency class c generated by d elements. Then $f(n, p, c, d)$ satisfies a linear recurrence relation with constant coefficients as n varies and p, c and d remain fixed.
- Secondly, the qualitative part of Newman and O'Brien's 'Conjecture P' [48] is confirmed.

Whilst the rationality of definable p-adic integrals is undoubtedly a significant theoretical advance, it is sadly of little help if one actually wishes to compute such an integral explicitly. This is due to the model-theoretic 'black-box' at the heart of the proof. For a set of cone integral data $\mathcal{D}$, the first author and Grunewald [17] considered the resolution of singularities attached to the polynomial $F = \prod_{i=0}^l f_i(\mathbf{x}) g_i(\mathbf{x})$. Using this approach they give an explicit expression for a cone integral $Z_{\mathcal{D}}(s, p)$ in terms of the data attached to the resolution. The resolution of F in some sense 'breaks it up' into irreducible smooth projective varieties E_i as i runs through some finite indexing set T. It is then proved that

$$Z_{\mathcal{D}}(s, p) = \sum_{I \subseteq T} c_{p,I} P_I(p, p^{-s}) \, , \tag{1.7}$$

where $c_{p,I}$ is the number of points mod p on all E_i for $i \in I$ and on no other E_i, and $P_I(p, p^{-s})$ are rational functions. It is then proved [17, Corollary 5.6] that if L is a ring additively isomorphic to $\mathbb{Z}^d$ for $d \in \mathbb{N}$, and $* \in \{\lhd, \le\}$, then

$$\zeta_{L,p}^*(s) = (1 - p^{-1})^{-d} Z_{\mathcal{D}^*}(s - d, p)$$

for suitable cone data $\mathcal{D}^*$; indeed we shall explicitly construct the polynomials comprising the cone data in Proposition 2.1.

The explicit expression (1.7) yields immediately another proof of the rationality of the local factors $\zeta^*_{L,p}(s)$. We mentioned above that the global zeta function $\zeta^*_L(s)$ of every Lie ring L additively isomorphic to $\mathbb{Z}^d$ has rational abscissa of convergence and always admits some analytic continuation beyond this abscissa, their proofs employ a variation of (1.7).

Expression (1.7) is also of interest when studying the uniformity the local zeta functions of a ring L. Since the factors $P_I(p, p^{-s})$ are uniform in p, the variation of $\zeta^*_{L,p}(s)$ as p varies is controlled by the variation of the coefficients $c_{p,I} \bmod p$. This of course raises the question of what varieties can be encoded by a nilpotent group. Some progress in answering this question has been made by Griffin [29].

The explicit expression is also of practical use in evaluating p-adic integrals. Guided by the resolution of singularities of the appropriate polynomial, the first author and Taylor compute in [24] the zeta function counting all subrings of the Lie ring $\mathfrak{sl}_2(\mathbb{Z})$. Numerous further such calculations have been performed by Taylor [57] and the second author [64] in their theses.

A further application of cone integrals is the conjecture due to the second author alluded to above and presented in Chap. 4. This conjecture is essentially a reciprocity conjecture involving cone integrals. It may be viewed as an attempt to generalise a theorem due to Denef and Meuser [8] on *Igusa-type* zeta functions, i.e. those defined by p-adic cone integrals for which $l = 0$ and $g_0(\mathbf{x}) = 1$.

1.4 Natural Boundaries of Euler Products

We mentioned above that local zeta functions of groups have meromorphic continuation to $\mathbb{C}$. However, the same is not true in general for the global zeta functions. In Chap. 5, we turn to the general analytic character of functions of the form

$$Z(s) = \prod_p W(p, p^{-s}) \tag{1.8}$$

defined as Euler products of two-variable polynomials. This includes the zeta functions of algebraic groups and many of the examples of zeta functions of groups and rings listed in Chap. 2.

We begin Chap. 5 by considering a particular example which arises in the zeta function of the algebraic group $\mathcal{G} = \mathrm{GSp}_6$ and a corresponding zeta function of type $\zeta^\wedge_{G,p}(s)$ for a nilpotent group. In particular we prove:

Proposition 1.9. *Let*

$$Z(s) = \prod_{p \; prime} Z_{\mathrm{GSp}_6,p}(s) .$$

Then $Z(s)$ (1) converges for $\Re(s) > 5$; (2) has meromorphic continuation to $\Re(s) > 4$; but (3) has a natural boundary at $\Re(s) = 4$ beyond which no further meromorphic continuation is possible.

As far as we can establish, this is the first place to document the failure of the global zeta function of an algebraic group to have meromorphic continuation to the whole complex plane. The proof of the natural boundary depends on showing that every point on the line $\Re(s) = 4$ can be realised as a limit point of zeros of the local factors $W(p, p^{-s})$ which all crucially lie on the right of $\Re(s) = 4$ (i.e. in the region of meromorphic continuation established in (2) above).

Throughout the remainder of Chap. 5, we generalise these ideas to prove a general result about the existence of natural boundaries for functions defined via Euler products of two-variable polynomials. This can be seen as contributing to a project begun by Estermann in the 1920s [25] and continued by Kurokawa [38, 39]. Estermann proved the following (see [25]):

Proposition 1.10. *Let $h(X) = 1 + a_1 X + \cdots a_d X^d = \prod(1 - \alpha_i X) \in \mathbb{Z}[X]$. Set $L(s) = \prod_p h(p^{-s})$ which converges for $\Re(s) > 1$. Then*

1. *$L(s)$ can be meromorphically continued to $\Re(s) > 0$.*
2. *If $|\alpha_i| = 1$ for $i = 1, \ldots, d$ (in which case we say that $h(X)$ is unitary) then $L(s)$ can be meromorphically continued to the whole complex plane. Otherwise $\Re(s) = 0$ is a natural boundary.*

In our case where we are dealing with polynomials in two variables, the following has been conjectured:

Conjecture 1.11. Let

$$W(X, Y) = 1 + \sum_{i=1}^{r} (a_{i0} + a_{i1} X + \cdots + a_{in_i} X^{n_i}) Y^i \in \mathbb{Z}[X, Y].$$

Set $L(s) = \prod_p W(p, p^{-s})$. Then $L(s)$ can be meromorphically continued to the whole complex plane if and only if for $i = 1, \ldots, n$ there exist unitary polynomials $g_i(Z)$ and integers b_i, c_i such that

$$W(X, Y) = g_1(X^{b_1} Y^{c_1})^{\pm 1} \cdots g_n(X^{b_n} Y^{c_n})^{\pm 1}.$$

One direction of the conjecture follows easily from Estermann's Theorem. We can view our result in Chap. 5 as a contribution to the other half of this conjecture. To explain our result we suppose firstly that any unitary factors of $W(X, Y)$ have been removed and that $W(X, Y) \neq 1$ (otherwise $Z(s)$ is meromorphic).

Let

$$\alpha = \max \left\{ \frac{n_k + 1}{k} : k = 1, \ldots, r \right\},$$

$$\beta = \max \left\{ \frac{n_k}{k} : k = 1, \ldots, r \right\}$$

and put

$$\widetilde{W_1}(X,Y) = \sum_{j/i=\beta} a_{ij} X^j Y^i \, .$$

This is one factor of something that we have called the *ghost* of $W(X,Y)$ (see [16] and [18]). We express $W(X,Y)$ as a unique cyclotomic expansion

$$W(X,Y) = \prod_{(n,m)\in\mathbb{N}^2} (1 - X^n Y^m)^{c_{n,m}} \tag{1.9}$$

with $c_{n,m} \in \mathbb{Z}$. Using this cyclotomic expansion we can prove:

Theorem 1.12. *$Z(s)$ converges on $\{\, s \in \mathbb{C} : \Re(s) > \alpha \,\}$ and can be meromorphically continued to $\{\, s \in \mathbb{C} : \Re(s) > \beta \,\}$.*

We conjecture that $\Re(s) = \beta$ will be the natural boundary for meromorphic continuation of $Z(s)$. We are able to prove the following:

Theorem 1.13. *Suppose that $W(X,Y) \neq 1$ and has no unitary factors. Suppose that either*

1. *$\widetilde{W_1}(X,Y)$ is not unitary; or*
2. *For each N there exists a prime $p > N$ and zeros of $W(p,Y)$ with $|Y| < p^\beta$, and there are finitely many pairs (n,m) with $c_{n,m} > 0$; or*
3. *For each N there exists a prime $p > N$ and zeros of $W(p,Y)$ with $|Y| < p^\beta$, and there are infinitely many pairs (n,m) with $c_{n,m} > 0$ and the Riemann Hypothesis holds.*

Then $\Re(s) = \beta$ is a natural boundary for $Z(s)$.

In case 1 we show that we are guaranteed local zeros on the right of $\Re(s) = \beta$. In cases 2 and 3 we must assume the existence of such zeros. As we shall explain, this actually covers the majority of polynomials. In case 2 we can get away without the Riemann Hypothesis, but in case 3 we must have some control over the zeros of the Riemann zeta function to be able to prove that their zeros won't kill the zeros of the local factors we will be using to realise our natural boundary.

A useful observation (see Corollaries 5.8 and 5.9) is that whenever β is an integer we can't be in case 3.

In Sect. 5.3 we explain some subcases of case 3 where we can avoid the Riemann Hypothesis by using current estimates for the number of zeros off the critical line. In Sect. 5.4 we speculate on some strategies for dealing with polynomials with all their local zeros to the left of $\Re(s) = \beta$. One case in which we are successful requires a strong assumption about the zeros of the Riemann zeta function:

Theorem 1.14. *Suppose that $W(X,Y) \neq 1$ and has no unitary factors. Suppose that there are an infinite number of pairs (n,m) with $c_{n,m} \neq 0$ and $(n + \frac{1}{2})/m > \beta$. Under the assumption that Riemann zeros are rationally independent (i.e. if $\rho = \tau + \sigma i$ and $\rho' = \tau' + \sigma' i$ are zeros of $\zeta(s)$ then $\sigma/\sigma' \notin \mathbb{Q}$) then $\Re(s) = \beta$ is a natural boundary for $Z(s)$.*

In Chap. 5 we introduce two hypotheses which can easily be checked in any individual case to determine whether the polynomial $W(X,Y)$ satisfies the conditions of Theorem 1.13.

In Chaps. 6 and 7 we return to the motivating examples of zeta functions of algebraic groups and nilpotent groups. All these examples satisfy the hypothesis of Theorem 1.13 that there exist local zeros to the right of the candidate natural boundary.

Let $\mathfrak{G}$ be one of the classical groups GO_{2l+1}, GSp_{2l} or GO_{2l}^+ of type B_l, C_l or D_l. Let W be the corresponding Weyl group and $\lambda(w)$ denote the length of an element $w \in W$, Φ the root system with fundamental roots $\alpha_1, \ldots, \alpha_l$, Φ^+ the set of positive roots of Φ, and a_i integers defined by

$$\prod_{\alpha \in \Phi^+} \alpha = \prod_{i=1}^{l} \alpha_i^{a_i}.$$

In [36], Igusa proved that $Z_{\mathfrak{G}}(s)$ could be expressed in terms of Riemann zeta functions and a function of type (1.8) where

$$W(X,Y) = P_{\mathfrak{G}}(X,Y) = \left(\sum_{w \in W} X^{-\lambda(w)} \prod_{\alpha_j \in w(\Phi^-)} X^{a_j} Y^{b_j} \right),$$

where b_i are integers defined by expressing the dominant weight of the natural representation in terms of the basis for the root system.

By analysing this explicit expressions of Igusa and the root systems in each particular case we apply in Chap. 6 the work of Chap. 5 to prove the following result which was first announced in [18]:

Theorem 1.15. *Let $\mathfrak{G}$ be one of the classical groups GO_{2l+1}, GSp_{2l} or GO_{2l}^+ of type B_l, C_l or D_l respectively. Then $Z_{\mathfrak{G}}(s)$ has abscissa of convergence $a_l + 1$ and has a natural boundary at $\Re(s) = \beta$ where*

1. *$\beta = l^2 - 1 = a_{l-1}$ if $\mathfrak{G} = GO_{2l+1}$,*
2. *$\beta = l(l+1)/2 - 2 = a_{l-2}/2 + 1$ if $\mathfrak{G} = GSp_{2l}$, and*
3. *$\beta = l(l-1)/2 - 2 = a_{l-2}/2$ if $\mathfrak{G} = GO_{2l}^+$.*

Here we are taking the natural representation in the definition of $Z_{\mathfrak{G}}(s)$. The proof of the Theorem in the case of GSp_{2l} and GO_{2l}^+ requires an application of a natural factorisation of the polynomial $P_{\mathfrak{G}}(X,Y)$ which we establish in Appendix B to remove various unitary factors which initially interfere with the analysis.

In Chap. 7 we consider the natural boundaries of the zeta functions of nilpotent groups presented in Chap. 2.

We mentioned earlier the concept of the ghost zeta function attached to $W(X, Y)$. This partly grew out of the analysis of Chap. 5. The philosophy of this book is that natural boundaries occur because the local zeros of $W(p, p^{-s})$ are shifted away from the candidate natural boundary but as p tends to infinity, these zeros tend to points on the boundary. The concept of the ghost polynomial grew out of this observation. The ghost polynomial $\widetilde{W}(X, Y)$ is defined so that its zeros are in some sense the limit of the zeros of $W(p, p^{-s})$ as p tends to infinity. In some philosophical sense $\widetilde{W}(X, Y)$ is the polynomial that $W(X, Y)$ is trying to be. This removes the first obstruction then to meromorphic continuation. So the interesting question is: does the zeta function defined by the ghost polynomial $\widetilde{W}(X, Y)$ have meromorphic continuation? If it does we say the ghost is *friendly*. For more details we refer the reader to [16] and [18] where the ghosts of the classical groups are proved to be friendly.

The ghost zeta functions attached to nilpotent groups are mostly friendly too. However there are a number that are unfriendly, in that they too fail to have meromorphic continuation to $\mathbb{C}$. In Chap. 7, we mention whether the ghosts of the zeta functions of nilpotent Lie rings calculated in Chap. 2 are friendly.

2

Nilpotent Groups: Explicit Examples

In this chapter we list some of the (now numerous) calculations of zeta functions of $\mathfrak{T}$-groups and Lie rings. The primary emphasis is on bringing into print explicit calculations that have yet to be published. However, we aim this chapter to be more than just a gallery of results. Hence we begin the chapter with some details about how these zeta functions have been calculated.

2.1 Calculating Zeta Functions of Groups

Zeta functions of groups have been calculated using a number of different methods. The first examples counted ideals in $\mathfrak{T}$-groups of class 2 and were calculated by Grunewald, Segal and Smith in [32]. A key part of their work is the formula [32, Lemma 6.1]

$$\zeta_{G,p}^{\triangleleft}(s) = \zeta_{\mathbb{Z}^n,p}(s) \sum_{B \leq A} |A : B|^{n-s} |G : X(B)|^{-s} , \qquad (2.1)$$

where $A = \gamma_2(G)$, $G/A \cong \mathbb{Z}^d$ and $X(B)/B = Z(G/B)$. Their calculations are made by evaluating (2.1) for each group in turn. Although there are a few general lemmas proved which help speed matters along, their methods are to some extent tailored to each group individually. Nonetheless, their methods suffice to calculate all but perhaps finitely many of the local factors $\zeta_{G,p}^{\triangleleft}(s)$ for every $\mathfrak{T}$-group G of class 2 and Hirsch length at most 6.

In [60], Voll uses (2.1) and the Bruhat-Tits building of $\mathrm{SL}_n(\mathbb{Q}_p)$ to compute normal zeta functions of $\mathfrak{T}$-groups whose centres are free abelian of rank 2 or 3. In particular, Voll computes the normal zeta function of all $\mathfrak{T}$-groups whose centre is of rank 2, and confirms the functional equation (1.5). This work is based on the classification of such groups by Grunewald and Segal [31]. For centres of rank 3, the geometry of the associated Pfaffian hypersurface comes into play. Provided the singularities of this hypersurface are in some sense not too severe, Voll gives a formula for the local normal zeta function of L

depending on the number of points on the Pfaffian hypersurface. A highlight
of this work is explicit expressions for the rational functions $P_1(X, Y)$ and
$P_2(X, Y)$ in the local normal zeta function of the 'elliptic curve example'
(1.6).

A more general approach is used by Voll in [61], where he considers the case
where the Pfaffian hypersurface has no lines. Indeed this occurs generically
if the abelianisation has rank greater than $4r - 10$, where r is the dimension
of the centre. Provided this Pfaffian is smooth and absolutely irreducible, the
functional equation (1.5) holds. Voll also gives in [61] an explicit formula for
the normal zeta functions of the class-2 nilpotent groups known as 'Grenham
groups', using a combinatorial formula for the number of points on flag varieties. This formula is also employed by Voll in [58], where he gives an explicit
formula for the local zeta functions counting all subgroups in the Grenham
groups.

One key assumption Voll makes in [61] is that the associated Pfaffian
hypersurface has no lines. A forthcoming paper by Paajanen [49] presents
the first step in overcoming this obstacle. She considers the normal zeta
function of a class-2 nilpotent group G_S which encodes the Segre surface
$S : x_1 x_4 - x_2 x_3 = 0$. In particular, she calculates that

$$\zeta_{G_S, p}^{\triangleleft}(s) = W_0(p, p^{-s}) + (p + 1)^2 W_1(p, p^{-s}) + 2(p + 1)W_2(p, p^{-s})$$

for explicit rational functions $W_i(p, p^{-s})$, $i = 0, 1, 2$. The coefficients $(p + 1)^2$
and $2(p + 1)$ arise from the geometry of S reduced mod p: being isomorphic
to $\mathbb{P}^1(\mathbb{F}_p) \times \mathbb{P}^1(\mathbb{F}_p)$ it has $(p + 1)^2$ points and $2(p + 1)$ lines.

Voll has also used combinatorial methods to yield an explicit expression
for the local normal zeta functions of the class-2 free nilpotent groups [62].
One key ingredient is an explicit expression for a sum of certain Hall polynomials. Whilst there seems to be no simple formula for the Hall polynomials
themselves, a polynomial expression for the sum has been known for some
time.

One approach common to the work of Voll and Paajanen is to decompose the local normal zeta function as a sum of rational functions with coefficients corresponding to invariants of a suitable algebraic variety. They are then
able to deduce functional equations by virtue of the fact that each individual
rational function with its coefficient satisfies the same functional equation.
In particular,

$$\zeta_{G_S, p}^{\triangleleft}(s)\big|_{p \to p^{-1}} = p^{28 - 12s} \zeta_{G_S, p}^{\triangleleft}(s) \,,$$

with the three rational functions above satisfying

$$W_0(X^{-1}, Y^{-1}) = X^{28} Y^{12} W_0(X, Y) \,,$$
$$W_1(X^{-1}, Y^{-1}) = X^{26} Y^{12} W_1(X, Y) \,,$$
$$W_2(X^{-1}, Y^{-1}) = X^{27} Y^{12} W_2(X, Y) \,.$$

The 'missing' powers of X are provided by the coefficients $(p+1)^2$ and $2(p+1)$.

2.2 Calculating Zeta Functions of Lie Rings

Most of the zeta functions presented in this chapter have been calculated by the method of Lie rings, p-adic integrals and ad-hoc resolutions of singularities. In particular, the zeta functions calculated in the theses of Taylor [57] and the second author [64] were calculated this way. In particular, we shall work with Lie rings instead of groups, and leave the reader to obtain the corresponding results concerning groups via the Mal'cev correspondence. We shall also make the assumption that our Lie rings are additively isomorphic to either $\mathbb{Z}^d$ or $\mathbb{Z}_p^d$, i.e. (additively) finitely generated and torsion-free.

Recall that $\zeta_{L,p}^*(s) = \zeta_{L\otimes\mathbb{Z}_p}^*(s)$. Given a $\mathbb{Z}_p$-Lie ring L with basis $\mathcal{B} = (\mathbf{e}_1, \ldots, \mathbf{e}_d)$ for L, calculating either of the zeta functions $\zeta_{L,p}^{\leq}$ or $\zeta_{L,p}^{\lhd}$ is essentially a four-stage calculation:

1. Constructing the cone integral.
2. Breaking the integral into a sum of monomial integrals.
3. Evaluating the monomial integrals.
4. Summing the resulting rational functions.

2.2.1 Constructing the Cone Integral

Let M be an upper-triangular $d \times d$ matrix $M = (m_{i,j})$ with entries in $\mathbb{Z}_p$. We may consider the rows $\mathbf{m}_1, \ldots, \mathbf{m}_d$ of this matrix to be additive generators of a submodule of L. This submodule will be a subring if

$$[\mathbf{m}_i, \mathbf{m}_j] \in \langle \mathbf{m}_1, \ldots, \mathbf{m}_d \rangle_{\mathbb{Z}_p} \text{ for all } 1 \leq i < j \leq d \tag{2.2}$$

and an ideal if

$$[\mathbf{e}_i, \mathbf{m}_j] \in \langle \mathbf{m}_1, \ldots, \mathbf{m}_d \rangle_{\mathbb{Z}_p} \text{ for all } 1 \leq i, j \leq d. \tag{2.3}$$

The following proposition and its proof gives us an explicit description of the *cone conditions*, i.e. the conditions of the form $v(f_i(\mathbf{x})) \leq v(g_i(\mathbf{x}))$ for $1 \leq i \leq l$. It is essentially Theorem 5.5 of [17].

Proposition 2.1. *Let L be a $\mathbb{Z}$-Lie ring with basis $\mathcal{B} = (\mathbf{e}_1, \ldots, \mathbf{e}_d)$. Let $V_p^{\lhd}$ be the set of all upper-triangular matrices over $\mathbb{Z}_p$ such that $\mathbb{Z}_p^d \cdot M \lhd L \otimes \mathbb{Z}_p$, and $V_p^{\leq}$ the set of such matrices such that $\mathbb{Z}_p^d \cdot M \leq L \otimes \mathbb{Z}_p$. Then $V_p^{\lhd}$ and $V_p^{\leq}$ are defined by the conjunction of polynomial divisibility conditions $v(f_i(\mathbf{x})) \leq v(g_i(\mathbf{x}))$ for $1 \leq i \leq l$. Furthermore, the conditions defining $V_p^{\lhd}$ satisfy $\deg f_i(\mathbf{x}) = \deg g_i(\mathbf{x})$, and those defining $V_p^{\leq}$ satisfy $\deg f_i(\mathbf{x}) + 1 = \deg g_i(\mathbf{x})$.*

Proof. Let $\mathbf{m}_1, \ldots, \mathbf{m}_d$ denote the rows of the matrix M, C_j the matrix whose rows are $\mathbf{c}_i = [\mathbf{e}_i, \mathbf{e}_j]$. Let M' denote the adjoint matrix of M and

$$M^\natural = M' \operatorname{diag}(m_{2,2}^{-1} \ldots m_{d,d}^{-1}, m_{3,3}^{-1} \ldots, m_{d,d}^{-1}, \ldots, m_{dd}^{-1}, 1) .$$

Since M is upper-triangular, the (i,k) entry of $M^\natural$ is a homogeneous polynomial of degree $k-1$ in the variables $m_{r,s}$ with $1 \leq r \leq s \leq k-1$.

The rows of M generate an ideal if we can solve, for each $1 \leq i, j \leq d$, the equation

$$\mathbf{m}_i C_j = (y_{i,j,1}, \ldots, y_{i,j,d}) M$$

for $(y_{i,j,1}, \ldots, y_{i,j,d}) \in \mathbb{Z}_p^d$. This rearranges to

$$\mathbf{m}_i C_j M^\natural = (m_{1,1} y_{i,j,1}, \ldots, m_{1,1} \ldots m_{d,d} y_{i,j,d})$$

for $(y_{i,j,1}, \ldots, y_{i,j,d}) \in \mathbb{Z}_p^d$. Set $g_{i,j,k}^{\triangleleft}(\mathbf{x})$ to be the kth entry of the d-tuple $\mathbf{m}_i C_j M^\natural$. $g_{i,j,k}^{\triangleleft}(\mathbf{x})$ is a homogeneous polynomial of degree k in the $m_{r,s}$, and if we set $f_{i,j,k}(\mathbf{x}) = m_{1,1} \ldots m_{k,k}$, we obtain the conditions $v(f_{i,j,k}(\mathbf{x})) \leq v(g_{i,j,k}^{\triangleleft}(\mathbf{x}))$ with $\deg(f_{i,j,k}(\mathbf{x})) = \deg(g_{i,j,k}^{\triangleleft}(\mathbf{x}))$.

Similarly, the rows of M generate a subring if we can solve, for $1 \leq i < j \leq d$,

$$\mathbf{m}_i \left(\sum_{r=j}^{d} m_{j,r} C_r \right) M^\natural = (m_{1,1} y_{i,j,1}, \ldots, m_{1,1} \ldots m_{d,d} y_{i,j,d})$$

for $(y_{i,j,1}, \ldots, y_{i,j,d}) \in \mathbb{Z}_p^d$. Again, we set $g_{i,j,k}^{\triangleleft}(\mathbf{x})$ to be the kth entry of the d-tuple $\mathbf{m}_i \left(\sum_{r=j}^{d} m_{j,r} C_r \right) M^\natural$. However, this time $g_{i,j,k}^{\triangleleft}(\mathbf{x})$ is a homogeneous polynomial of degree $k+1$, so we obtain conditions $v(f_{i,j,k}(\mathbf{x})) \leq v(g_{i,j,k}^{\triangleleft}(\mathbf{x}))$. Furthermore, $\deg(f_{i,j,k}(\mathbf{x})) + 1 = \deg(g_{i,j,k}^{\triangleleft}(\mathbf{x}))$. □

Whilst every subring or ideal H has a matrix M whose rows additively generate H, these matrices are by no means unique. Multiplying a row by a p-adic unit or adding a multiple of a row to another row above it may change the matrix but does not alter the subring additively generated by the rows. Each diagonal entry $m_{i,i}$ is unique up to multiplication by p-adic units, hence the measure of values it can take is $(1 - p^{-1})|m_{i,i}|_p$. Each off-diagonal entry $m_{i,j}$ is only unique modulo $|m_{j,j}|_p^{-1}$. Hence the measure of upper-triangular matrices generating H is $(1 - p^{-1})^d |m_{1,1}|_p |m_{2,2}|_p^2 \ldots |m_{d,d}|_p^d$. Note that although $m_{i,i}$ may vary, $|m_{i,i}|_p$ is uniquely determined by H.

Finally, we note that the index of H is $|m_{1,1} m_{2,2} \ldots m_{d,d}|_p^{-1}$. Hence we may write

$$\zeta_{L,p}^*(s) = (1 - p^{-1})^{-d} \int_{V_p^*} |m_{1,1} \ldots m_{d,d}|_p^s |m_{1,1}^1 \ldots m_{d,d}^d|_p^{-1} \, d\mu , \qquad (2.4)$$

or

$$\zeta^*_{L,p}(s+d) = (1-p^{-1})^{-d} \int_{V^*_p} |m_{1,1} \ldots m_{d,d}|^s_p |m^{d-1}_{1,1} \ldots m^1_{d-1,d-1}|_p \, d\mu \ . \quad (2.5)$$

Note that the translation in (2.5) is necessary. Equation (2.4) is not a cone integral since the constant (independent of s) term in the integrand has a negative exponent. We complete the set of cone data by setting $f_0(\mathbf{x}) = m_{1,1} \ldots m_{d,d}$, $g_0(\mathbf{x}) = m^{d-1}_{1,1} \ldots m_{d-1,d-1}$ and $\mathcal{D} = \{f_0(\mathbf{x}), g_0(\mathbf{x}), \ldots, f_l(\mathbf{x}), g_l(\mathbf{x})\}$. We therefore obtain the following result.

Proposition 2.2. *Let L be a Lie ring additively isomorphic to $\mathbb{Z}^d$, $* \in \{\leq, \triangleleft\}$. There exists a set of cone integral data $\mathcal{D} = \{f_0, g_0, \ldots, f_l, g_l$ such that, for all primes p,*

$$\zeta^*_{L,p}(s+d) = (1-p^{-1})^{-d} Z_{\mathcal{D}}(s,p) \ .$$

Furthermore, $\deg f_0 = d$, $\deg g_0 = \binom{d}{2}$.

2.2.2 Resolution

Once we have constructed the cone integral, the next step is to break the integral into a sum of integrals with monomial conditions. As mentioned in the Introduction, resolution of singularities gives us one way of doing this, and more importantly guarantees that this can always be done. Hironaka's proof of resolution of singularities of any singular variety defined over a field of characteristic 0 has been refined by Villamayor, Encinas, Bierstone and Milman, and Hauser amongst others to produce an explicit constructive procedure. In particular, Bodnár and Schicho have implemented a computer program to calculate resolutions. We refer the reader wanting to know more to Hauser's accessible article on resolution [34] and its comprehensive bibliography.

However, we shall not use resolution of singularities, for a number of reasons. Firstly, the computer program of Bodnár and Schicho works best in small dimensions, and we shall typically require resolutions of a polynomial with a large number of variables. Secondly, we shall find that we do not need to resolve all the singularities of the polynomial $F = \prod_{i=0}^l f_i(\mathbf{x}) g_i(\mathbf{x})$. Singularities lying outside V^*_p do not need to be resolved. Thirdly, there are 'tricks' that can be applied to simplify the polynomial conditions and speed up the process of decomposing the integral as a sum of monomial integrals. Some of these will take advantage of the fact we are working over $\mathbb{Q}_p$, whereas resolution is a general procedure for arbitrary fields of characteristic 0. A further disadvantage of resolution is the highly technical language it is most rigorously formulated in. We do not wish to alienate readers unfamiliar with this advanced machinery.

Therefore, we resolve singularities in an elementary and 'ad-hoc' manner. A collection of 'tricks' are used to simplify the conditions under the integral, and when the conditions can be simplified no further we bisect the integral. This bisection is achieved by choosing a pair of variables and splitting the domain of integration into two parts depending on which variable has the larger valuation. Further 'tricks' and bisections may then be necessary to reduce the integral into smaller and smaller pieces until all the pieces become monomial.

The idea of bisecting the integral as described above has its origins in the concept of a *blow-up*, an operation fundamental to the process of resolution of singularities. Indeed, we shall refer to our bisections as 'blow-ups'. Furthermore, we can use ideas originating from algebraic geometry to provide motivation for our choices of blow-ups. For example, suppose a non-monomial factor of one of the cone conditions is of the form $Px_j + Qx_k$ for variables x_j and x_k and nonzero polynomials P and Q. Let us also assume x_j and x_k have nontrivial integrand exponent or feature somewhere in a monomial condition. The polynomial F, being the product of all the cone data polynomials, has the factors x_j, x_k and $Px_j + Qx_k$, and therefore has a singularity with non-normal crossings at $x_j = x_k = 0$. A blow-up involving x_j and x_k will then replace this polynomial factor with $x_j(P + Qx'_k)$ (where $x_k = x_j x'_k$) or $x_k(Px'_j + Q)$ (where $x_j = x'_j x_k$) on the two sides of the blow-up. If P and Q are both independent of x_j and x_k, then this trick reduces the sum of the total degrees of the terms of the non-monomial factor. This trick is even more useful when one of x_j and x_k divides the other side of the condition, since the monomial factor x_j or x_k introduced above will cancel out. Algebraic geometry therefore provides inspiration for our method, but we do not totally rely on it.

Initially, the integrand and the left-hand side of each condition $v(f_i(\mathbf{x})) \le v(g_i(\mathbf{x}))$ is monomial, and this is something we preserve. For brevity we also write $f_i(\mathbf{x}) \mid g_i(\mathbf{x})$ instead of $v(f_i(\mathbf{x})) \le v(g_i(\mathbf{x}))$.

Examples of 'Resolution'

To illustrate the concepts in the previous section, we present two example calculation, where we construct the p-adic integral corresponding to a Lie ring and in each case apply some 'tricks' and blow-ups to split it into monomial integrals. The first example will illustrate the basic ideas, with some more unusual and less obvious tricks employed in the second.

For the first example, we shall choose to count all subrings of the Lie ring

$$L = \langle x_1, x_2, x_3, x_4, y_1, y_2 : [x_1, x_2] = y_1, [x_1, x_3] = y_2, [x_2, x_4] = y_2 \rangle .$$

In this case, the set $V_p^{\le}$ is given by

$$V_p^{\le} = \{ (m_{1,1}, m_{1,2} \ldots, m_{6,6}) \in \mathbb{Z}_p^{21} : f_i(\mathbf{x}) \le g_i(\mathbf{x}) \text{ for } 1 \le i \le 6 \} ,$$

where the six[1] conditions $f_i(\mathbf{x}) \mid g_i(\mathbf{x})$ are listed below:

---

[1] It is mere coincidence that there are six conditions in this case. Generally the number of conditions obtained bears no relation to the rank of the underlying Lie ring.

$$m_{5,5} \mid m_{1,1}m_{2,2} \,,$$

$$m_{6,6} \mid m_{1,2}m_{4,4} \,,$$

$$m_{6,6} \mid m_{2,2}m_{3,4} \,,$$

$$m_{6,6} \mid m_{2,2}m_{4,4} \,,$$

$$m_{6,6} \mid m_{1,1}m_{3,3} + m_{1,2}m_{3,4} \,,$$

$$m_{5,5}m_{6,6} \mid m_{1,1}m_{2,2}m_{5,6} - m_{1,1}m_{2,3}m_{5,5} - m_{1,2}m_{2,4}m_{5,5} + m_{1,4}m_{2,2}m_{5,5} \,.$$

These conditions are independent of $m_{1,3}$ and $m_{i,j}$ for $1 \le i \le 4$, $5 \le j \le 6$. For the sake of clarity, we shall relabel the remaining 12 variables as $a, b, \ldots, l$. Thus,

$$\zeta_{\bar{L},p}^{\le}(s) = (1 - p^{-1})^{-6} I \,,$$

where

$$I = \int_W |a|_p^{s-1} |d|_p^{s-2} |g|_p^{s-3} |i|_p^{s-4} |j|_p^{s-5} |l|_p^{s-6} \, \mathrm{d}\mu$$

and W is the subset of $(a, b, \ldots, l) \in \mathbb{Z}_p^{12}$ defined by the conditions

$$j \mid ad \,, \quad l \mid bi \,, \quad l \mid dh \,, \quad l \mid di \,, \quad l \mid ag + bh \,, \quad jl \mid adk - aej - bfj + cdj \,.$$

We perform a blow-up with l and d to remove the variable c. On one side of the blow-up it disappears altogether, on the other its coefficient dj divides the sum of the other terms of the polynomial:

1. $v(l) \le v(d)$: set $d = d'l$. The conditions $l \mid dh$ and $l \mid di$ become trivially true, and we can also remove the term $cd'jl$ from the last condition. Thus

$$I_1 = \int_{\substack{j \mid ad'l \\ l \mid bi \\ l \mid ag+bh \\ jl \mid ad'kl - aej - bfj}} |a|_p^{s-1} |d'|_p^{s-2} |g|_p^{s-3} |i|_p^{s-4} |j|_p^{s-5} |l|_p^{2s-7} \, \mathrm{d}\mu \,.$$

Note that the exponent of $|l|_p$ is $2s - 7$, as opposed to $2s - 8 = (s - 2) + (s - 6)$. The discrepancy is caused by the dilation of the measure that the change $d = d'l$ brings about. By dividing the l out of d, we have allowed d' to take a greater measure of values in $\mathbb{Z}_p$ than d. Hence we introduce a Jacobean $|l|_p$ into the integrand to balance out the dilation.

2. $v(l) > v(d)$: set $l = dl'$ with $v(l') \ge 1$. This then implies $l' \mid h$ and $l' \mid i$. To remove these two variable-divides-variable conditions, set $h = h'l'$ and $i = i'l'$.

$$I_2 = \int_{\substack{j \mid ad \\ d \mid bi' \\ dl' \mid ag+bh'l' \\ djl' \mid adk - aej - bfj + cdj \\ v(l') \ge 1}} |a|_p^{s-1} |d|_p^{2s-7} |g|_p^{s-3} |i|_p^{s-4} |j|_p^{s-5} |l'|_p^{s-4} \, \mathrm{d}\mu \,.$$

The last condition implies

$$dj \mid adk - aej - bfj \tag{2.6}$$

and thus $l \mid c + (adk - aej - bfj)/dj$, so we shall set $c = c' - (adk - aej - bfj)/dj$. After this substitution, the conditions no longer imply (2.6), so to avoid altering the value of the integral, we must explicitly enforce (2.6). We can also set $c' = c''l$ to remove the condition $l \mid c'$. Hence

$$I_2 = \int_{\substack{j \mid ad \\ d \mid bi' \\ dl' \mid ag + bh'l' \\ dj \mid adk - aej - bfj \\ v(l') \geq 1}} |a|_p^{s-1} |d|_p^{2s-7} |g|_p^{s-3} |i'|_p^{s-4} |j|_p^{s-5} |l'|_p^{2s-7} \, d\mu \ .$$

In both cases we have removed c or c'' from the conditions and the number of terms in the last condition has dropped from 4 to 3.

We play a similar trick on I_1 and I_2 to remove f. By a stroke of luck it turns out to also eliminate h from I_1 and h' from I_2:

1.1. $v(l) \leq v(b)$: set $b = b'l$. Terms $b'hl$ and $-b'fjl$ disappear from the last two conditions:

$$I_{1.1} = \int_{\substack{j \mid ad'l \\ l \mid ag \\ jl \mid a(d'kl - ej)}} |a|_p^{s-1} |d'|_p^{s-2} |g|_p^{s-3} |i|_p^{s-4} |j|_p^{s-5} |l|_p^{2s-6} \, d\mu \ .$$

1.2. $v(l) > v(b)$: set $l = bl'$ with $v(l') \geq 1$, and $i = i'l'$. Now $b \mid ag$ and $bj \mid a(bd'kl' - ej)$ are implied by the last two conditions, so we set $h = h'l - ag/b$ and $f = f'l + a(bd'kl' - ej)/bj$. Again, we must introduce explicitly the implied conditions.

$$I_{1.2} = \int_{\substack{j \mid abd'l' \\ b \mid ag \\ bj \mid a(bd'kl' - ej) \\ v(l') \geq 1}} |a|_p^{s-1} |b|_p^{2s-6} |d'|_p^{s-2} |g|_p^{s-3} |i'|_p^{s-4} |j|_p^{s-5} |l'|_p^{3s-8} \, d\mu \ .$$

2.1. $v(d) \leq v(b)$: set $b = b'd$:

$$I_{2.1} = \int_{\substack{j \mid ad \\ dl' \mid ag \\ dj \mid a(dk - ej) \\ v(l') \geq 1}} |a|_p^{s-1} |d|_p^{2s-6} |g|_p^{s-3} |i'|_p^{s-4} |j|_p^{s-5} |l'|_p^{2s-7} \, d\mu \ .$$

2.2. $v(d) > v(b)$: set $d = bd'$ with $v(d') \geq 1$, $i' = d'i''$. Also $bl' \mid ag$ and $bj \mid a(bd'k - ej)$, so we can set $h' = d'h' - ag/bl'$ and $f = df' + a(bd'k - ej)/bj$:

$$I_{2.2} = \int\limits_{\substack{j \mid abd' \\ bl' \mid ag \\ bj \mid a(bd'k - ej) \\ v(l') \geq 1 \\ v(d') \geq 1}} |a|_p^{s-1} |b|_p^{2s-6} |d'|_p^{3s-8} |g|_p^{s-3} |i''|_p^{s-4} |j|_p^{s-5} |l'|_p^{2s-7} \, d\mu \; .$$

All four of these integrals are very similar, and can be reduced to monomials in the same way. For simplicity we shall consider only $I_{1.1}$.

1.1.1. $v(j) \leq v(d'kl)$: in this case, $d'kl/j$ is an integer, so we may set $e = e' + d'kl/j$:

$$I_{1.1.1} = \int\limits_{\substack{j \mid ad'l \\ l \mid ag \\ j \mid d'kl \\ l \mid ae'}} |a|_p^{s-1} |d'|_p^{s-2} |g|_p^{s-3} |i|_p^{s-4} |j|_p^{s-5} |l|_p^{2s-6} \, d\mu \; .$$

1.1.2. $v(j) > v(d'kl)$: set $j = j'd'kl$ with $v(j') \geq 1$:

$$I_{1.1.2} = \int\limits_{\substack{j'k \mid a \\ l \mid ag \\ j'l \mid a(1 - ej') \\ v(j') \geq 1}} |a|_p^{s-1} |d'|_p^{2s-6} |g|_p^{s-3} |i|_p^{s-4} |j'|_p^{s-5} |k|_p^{s-4} |l|_p^{3s-10} \, d\mu \; .$$

Since $v(j') \geq 1$, $v(1 - ej') = 0$. Thus

$$I_{1.1.2} = \int\limits_{\substack{j'k \mid a \\ l \mid ag \\ j'l \mid a \\ v(j') \geq 1}} |a|_p^{s-1} |d'|_p^{2s-6} |g|_p^{s-3} |i|_p^{s-4} |j'|_p^{s-5} |k|_p^{s-4} |l|_p^{3s-10} \, d\mu \; .$$

In this case we can break up the initial integral into eight monomial integrals, however larger examples may need to be broken up into many more integrals. Evaluating these monomial integrals and summing gives us the local zeta function counting all subrings in $\mathfrak{g}_{6,4}$, which can be found below on p. 44.

The second example is more involved, and demonstrates some other tricks which sometimes come in useful. We count ideals in the free class-3 2-generator nilpotent Lie ring $F_{3,2}$. This has presentation

$$\langle x_1, x_2, y, z_1, z_2 : [x_1, x_2] = y, [x_1, y] = z_1, [x_2, y] = z_2 \rangle \; .$$

Now

$$I := \zeta^{\triangleleft}_{F_{3,2},p}(s) = (1 - p^{-1})^{-5} \int_W |m_{1,1}|_p^{s-1} \dots |m_{5,5}|_p^{s-5} \, d\mu \,,$$

where W is defined by the conjunction of the following conditions:

$$m_{3,3} \mid m_{1,1}\,, \quad m_{3,3} \mid m_{1,2}\,, \quad m_{3,3} \mid m_{2,2}\,, \quad m_{4,4} \mid m_{1,1}\,, \quad m_{4,4} \mid m_{3,3}\,,$$
$$m_{5,5} \mid m_{2,2}\,, \quad m_{5,5} \mid m_{2,3}\,, \quad m_{5,5} \mid m_{3,3}\,, \quad m_{3,3}m_{4,4} \mid m_{1,1}m_{3,4}\,,$$
$$m_{4,4}m_{5,5} \mid m_{3,3}m_{4,5}\,, \quad m_{3,3}m_{4,4} \mid m_{1,2}m_{3,4} - m_{1,3}m_{3,3}\,,$$
$$m_{3,3}m_{4,4} \mid m_{2,2}m_{3,4} - m_{2,3}m_{3,3}\,, \quad m_{4,4}m_{5,5} \mid m_{1,1}m_{4,5} - m_{1,2}m_{4,4}\,,$$
$$m_{3,3}m_{4,4}m_{5,5} \mid m_{1,2}m_{3,4}m_{4,5} - m_{1,2}m_{3,5}m_{4,4} - m_{1,3}m_{3,3}m_{4,5}\,,$$
$$m_{3,3}m_{4,4}m_{5,5} \mid m_{2,2}m_{3,4}m_{4,5} - m_{2,2}m_{3,5}m_{4,4} - m_{2,3}m_{3,3}m_{4,5}\,,$$
$$m_{3,3}m_{4,4}m_{5,5} \mid m_{1,1}m_{3,4}m_{4,5} - m_{1,1}m_{3,5}m_{4,4} - m_{1,3}m_{3,3}m_{4,4}\,.$$

We start by setting $m_{1,1} = m'_{1,1}m_{3,3}$, $m_{1,2} = m'_{1,2}m_{3,3}$, $m_{2,2} = m'_{2,2}m_{3,3}$, $m_{3,3} = m'_{3,3}m_{4,4}$ and $m_{2,3} = m'_{2,3}m_{5,5}$. Doing so 'uses up' five of the first eight conditions. These conditions, and the changes that eliminate them, are typical when calculating local ideal zeta functions. Variables $m_{1,4}$, $m_{1,5}$, $m_{2,4}$ and $m_{2,5}$ don't feature among the above conditions. Relabelling the remainder from a to k tells us that

$$I = (1 - p^{-1})^{-5} \int_{W'} |a|_p^{s-1} |d|_p^{s-2} |f|_p^{3s-3} |i|_p^{4s-6} |k|_p^{s-4} \, d\mu \,,$$

where W is the subset of all $(a, \dots, k) \in \mathbb{Z}_p^{11}$ satisfying

$$i \mid ag\,, \quad k \mid fi\,, \quad k \mid fj\,, \quad i \mid bg - c\,, \quad i \mid dg - ek\,, \quad ik \mid agj - ahi - ci\,,$$
$$ik \mid bgj - bhi - cj\,, \quad ik \mid dgj - dhi - ekj\,.$$

Our focus is on the conditions and how to perform blow-ups to reduce the conditions to monomials. We shall therefore neglect to track the changes to the integrand.

We started the last calculation by aiming to remove a variable from the integral. We cannot do the same here. Instead, we choose a blow-up between i and j. Note that each term of the right-hand side of each of the last three conditions above contains an i or a j. Where $v(i) \leq v(j)$, we set $i = i'j$ and then $h = h' + gj'$ to obtain that

$$W_1 := \left\{ (a, \dots, k) \in \mathbb{Z}_p^{11} : \begin{array}{llll} i \mid ag\,, & k \mid fi\,, & k \mid dh'\,, & i \mid bg - c\,, \\ i \mid dg - ek\,, & k \mid ah' + c\,, & k \mid bh' + cj' \end{array} \right\}.$$

A blow-up with k and c is the thing to do here. Where $v(k) \leq v(c)$, two of the binomial conditions drop to monomial and a blow-up with i and k will suffice to reduce to monomials. However, more interesting things happen when $v(k) > v(c)$. Firstly, let's set $k = ck'$ with $v(k') \geq 1$, and then set $j' = j''k - bh'/c$:

$$W_{1.2} := \left\{ (a, \ldots, k') \in \mathbb{Z}_p^{11} : \begin{array}{c} c \mid bh', \quad i \mid ag, \quad ck' \mid fi, \quad ck' \mid dh', \\ i \mid bg - c, \quad i \mid dg - eck', \quad ck' \mid ah' + c, \\ v(k') \geq 1 \end{array} \right\}.$$

Consider the last condition, $ck' \mid ah' + c$. Since $v(k') \geq 1$, $v(ck') > v(c)$. This implies that $v(ah') = v(c)$, so that $ah' \mid c$. Set $c = ac'h'$:

$$W_{1.2} = \left\{ (a, \ldots, k') \in \mathbb{Z}_p^{11} : \begin{array}{c} a \mid b, \quad i \mid ag, \quad ac'h'k' \mid fi, \quad ac'k' \mid d, \\ i \mid bg - ac'h', \quad i \mid dg - ac'eh'k', \\ c'k' \mid 1 + c', \quad v(k') \geq 1 \end{array} \right\}.$$

$c'k' \mid 1 + c'$ and $v(k') \geq 1$ imply that $c' \equiv -1 \pmod{p}$, in particular c' is a unit. We set $c' = c''k - 1$ as well as $b = ab'$ and $d = ad'k'$. After some tidying, we end with the following monomial conditions:

$$W_{1.2} = \left\{ (a, \ldots, k') \in \mathbb{Z}_p^{11} : i \mid ag, \quad ah'k' \mid fi, \quad i \mid ah', \quad v(k') \geq 1 \right\}.$$

We now return to the second half of the initial blow-up. We have

$$W_2 := \left\{ (a, \ldots, k) \in \mathbb{Z}_p^{11} : \begin{array}{c} k \mid fj, \quad i'j \mid ag, \quad i'j \mid bg - c, \quad j \mid dh - e''k, \\ k \mid d(g - hi'), \quad i'k \mid bg - bhi' - c, \\ i'k \mid ag - ahi' - ci', \quad v(i') \geq 1 \end{array} \right\}.$$

It is best not to do a blow-up at this point. Instead, we do a couple of changes of variable. Firstly, we set $g = g' + hi'$. Note that this change will make two conditions longer. Setting $c = c' + bg'$ and then $c' = c''i'k$ gives us the binomial conditions

$$W_2 = \left\{ (a, \ldots, k) \in \mathbb{Z}_p^{11} : \begin{array}{c} k \mid dg', \quad k \mid fj, \quad j \mid bh - c''k, \\ j \mid dh - e''k, \quad i'j \mid a(g' + hi'), \\ i'k \mid g'(a - bi'), \quad v(i') \geq 1 \end{array} \right\}.$$

A blow-up between j and k will remove the first two binomial conditions. It is then routine (although not trivial) to split the two parts into monomials. Evaluating the resulting monomial integrals and summing yields $\zeta_{F_{3,2},p}^{\triangleleft}(s)$, on p. 51.

2.2.3 Evaluating Monomial Integrals

A p-adic cone integral with monomial conditions can be expressed as a sum of integral points within a polyhedral cone in $\mathbb{R}^n$, and there are algorithms for evaluating such sums. One such example is the Elliott–MacMahon algorithm described in [54]. However, the second author considered an alternative approach, which appears to be more efficient for the monomial cone integrals arising from zeta functions of Lie rings, but is not guaranteed to terminate.

This approach is to continue applying 'blow-ups' to further decompose the monomial integrals until the conditions become trivial. One strategy for choosing blow-ups is to choose the two variables which appear most frequently on opposite sides of conditions without appearing on the same side. It is not difficult to automate this strategy, and in practice it has worked well, but it is not difficult to construct integrals for which this strategy will fail.

Most of the 'tricks' described in the previous section are aimed at reducing non-monomial conditions to monomials and so cannot be applied. The exception is that any conditions $f_i(\mathbf{x}) \mid g_i(\mathbf{x})$ where $g_i(\mathbf{x})$ is a single variable x_j can be removed by setting $x_j = x'_j f_i(\mathbf{x})$.

2.2.4 Summing the Rational Functions

The final stage is to sum the rational functions resulting from the trivial integrals. Whilst being the most elementary, it can also be the most computationally intensive. Given a perhaps large collection of rational functions in two variables, we must add them up. This sort of summation can easily be performed by a computer algebra system such as Maple or Magma. Indeed this is the approach used by Taylor [57]. However, we can make use of the fact that these rational functions are of the form

$$\frac{P(X,Y)}{\prod_{i=1}^{r}(1 - X^{a_i}Y^{b_i})}$$

for some bivariate polynomial $P(X,Y)$ with $a_i, b_i \in \mathbb{N}$. Typically, many of the factors of the denominator will cancel out once all the terms have been summed. If there are a large number of rational functions, it is advantageous to pick factors we believe will cancel, sum all the rational functions with this factor in the denominator and then hope that the factor cancels in this partial sum. We may then replace the rational functions we summed with the partial sum and continue. With less factors in the denominator, the remaining rational functions should sum more quickly.

2.3 Explicit Examples

For the rest of this chapter we give explicit expressions for the local zeta functions of many Lie rings. We also list the functional equation satisfied by these local zeta functions (where applicable), and the abscissae of convergence of the corresponding global zeta functions. We also give the order of the pole on the abscissa of convergence when it is not a simple pole. Unless we state otherwise, the local zeta functions we present are uniform, i.e. are given by the same rational function in p and p^{-s} for all primes p.

It may be noted that there are more zeta functions counting ideals than all subrings. There are usually more conditions under a p-adic integral counting ideals than under one counting all subrings, but the cone conditions for counting ideals are simpler.

The calculations involved are frequently long and tedious and were often performed with computer assistance. Therefore we shall not provide proofs of the calculations. This contrasts with the approach of Taylor [57], who does provide proofs of his calculations in his thesis. One such proof runs to 40 pages. There are several zeta functions of comparable or greater complexity presented in this chapter, and we simply don't have the space to present the proofs. Nonetheless we believe that all the zeta functions listed below are correct. In particular, there shouldn't have been any errors in transcription since the LATEX source for each zeta function was generated from the computer calculations.

The advent of computer calculations has also led to zeta functions with the numerator and denominator of large degree. We have confined some of the larger numerator polynomials to Appendix A. However, there are four excessively large polynomials which we have chosen not to include since we do not feel the extra 23 pages they would require would be justified. Further details may be obtained from the authors on request.

Many of the examples will satisfy a functional equation of the form

$$\zeta^*_{L,p}(s)\big|_{p \to p^{-1}} = (-1)^c p^{b-as} \zeta^*_{L,p}(s) \tag{2.7}$$

for all but perhaps finitely many primes p. However, there are a small number that don't. When we say that a local zeta function 'satisfies no functional equation', we mean that it satisfies no functional equation of the form (2.7).

The Lie rings we shall be considering can be presented conveniently by giving a basis and the nontrivial Lie brackets of the basis elements. Most of these Lie brackets will be zero, so we make the convention that, up to antisymmetry, any Lie bracket not listed is zero.

2.4 Free Abelian Lie Rings

Let $L = \mathbb{Z}^d$, the free abelian Lie ring of rank d. Then

$$\zeta^{\triangleleft}_L(s) = \zeta^{\leq}_L(s) = \prod_{i=0}^{d-1} \zeta(s-i) \, ,$$

where $\zeta(s)$ is the Riemann zeta function. Hence this function is meromorphic on the whole of $\mathbb{C}$. In particular, the Tauberian Theorem (Theorem 1.8) mentioned in the Introduction allows us to deduce that if a_n is the number of subgroups of index n in $\mathbb{Z}^2$, then

$$\sum_{i=1}^{n} a_i \sim \frac{\pi^2}{12} n^2 \, ,$$

a result which seems remarkably difficult to obtain without the machinery of zeta functions.

In [22] it is shown that for any finite extension G of the free abelian group $\mathbb{Z}^d$, the zeta functions $\zeta_G^*(s)$ are all meromorphic. This is proved by relating the zeta functions to classical L-functions that arise in the work of Solomon, Bushnell and Reiner. The zeta functions of the 17 plane crystallographic groups, also known as the 'wallpaper groups', were calculated by McDermott and are listed in [22].

We shall see that many of the zeta functions have a factor similar to the local factor of $\zeta_{\mathbb{Z}^d}(s)$. It is therefore convenient to use the notation

$$\zeta_{\mathbb{Z}^n,p}(s) = \prod_{i=0}^{n-1} \zeta_p(s-i) , \tag{2.8}$$

where $\zeta_p(s) = (1-p^{-s})^{-1}$ is the p-factor of the Riemann zeta function.

2.5 Heisenberg Lie Ring and Variants

Let $\mathcal{H}$ be the free class two, two generator nilpotent Lie ring. This is the Lie ring of strictly upper-triangular matrices

$$U_3(\mathbb{Z}) = \begin{pmatrix} 0 & \mathbb{Z} & \mathbb{Z} \\ 0 & 0 & \mathbb{Z} \\ 0 & 0 & 0 \end{pmatrix} .$$

It is given by the presentation

$$\mathcal{H} = \langle x, y, z : [x,y] = z \rangle ,$$

where, as mentioned above, $[x,z] = [y,z] = 0$. For $n \geq 2$, let $\mathcal{H}^n$ denote the direct product of n copies of the Heisenberg Lie ring.

Theorem 2.3 ([32]).

$$\zeta_{\mathcal{H},p}^{\triangleleft}(s) = \zeta_{\mathbb{Z}^2,p}(s)\zeta_p(3s-2) ,$$
$$\zeta_{\mathcal{H},p}^{\leq}(s) = \zeta_{\mathbb{Z}^2,p}(s)\zeta_p(2s-2)\zeta_p(2s-3)\zeta_p(3s-3)^{-1} .$$

These zeta functions satisfy the functional equations

$$\zeta_{\mathcal{H},p}^{\triangleleft}(s)\Big|_{p\to p^{-1}} = -p^{3-5s}\zeta_{\mathcal{H},p}^{\triangleleft}(s) ,$$
$$\zeta_{\mathcal{H},p}^{\leq}(s)\Big|_{p\to p^{-1}} = -p^{3-3s}\zeta_{\mathcal{H},p}^{\leq}(s) .$$

The corresponding global zeta functions have abscissa of convergence $\alpha_{\mathcal{H}}^{\triangleleft} = \alpha_{\mathcal{H}}^{\leq} = 2$, with $\zeta_{\mathcal{H}}^{\leq}(s)$ having a double pole at $s = 2$.

Theorem 2.4 ([32, 57]).

$$\zeta_{\mathcal{H}^2,p}^{\lhd}(s) = \zeta_{\mathbb{Z}^4,p}(s)\zeta_p(3s-4)^2\zeta_p(5s-5)\zeta_p(5s-4)^{-1},$$

$$\zeta_{\mathcal{H}^2,p}^{\leq}(s) = \zeta_{\mathbb{Z}^4,p}(s)\zeta_p(2s-4)^2\zeta_p(2s-5)^2\zeta_p(3s-5)\zeta_p(3s-7)\zeta_p(3s-8)$$
$$\times\, W_{\mathcal{H}^2}^{\leq}(p,p^{-s}),$$

where $W_{\mathcal{H}^2}^{\leq}(X,Y)$ is

$$1 - X^4Y^3 - 3X^5Y^3 - X^7Y^3 + X^5Y^4 - X^9Y^4 - X^8Y^5 + 3X^9Y^5 - 2X^{11}Y^5$$
$$+ X^{10}Y^6 + 3X^{11}Y^6 + 3X^{12}Y^6 + 2X^{13}Y^6 + X^{14}Y^6 - X^{14}Y^7 + X^{15}Y^7$$
$$- X^{14}Y^8 + X^{15}Y^8 - X^{15}Y^9 - 2X^{16}Y^9 - 3X^{17}Y^9 - 3X^{18}Y^9 - X^{19}Y^9$$
$$+ 2X^{18}Y^{10} - 3X^{20}Y^{10} + X^{21}Y^{10} + X^{20}Y^{11} - X^{24}Y^{11} + X^{22}Y^{12}$$
$$+ 3X^{24}Y^{12} + X^{25}Y^{12} - X^{29}Y^{15}.$$

These zeta functions satisfy the functional equations

$$\zeta_{\mathcal{H}^2,p}^{\lhd}(s)\Big|_{p\to p^{-1}} = p^{15-10s}\zeta_{\mathcal{H}^2,p}^{\lhd}(s),$$

$$\zeta_{\mathcal{H}^2,p}^{\leq}(s)\Big|_{p\to p^{-1}} = p^{15-6s}\zeta_{\mathcal{H}^2,p}^{\leq}(s).$$

The corresponding global zeta functions have abscissa of convergence $\alpha_{\mathcal{H}^2}^{\lhd} = \alpha_{\mathcal{H}^2}^{\leq} = 4$.

Theorem 2.5 ([57]).

$$\zeta_{\mathcal{H}^3,p}^{\lhd}(s) = \zeta_{\mathbb{Z}^6,p}(s)\zeta_p(3s-6)^3\zeta_p(5s-7)\zeta_p(7s-8)\zeta_p(8s-14)W_{\mathcal{H}^3}^{\lhd}(p,p^{-s}),$$

where $W_{\mathcal{H}^3}^{\lhd}(X,Y)$ is

$$1 - 3X^6Y^5 + 2X^7Y^5 + X^6Y^7 - 2X^7Y^7 + X^{12}Y^8 - 2X^{13}Y^8 + 2X^{13}Y^{12}$$
$$- X^{14}Y^{12} + 2X^{19}Y^{13} - X^{20}Y^{13} - 2X^{19}Y^{15} + 3X^{20}Y^{15} - X^{26}Y^{20}.$$

This zeta function satisfies the functional equation

$$\zeta_{\mathcal{H}^3,p}^{\lhd}(s)\Big|_{p\to p^{-1}} = -p^{36-15s}\zeta_{\mathcal{H}^3,p}^{\lhd}(s).$$

The corresponding global zeta function has abscissa of convergence $\alpha_{\mathcal{H}^3}^{\lhd} = 6$.

Theorem 2.6 ([64]).

$$\zeta_{\mathcal{H}^4,p}^{\lhd}(s) = \zeta_{\mathbb{Z}^8,p}(s)\zeta_p(3s-8)^4\zeta_p(5s-9)\zeta_p(7s-10)\zeta_p(8s-18)\zeta_p(9s-11)$$
$$\times\, \zeta_p(10s-20)\zeta_p(11s-27)W_{\mathcal{H}^4}^{\lhd}(p,p^{-s}),$$

where the polynomial $W_{\mathcal{H}^4}^{\triangleleft}(X, Y)$ is given in Appendix A on p. 179. This zeta function satisfies the functional equation

$$\zeta_{\mathcal{H}^4, p}^{\triangleleft}(s)\Big|_{p \to p^{-1}} = p^{66-20s} \zeta_{\mathcal{H}^4, p}^{\triangleleft}(s) \ .$$

The corresponding global zeta function has abscissa of convergence $\alpha_{\mathcal{H}^4}^{\triangleleft} = 8$.

Theorem 2.7. *Let $(K : \mathbb{Q}) = 2$, R be the ring of integers of K and $L = U_3(R)$. Then*

1. *If p is inert (of which there are possibly infinitely many) then*

$$\zeta_{L,p}^{\triangleleft}(s) = \zeta_{\mathbb{Z}^4, p}(s)\zeta_p(5s - 5)\zeta_p(6s - 8)(1 + p^{4-5s}) \ .$$

2. *If p is ramified (of which there are only finitely many) then*

$$\zeta_{L,p}^{\triangleleft}(s) = \zeta_{\mathbb{Z}^4, p}(s)\zeta_p(3s - 4)\zeta_p(5s - 5) \ .$$

3. *If p is split then $U_3(R \otimes \mathbb{Z}_p) = U_3(\mathbb{Z}_p) \times U_3(\mathbb{Z}_p)$ and we already have a calculation of this factor from Theorem 2.4 above.*

For all split or inert primes p, this zeta function satisfies the functional equation

$$\zeta_{L,p}^{\triangleleft}(s)\Big|_{p \to p^{-1}} = p^{15-10s} \zeta_{L,p}^{\triangleleft}(s) \ ,$$

whereas for p ramified,

$$\zeta_{L,p}^{\triangleleft}(s)\Big|_{p \to p^{-1}} = p^{15-12s} \zeta_{L,p}^{\triangleleft}(s) \ .$$

The corresponding global zeta function has abscissa of convergence $\alpha_L^{\triangleleft} = 4$.

Taking the Euler product of all these factors we can represent the global zeta function in terms of the Riemann zeta function and the Dedekind zeta function $\zeta_K(s)$ of the underlying quadratic number field K (as observed in Corollary 8.2 of [32]):

Corollary 2.8.

$$\zeta_L^{\triangleleft}(s) = \zeta_{\mathbb{Z}^4}(s)\zeta(5s - 4)\zeta(5s - 5)\zeta_K(3s - 4)/\zeta_K(5s - 4) \ . \tag{2.9}$$

Theorem 2.9 ([32, 57]). *Let $L = U_3(R_3)$ be the Lie ring of 3×3 upper triangular matrices over the ring of integers R_3 of a algebraic number field K of degree 3 over $\mathbb{Q}$.*

1. *If p is inert in R_3, then*

$$\zeta_{L,p}^{\triangleleft}(s) = \zeta_{\mathbb{Z}^6, p}(s)\zeta_p(7s - 8)\zeta_p(8s - 14)\zeta_p(9s - 18)W_{L,\text{in}}^{\triangleleft}(p, p^{-s})$$

where

$$W_{L,\text{in}}^{\triangleleft}(X, Y) = 1 + X^6Y^7 + X^7Y^7 + X^{12}Y^8 + X^{13}Y^8 + X^{19}Y^{15} \ .$$

2. If p ramifies completely in R_3 (i.e. if $(p) = \mathfrak{p}^3$ for some prime ideal $\mathfrak{p}$), then

$$\zeta_{L,p}^{\triangleleft}(s) = \zeta_{\mathbb{Z}^6,p}(s)\zeta_p(3s - 6)\zeta_p(7s - 8)\zeta_p(8s - 14)(1 + p^{7-5s}) .$$

3. If p ramifies partially in R_3 (i.e. if $(p) = \mathfrak{p}^2\mathfrak{q}$ for prime ideals $\mathfrak{p} \neq \mathfrak{q}$),

$$\zeta_{L,p}^{\triangleleft}(s) = \zeta_{\mathbb{Z}^6,p}(s)\zeta_p(3s - 6)^2\zeta_p(5s - 7)\zeta_p(7s - 8)\zeta_p(8s - 14)W_{L,\mathrm{rp}}^{\triangleleft}(p, p^{-s}),$$

where

$$\begin{aligned}W_{L,\mathrm{rp}}^{\triangleleft}(X, Y) = {}& 1 - X^6Y^5 + X^7Y^5 - X^7Y^7 - X^{13}Y^8 + X^{13}Y^{10} \\ & - X^{14}Y^{10} + X^{20}Y^{15} .\end{aligned}$$

4. If p splits completely in R_3:

$$\zeta_{L,p}^{\triangleleft}(s) = \zeta_{\mathbb{Z}^6,p}(s)\zeta_p(3s - 6)^3\zeta_p(5s - 7)\zeta_p(7s - 8)\zeta_p(8s - 14)W_{L,\mathrm{sc}}^{\triangleleft}(p, p^{-s}),$$

where $W_{L,\mathrm{sc}}^{\triangleleft} = W_{\mathcal{H}^3}^{\triangleleft}(X, Y)$ given above on p. 35.

5. If p splits partially in R_3 (i.e. $(p) = \mathfrak{p}\mathfrak{q}$ for prime ideals $\mathfrak{p} \neq \mathfrak{q}$):

$$\zeta_{L,p}^{\triangleleft}(s) = \zeta_{\mathbb{Z}^6,p}(s)\zeta_p(3s - 6)\zeta_p(5s - 7)\zeta_p(7s - 8)\zeta_p(6s - 12)\zeta_p(8s - 14)$$
$$\times W_{L,\mathrm{sp}}^{\triangleleft}(p, p^{-s}) ,$$

where

$$\begin{aligned}W_{L,\mathrm{sp}}^{\triangleleft}(X, Y) = {}& 1 + X^6Y^5 - X^6Y^7 - X^{12}Y^8 - X^{14}Y^{12} - X^{20}Y^{13} \\ & + X^{20}Y^{15} + X^{26}Y^{20} .\end{aligned}$$

For all primes that do not ramify, this zeta function satisfies the functional equation

$$\zeta_{L,p}^{\triangleleft}(s)\Big|_{p \to p^{-1}} = -p^{36-15s}\zeta_{L,p}^{\triangleleft}(s) .$$

The corresponding global function has abscissa of convergence $\alpha_L^{\triangleleft} = 6$.

Remark 2.10.   1. Cases 3 and 5 can only occur if the field K is not a normal extension of $\mathbb{Q}$.

2. As with the case with a quadratic number field, the p-local normal zeta function does satisfy a functional equation even when p ramifies. If f_p is the ramification degree of p in K, then

$$\zeta_{L,p}^{\triangleleft}(s)\Big|_{p \to p^{-1}} = -p^{36-(13+2f_p)s}\zeta_{L,p}^{\triangleleft}(s)$$

for all primes p.

It is possible to write the global zeta function of L in terms of Riemann zeta functions, the zeta function of the number field and Euler products of these two variable polynomials. However, the end result is not as neat as (2.9):

Proposition 2.11. *If $(K : \mathbb{Q}) = 3$, R is the ring of integers of K and $L = U_3(R)$ then*

$$\zeta_L^\triangleleft(s) = \zeta_{\mathbb{Z}^6}(s)\zeta(5s - 7)\zeta(7s - 8)\zeta(8s - 14)\zeta_K(3s - 6)\prod_p W_{L,p}^\triangleleft(p, p^{-s}) \,,$$

where

$$W_{L,p}^\triangleleft(X,Y) = \begin{cases} W_{L,\mathrm{in}}^\triangleleft(X,Y)(1 - X^7Y^5) & \textit{if } p \textit{ is inert in } R, \\ 1 - X^{14}Y^{10} & \textit{if } p \textit{ ramifies completely in } R, \\ W_{L,\mathrm{rp}}^\triangleleft(X,Y) & \textit{if } p \textit{ ramifies partially in } R, \\ W_{L,\mathrm{sc}}^\triangleleft(X,Y) & \textit{if } p \textit{ splits completely in } R, \\ W_{L,\mathrm{sp}}^\triangleleft(X,Y) & \textit{if } p \textit{ splits partially in } R. \end{cases}$$

2.6 Grenham's Lie Rings

The next examples are calculations made by Grenham in his D.Phil. thesis [28] of zeta functions of Lie rings $\mathcal{G}_n$ with the following presentation:

$$\mathcal{G}_n = \langle z, x_1, \dots, x_{n-1}, y_1, \dots, y_{n-1} : [z, x_i] = y_i \ (1 \le i \le n - 1)\rangle \,.$$

These Lie rings are class-2 nilpotent. $\mathcal{G}_2 \cong \mathcal{H}$, the Heisenberg Lie ring again. Grenham calculated $\zeta_{\mathcal{G}_n,p}^\triangleleft(s)$ and $\zeta_{\mathcal{G}_n,p}^{\le}(s)$ for $n \le 5$. They all have the form of products of local Riemann zeta functions together with one of the palindromic polynomials.

Theorem 2.12 ([32, 28]).

$$\zeta_{\mathcal{G}_3,p}^\triangleleft(s) = \zeta_{\mathbb{Z}^3,p}(s)\zeta_p(3s - 3)^2\zeta_p(3s - 4)\zeta_p(5s - 6)\zeta_p(6s - 6)^{-1} \,,$$
$$\zeta_{\mathcal{G}_3,p}^{\le}(s) = \zeta_{\mathbb{Z}^3,p}(s)\zeta_p(2s - 4)\zeta_p(2s - 5)\zeta_p(3s - 6)W_{\mathcal{G}_3}^{\le}(p, p^{-s}) \,,$$

where

$$W_{\mathcal{G}_3}^{\le}(X,Y) = 1 + X^3Y^2 + X^4Y^2 - X^4Y^3 - X^5Y^3 - X^8Y^5 \,.$$

These zeta functions satisfy the functional equations

$$\zeta_{\mathcal{G}_3,p}^\triangleleft(s)\Big|_{p \to p^{-1}} = -p^{10-8s}\zeta_{\mathcal{G}_3,p}^\triangleleft(s) \,,$$
$$\zeta_{\mathcal{G}_3,p}^{\le}(s)\Big|_{p \to p^{-1}} = -p^{10-5s}\zeta_{\mathcal{G}_3,p}^{\le}(s) \,.$$

The corresponding global zeta functions have abscissa of convergence $\alpha_{\mathcal{G}_3}^\triangleleft = \alpha_{\mathcal{G}_3}^{\le} = 3$, with $\zeta_{\mathcal{G}_3}^{\le}(s)$ having a double pole at $s = 3$.

Theorem 2.13 ([28]).

$$\zeta_{\mathcal{G}_4,p}^{\triangleleft}(s) = \zeta_{\mathbb{Z}^4,p}(s)\zeta_p(3s-6)\zeta_p(5s-10)\zeta_p(7s-12)W_{\mathcal{G}_4}^{\triangleleft}(p,p^{-s}),$$

where

$$W_{\mathcal{G}_4}^{\triangleleft}(X,Y) = 1 + X^4Y^3 + X^5Y^3 + X^8Y^5 + X^9Y^5 + X^{13}Y^8,$$

and

$$\zeta_{\mathcal{G}_4,p}^{\leq}(s) = \zeta_{\mathbb{Z}^4,p}(s)\zeta_p(2s-5)\zeta_p(2s-6)\zeta_p(2s-7)\zeta_p(3s-10)\zeta_p(4s-12)$$
$$\times\, W_{\mathcal{G}_4}^{\leq}(p,p^{-s})$$

where $W_{\mathcal{G}_4}^{\leq}(X,Y)$ is

$$1 + X^4Y^2 + X^5Y^2 + X^6Y^2 - X^5Y^3 - X^6Y^3 - X^7Y^3 + X^8Y^3 + X^9Y^3$$
$$- X^9Y^4 - X^{10}Y^4 - X^{11}Y^4 - X^{14}Y^6 - X^{15}Y^6 - X^{16}Y^6 + X^{16}Y^7 + X^{17}Y^7$$
$$- X^{18}Y^7 - X^{19}Y^7 - X^{20}Y^7 + X^{19}Y^8 + X^{20}Y^8 + X^{21}Y^8 + X^{25}Y^{10}.$$

These zeta functions satisfy the functional equations

$$\zeta_{\mathcal{G}_4,p}^{\triangleleft}(s)\Big|_{p\to p^{-1}} = -p^{21-11s}\zeta_{\mathcal{G}_4,p}^{\triangleleft}(s),$$
$$\zeta_{\mathcal{G}_4,p}^{\leq}(s)\Big|_{p\to p^{-1}} = -p^{21-7s}\zeta_{\mathcal{G}_4,p}^{\leq}(s).$$

The corresponding global zeta functions have abscissa of convergence $\alpha_{\mathcal{G}_4}^{\triangleleft} = \alpha_{\mathcal{G}_4}^{\leq} = 4$, with $\zeta_{\mathcal{G}_4}^{\leq}(s)$ having a double pole at $s = 4$.

Theorem 2.14 ([28]).

$$\zeta_{\mathcal{G}_5,p}^{\triangleleft}(s) = \zeta_{\mathbb{Z}^5,p}(s)\zeta_p(3s-8)\zeta_p(5s-14)\zeta_p(7s-18)\zeta_p(9s-20)W_{\mathcal{G}_5}^{\triangleleft}(p,p^{-s})$$

where $W_{\mathcal{G}_5}^{\triangleleft}(X,Y)$ is

$$1 + X^5Y^3 + X^6Y^3 + X^7Y^3 + X^{10}Y^5 + X^{11}Y^5 + 2X^{12}Y^5 + X^{13}Y^5 + X^{15}Y^7$$
$$+ X^{16}Y^7 + X^{17}Y^7 + X^{17}Y^8 + X^{18}Y^8 + X^{19}Y^8 + X^{21}Y^{10} + 2X^{22}Y^{10}$$
$$+ X^{23}Y^{10} + X^{24}Y^{10} + X^{27}Y^{12} + X^{28}Y^{12} + X^{29}Y^{12} + X^{34}Y^{15},$$

and

$$\zeta_{\mathcal{G}_5,p}^{\leq}(s) = \zeta_{\mathbb{Z}^5,p}(s)\zeta_p(2s-6)\zeta_p(2s-8)\zeta_p(2s-9)\zeta_p(3s-14)\zeta_p(4s-18)$$
$$\times\, \zeta_p(5s-20)\zeta_p(s-2)^{-1}W_{\mathcal{G}_5}^{\leq}(p,p^{-s}),$$

where $W_{\mathcal{G}_5}^{\leq}(X,Y)$ is

$$1 + X^2Y + X^4Y^2 + X^5Y^2 + X^6Y^2 + 2X^7Y^2 + X^8Y^2 + X^9Y^3 + 2X^{10}Y^3$$
$$+ X^{11}Y^3 + 2X^{12}Y^3 + X^{13}Y^3 + X^{12}Y^4 + 2X^{14}Y^4 + 2X^{15}Y^4 + X^{16}Y^4$$
$$+ X^{17}Y^4 + 2X^{17}Y^5 + X^{18}Y^5 + 2X^{19}Y^5 + X^{20}Y^5 - X^{18}Y^6 - X^{20}Y^6$$
$$+ X^{21}Y^6 + 2X^{22}Y^6 + 2X^{23}Y^6 + 2X^{24}Y^6 + X^{25}Y^6 - X^{22}Y^7 - 2X^{23}Y^7$$
$$- 2X^{24}Y^7 - 2X^{25}Y^7 - X^{26}Y^7 + X^{27}Y^7 + X^{29}Y^7 - X^{27}Y^8 - 2X^{28}Y^8$$
$$- X^{29}Y^8 - 2X^{30}Y^8 - X^{30}Y^9 - X^{31}Y^9 - 2X^{32}Y^9 - 2X^{33}Y^9 - X^{35}Y^9$$
$$- X^{34}Y^{10} - 2X^{35}Y^{10} - X^{36}Y^{10} - 2X^{37}Y^{10} - X^{38}Y^{10} - X^{39}Y^{11}$$
$$- 2X^{40}Y^{11} - X^{41}Y^{11} - X^{42}Y^{11} - X^{43}Y^{11} - X^{45}Y^{12} - X^{47}Y^{13} \ .$$

These zeta functions satisfy the functional equations

$$\zeta^{\triangleleft}_{\mathcal{G}_5,p}(s)\Big|_{p \to p^{-1}} = -p^{36-14s}\zeta^{\triangleleft}_{\mathcal{G}_5,p}(s) \ ,$$
$$\zeta^{\leq}_{\mathcal{G}_5,p}(s)\Big|_{p \to p^{-1}} = -p^{36-9s}\zeta^{\leq}_{\mathcal{G}_5,p}(s) \ .$$

The corresponding global zeta functions have abscissa of convergence $\alpha^{\triangleleft}_{\mathcal{G}_5} = \alpha^{\leq}_{\mathcal{G}_5} = 5$, *with* $\zeta^{\leq}_{\mathcal{G}_5}(s)$ *having a triple pole at* $s = 5$.

In [61], Voll has given an explicit expression for $\zeta^{\triangleleft}_{\mathcal{G}_n,p}(s)$, and in a forthcoming paper, gives a similar expression for $\zeta^{\leq}_{\mathcal{G}_n,p}(s)$. In particular, he proves that

Theorem 2.15. *Let* $n > 1$. *Then for all primes* p, $\zeta^{\triangleleft}_{\mathcal{G}_n,p}(s)$ *and* $\zeta^{\leq}_{\mathcal{G}_n,p}(s)$ *satisfy the functional equations*

$$\zeta^{\triangleleft}_{\mathcal{G}_n,p}(s)\Big|_{p \to p^{-1}} = -p^{\binom{2n-1}{2}-(3n-1)s}\zeta^{\triangleleft}_{\mathcal{G}_n,p}(s) \ ,$$
$$\zeta^{\leq}_{\mathcal{G}_n,p}(s)\Big|_{p \to p^{-1}} = -p^{\binom{2n-1}{2}-(2n-1)s}\zeta^{\leq}_{\mathcal{G}_n,p}(s) \ .$$

Grenham proved that the abscissa of convergence of $\zeta^{\triangleleft}_{\mathcal{G}_n}(s)$ is n. Voll gives in [61] an expression for the abscissa of convergence of $\zeta^{\leq}_{\mathcal{G}_n}(s)$, which agrees with an expression previously derived by Paajanen. In particular, $\alpha^{\leq}_{\mathcal{G}_6}(s) = 19/3$.

2.7 Free Class-2 Nilpotent Lie Rings

Let $F_{2,n}$ denote the free nilpotent Lie ring of class two on n generators. $F_{2,2}$ is the Heisenberg Lie ring once again.

2.7.1 Three Generators

Theorem 2.16 ([32, 57]). *Let the Lie ring* $F_{2,3}$ *have presentation*

$$\langle x_1, x_2, x_3, y_1, y_2, y_3 : [x_1, x_2] = y_1, [x_1, x_3] = y_2, [x_2, x_3] = y_3 \rangle \ .$$

Then

$$\zeta^{\triangleleft}_{F_{2,3},p}(s) = \zeta_{\mathbb{Z}^3,p}(s)\zeta_p(3s - 5)\zeta_p(5s - 8)\zeta_p(6s - 9)W^{\triangleleft}_{F_{2,3}}(p, p^{-s}) \ ,$$

where

$$W^{\triangleleft}_{F_{2,3}}(X, Y) = 1 + X^3Y^3 + X^4Y^3 + X^6Y^5 + X^7Y^5 + X^{10}Y^8 \ ,$$

and

$$\zeta^{\leq}_{F_{2,3},p}(s) = \zeta_{\mathbb{Z}^3,p}(s)\zeta_p(2s - 4)\zeta_p(2s - 5)\zeta_p(2s - 6)\zeta_p(3s - 6)\zeta_p(3s - 7)$$
$$\times \zeta_p(3s - 8)\zeta_p(4s - 8)^{-1}W^{\leq}_{F_{2,3}}(p, p^{-s}) \ ,$$

where $W^{\leq}_{F_{2,3}}(X, Y)$ is

$$1 + X^3Y^2 + X^4Y^2 + X^5Y^2 - X^4Y^3 - X^5Y^3 - X^6Y^3 - X^7Y^4 - X^9Y^4$$
$$- X^{10}Y^5 - X^{11}Y^5 - X^{12}Y^5 + X^{11}Y^6 + X^{12}Y^6 + X^{13}Y^6 + X^{16}Y^8 \ .$$

These zeta functions satisfy the functional equations

$$\zeta^{\triangleleft}_{F_{2,3},p}(s)\Big|_{p \to p^{-1}} = p^{15-9s}\zeta^{\triangleleft}_{F_{2,3},p}(s) \ ,$$
$$\zeta^{\leq}_{F_{2,3},p}(s)\Big|_{p \to p^{-1}} = p^{15-6s}\zeta^{\leq}_{F_{2,3},p}(s) \ .$$

The corresponding global zeta functions have abscissa of convergence $\alpha^{\triangleleft}_{F_{2,3}} = 3$, $\alpha^{\leq}_{F_{2,3}} = 7/2$.

The zeta function counting all subrings is interesting since the abscissa of convergence is not an integer and is strictly greater than the rank of the abelianisation of G. This was the first such example calculated at nilpotency class 2.

2.7.2 n Generators

In [62], Voll gives an explicit formulae for the local ideal zeta functions of $F_{2,n}$ for all n. We shall not replicate Voll's explicit formulae for these functions, but we shall state some corollaries he deduces. Put $h(n) = \frac{1}{2}n(n + 1)$, the rank of $F_{2,n}$.

Corollary 2.17. *The local zeta functions $\zeta^{\triangleleft}_{F_{2,n},p}(s)$ are uniform, i.e. are given by the same rational function in p and p^{-s} for all primes p.*

Corollary 2.18. *The local ideal zeta function of $F_{2,n}$ satisfies the local functional equation*

$$\zeta^{\triangleleft}_{F_{2,n},p}(s)\Big|_{p\to p^{-1}} = (-1)^{h(n)}p^{\binom{h(n)}{2}-(h(n)+n)s}\zeta^{\triangleleft}_{F_{2,n},p}(s)$$

for all primes p.

Corollary 2.19. *The abscissa of convergence of $\zeta^{\triangleleft}_{F_{2,n}}(s)$ is*

$$\alpha^{\triangleleft}_{F_{2,n}} = \max\left\{n, \frac{\left(\binom{n}{2}-j\right)(n+j)+1}{h(n)-j} \,\middle|\, j \in \{1,\ldots,\binom{n}{2}-1\}\right\}$$

and $\zeta^{\triangleleft}_{F_{2,n}}(s)$ has a simple pole at $s = \alpha^{\triangleleft}_{F_{2,n}}$.

In particular, $F_{2,5}$ has abscissa of convergence $\alpha^{\triangleleft}_{F_{2,5}} = 51/10$. Indeed, this is the first Lie ring whose local ideal zeta function is known to have abscissa of convergence strictly greater than the rank of the abelianisation.

2.8 The 'Elliptic Curve Example'

Theorem 2.20 ([60]). *Let E denote the elliptic curve $y^2 = x^3 - x$. Define the nilpotent Lie ring L_E by the presentation*

$$L_E = \left\langle x_1,\ldots,x_6,y_1,y_2,y_3 : \begin{array}{l} [x_1,x_4]=y_3, [x_1,x_5]=y_1, [x_1,x_6]=y_2, \\ [x_2,x_4]=y_1, [x_2,x_5]=y_3, \\ [x_3,x_4]=y_2, [x_3,x_6]=y_1 \end{array} \right\rangle.$$

Then, for all but finitely many primes p, the local zeta function of L_E is given by

$$\zeta^{\triangleleft}_{L_E,p}(s) = \zeta_{\mathbb{Z}^6,p}(s)\zeta_p(5s-7)\zeta_p(7s-8)\zeta_p(9s-18)\zeta_p(8s-14)$$
$$\times (P_1(p,p^{-s}) + |E(\mathbb{F}_p)|P_2(p,p^{-s})),$$

where

$$|E(\mathbb{F}_p)| = \left|\left\{ (x:y:z) \in \mathbb{P}^2(\mathbb{F}_p) : y^2z = x^3 - xz^2 \right\}\right|,$$
$$P_1(X,Y) = (1 + X^6Y^7 + X^7Y^7 + X^{12}Y^8 + X^{13}Y^8 + X^{19}Y^{15})(1 - X^7Y^5),$$
$$P_2(X,Y) = X^6Y^5(1 - Y^2)(1 + X^{13}Y^8).$$

In [13] it was shown that this zeta function is not finitely uniform, thus answering in the negative a question posed by Grunewald, Segal and Smith in [32] that seemed 'plausible'. However, there was some doubt as to whether this zeta function would satisfy a functional equation similar to that satisfied by other local ideal zeta functions of Lie rings of class 2. The dependency on the number of points mod p on an elliptic curve did cast some doubt on this. However, it can easily be checked that

$$P_1(X^{-1}, Y^{-1}) = X^{-26} Y^{-20} P_1(X, Y) \,,$$
$$P_2(X^{-1}, Y^{-1}) = X^{-25} Y^{-20} P_2(X, Y) \,.$$

Together with the functional equation of the Weil zeta function applied to $|E(\mathbb{F}_p)|$, this yields

Corollary 2.21 (Voll [60]). *For all but finitely many primes p,*

$$\zeta_{L_E, p}^{\triangleleft}(s)\Big|_{p \to p^{-1}} = -p^{36 - 15s} \zeta_{L_E, p}^{\triangleleft}(s) \,.$$

2.9 Other Class Two Examples

We start with a number of Lie rings which appear in [32].

Theorem 2.22 ([32]). *Let $G(m, r)$ denote the direct product of $\mathbb{Z}^r$ with the central product of m copies of the Heisenberg Lie ring $\mathcal{H}$. Then $G(m, r)$ has Hirsch length $2m + r + 1$.*

$$\zeta_{G(m,r), p}^{\triangleleft}(s) = \zeta_{\mathbb{Z}^{2m+r}, p}(s) \zeta_p((2m + 1)s - (2m + r)) \,.$$

For $m \leq 2$,

$$\zeta_{G(1,r), p}^{\leq}(s) = \zeta_{\mathbb{Z}^{r+2}, p}(s) \zeta_p(2s - (r + 2)) \zeta_p(2s - (r + 3)) \zeta_p(3s - (r + 3))^{-1} \,,$$

$$\zeta_{G(2,r), p}^{\leq}(s) = \zeta_{\mathbb{Z}^{r+4}, p}(s) \zeta_p(3s - (r + 4)) \zeta_p(3s - (r + 6)) \zeta_p(3s - (r + 7))$$
$$\times W_{G(2,r)}^{\leq}(p, p^{-s}) \,,$$

where

$$W_{G(2,r)}^{\leq}(X, Y) = 1 + X^{r+5} Y^3 - X^{r+5} Y^4 - X^{r+6} Y^4 - X^{r+7} Y^4 - X^{r+8} Y^4$$
$$+ X^{r+8} Y^5 + X^{2r+13} Y^8 \,.$$

These zeta functions satisfy the functional equations

$$\zeta_{G(m,r), p}^{\triangleleft}(s)\Big|_{p \to p^{-1}} = (-1)^{2m+r+1} p^{\binom{2m+r+1}{2} - (4m+r+1)s} \zeta_{G(m,r), p}^{\triangleleft}(s) \,,$$

$$\zeta_{G(m,r), p}^{\leq}(s)\Big|_{p \to p^{-1}} = (-1)^{2m+r+1} p^{\binom{2m+r+1}{2} - (2m+r+1)s} \zeta_{G(m,r), p}^{\leq}(s) \; (m = 1, 2) \,.$$

The corresponding global zeta functions have abscissa of convergence $\alpha_{G(m,r)}^{\triangleleft} = 2m + r$ for all $m \in \mathbb{N}_{>0}$, $r \in \mathbb{N}$ and $\alpha_{G(m,r)}^{\leq} = 2m + r$ for $m \in \{1, 2\}$, $r \in \mathbb{N}$.

Theorem 2.23 ([32]). *For $r \in \mathbb{N}$,*

$$\zeta^{\triangleleft}_{\mathcal{G}_3 \times \mathbb{Z}^r, p}(s) = \zeta_{\mathbb{Z}^{r+3}_p}(s)\zeta_p(3s - (r + 4))\zeta_p(5s - (2r + 6))(1 + p^{r+3-3s}) ,$$

$$\zeta^{\leq}_{\mathcal{G}_3 \times \mathbb{Z}^r, p}(s) = \zeta_{\mathbb{Z}^{r+3}_p}(s)\zeta_p(2s - (r + 4))\zeta_p(2s - (r + 5))\zeta_p(3s - (2r + 6))$$
$$\times W^{\leq}_{\mathcal{G}_3 \times \mathbb{Z}^r}(p, p^{-s}) ,$$

where

$$W^{\leq}_{\mathcal{G}_3 \times \mathbb{Z}^r}(p, p^{-s}) = 1 + X^{r+3}Y^2 + X^{r+4}Y^2 - X^{r+4}Y^3 - X^{r+5}Y^3 - X^{2r+8}Y^5 .$$

These zeta functions satisfy the functional equations

$$\zeta^{\triangleleft}_{\mathcal{G}_3 \times \mathbb{Z}^r, p}(s)\Big|_{p \to p^{-1}} = (-1)^{r+5}p^{\binom{r+5}{2} - (r+8)s}\zeta^{\triangleleft}_{\mathcal{G}_3 \times \mathbb{Z}^r, p}(s) ,$$

$$\zeta^{\leq}_{\mathcal{G}_3 \times \mathbb{Z}^r, p}(s)\Big|_{p \to p^{-1}} = (-1)^{r+5}p^{\binom{r+5}{2} - (r+5)s}\zeta^{\leq}_{\mathcal{G}_3 \times \mathbb{Z}^r, p}(s) .$$

The corresponding global zeta functions have abscissa of convergence $\alpha^{\triangleleft}_{\mathcal{G}_3 \times \mathbb{Z}^r} = \alpha^{\leq}_{\mathcal{G}_3 \times \mathbb{Z}^r} = r + 3$.

The calculations of the ideal zeta functions were made by Grunewald, Segal and Smith in [32]. Note that they use the more cumbersome notation $F_{2,3}/\langle z \rangle$ in place of $\mathcal{G}_3$.

Theorem 2.24 ([32, 64]). *Let*

$$\mathfrak{g}_{6,4} = \langle x_1, x_2, x_3, x_4, y_1, y_2 : [x_1, x_2] = y_1, [x_1, x_3] = y_2, [x_2, x_4] = y_2 \rangle .$$

Then

$$\zeta^{\triangleleft}_{\mathfrak{g}_{6,4}, p}(s) = \zeta_{\mathbb{Z}^4, p}(s)\zeta_p(3s - 4)\zeta_p(5s - 5)\zeta_p(6s - 9)\zeta_p(8s - 9)^{-1} ,$$

$$\zeta^{\leq}_{\mathfrak{g}_{6,4}, p}(s) = \zeta_{\mathbb{Z}^4, p}(s)\zeta_p(2s - 5)\zeta_p(3s - 5)\zeta_p(3s - 7)\zeta_p(3s - 8)\zeta_p(4s - 9)$$
$$\times \zeta_p(4s - 11)\zeta_p(5s - 12)W^{\leq}_{\mathfrak{g}_{6,4}}(p, p^{-s}) ,$$

where $W^{\leq}_{\mathfrak{g}_{6,4}}(X, Y)$ is given in Appendix A on p. 180. These zeta functions satisfy the functional equations

$$\zeta^{\triangleleft}_{\mathfrak{g}_{6,4}, p}(s)\Big|_{p \to p^{-1}} = p^{15-10s}\zeta^{\triangleleft}_{\mathfrak{g}_{6,4}, p}(s) ,$$

$$\zeta^{\leq}_{\mathfrak{g}_{6,4}, p}(s)\Big|_{p \to p^{-1}} = p^{15-6s}\zeta^{\leq}_{\mathfrak{g}_{6,4}, p}(s) .$$

The corresponding global zeta functions have abscissa of convergence $\alpha^{\triangleleft}_{\mathfrak{g}_{6,4}} = \alpha^{\leq}_{\mathfrak{g}_{6,4}} = 4$.

In [32], this Lie ring is given the more cumbersome name $F_{2,3}/\langle z\rangle \cdot \mathbb{Z}$. For brevity we have changed the name. The new name is borrowed from the classification of nilpotent Lie algebras of dimension 6 mentioned in Sect. 2.14 below.

Let T_n denote the maximal class-two quotient of the Lie ring of unitriangular $n \times n$ matrices. T_n has presentation

$$\langle x_1, \ldots, x_n, y_1, \ldots, y_{n-1} : [x_i, x_{i+1}] = y_i \text{ for } 1 \le i \le n-1\rangle \ .$$

T_2 is the Heisenberg Lie ring once again, and $T_3 \cong \mathcal{G}_3$, whose zeta functions are given in Sect. 2.6.

Theorem 2.25 ([57, 64]).

$$\zeta_{T_4,p}^{\triangleleft}(s) = \zeta_{\mathbb{Z}^4,p}(s)\zeta_p(3s-5)^2\zeta_p(5s-6)\zeta_p(5s-8)\zeta_p(6s-10)\zeta_p(7s-12)$$
$$\times W_{T_4}^{\triangleleft}(p,p^{-s}) \ ,$$

where $W_{T_4}^{\triangleleft}(X,Y)$ is

$$1 + X^4Y^3 - X^5Y^5 + X^8Y^5 - X^8Y^6 - X^9Y^6 - X^{10}Y^8 - X^{12}Y^8 - X^{13}Y^9$$
$$+ X^{13}Y^{10} - 2X^{14}Y^{10} + X^{14}Y^{11} + X^{15}Y^{11} - X^{16}Y^{11} - X^{17}Y^{11} + 2X^{17}Y^{12}$$
$$- X^{18}Y^{12} + X^{18}Y^{13} + X^{19}Y^{14} + X^{21}Y^{14} + X^{22}Y^{16} + X^{23}Y^{16} - X^{23}Y^{17}$$
$$+ X^{26}Y^{17} - X^{27}Y^{19} - X^{31}Y^{22} \ ,$$

and

$$\zeta_{T_4,p}^{\le}(s) = \zeta_{\mathbb{Z}^4,p}(s)\zeta_p(2s-5)^2\zeta_p(2s-6)^2\zeta_p(3s-6)\zeta_p(3s-8)^2\zeta_p(3s-9)$$
$$\times \zeta_p(4s-12)\zeta_p(5s-14)W_{T_4}^{\le}(p,p^{-s}) \ ,$$

where the polynomial $W_{T_4}^{\le}(X,Y)$ is given in Appendix A on p. 180. These zeta functions satisfy the functional equations

$$\zeta_{T_4,p}^{\triangleleft}(s)\Big|_{p\to p^{-1}} = -p^{21-11s}\zeta_{T_4,p}^{\triangleleft}(s) \ ,$$
$$\zeta_{T_4,p}^{\le}(s)\Big|_{p\to p^{-1}} = -p^{21-7s}\zeta_{T_4,p}^{\le}(s) \ .$$

The corresponding global zeta functions have abscissa of convergence $\alpha_{T_4}^{\triangleleft} = \alpha_{T_4}^{\le} = 4$.

2.10 The Maximal Class Lie Ring M_3 and Variants

The most well-understood zeta functions of Lie rings are those for Lie rings of nilpotency class 2. However, as we move to higher nilpotency classes, there is much less in the way of theory to help us. In particular, as we mentioned in Chap. 1, the Mal'cev correspondence can be avoided for nilpotency class 2. There is no such shortcut in higher nilpotency classes.

Taylor [57] was the first to calculate the zeta functions of a class-3-nilpotent Lie ring, and since then the second author has greatly enlarged the stock of examples at class 3.

In some sense, the 'simplest' Lie rings of nilpotency class n are the Lie rings M_n, with presentation

$$M_n = \langle z, x_1, x_2, \ldots, x_n : [z, x_i] = x_{i+1} \text{ for } i = 1, \ldots, n-1 \rangle .$$

In particular, $\mathcal{H} = M_2$. We now consider M_3 and some variations.

Theorem 2.26. *For* $r \in \mathbb{Z}$,

$$\zeta^{\triangleleft}_{M_3 \times \mathbb{Z}^r, p}(s) = \frac{\zeta_{\mathbb{Z}^{r+2}, p}(s)\zeta_p(3s - (r+2))\zeta_p(4s - (r+2))\zeta_p(5s - (r+3))}{\zeta_p(5s - (r+2))} ,$$

and

$$\zeta^{\leq}_{M_3 \times \mathbb{Z}^r, p}(s) = \zeta_{\mathbb{Z}^{r+2}, p}(s)\zeta_p(2s - (r+3))\zeta_p(3s - (2r+4))$$
$$\times \zeta_p(4s - (2r+6))W^{\leq}_{M_3 \times \mathbb{Z}^r}(p, p^{-s}) ,$$

where

$$W^{\leq}_{M_3 \times \mathbb{Z}^r}(p, p^{-s}) = 1 + X^{r+2}Y^2 + X^{r+3}Y^2 - X^{r+3}Y^3 - X^{r+5}Y^4 + X^{2r+6}Y^4$$
$$- 2X^{2r+6}Y^5 - 2X^{2r+7}Y^5 + X^{2r+7}Y^6 - X^{3r+8}Y^6$$
$$- X^{3r+10}Y^7 + X^{3r+10}Y^8 + X^{3r+11}Y^8 + X^{4r+13}Y^{10} .$$

These zeta functions satisfy the functional equations

$$\zeta^{\triangleleft}_{M_3 \times \mathbb{Z}^r, p}(s)\Big|_{p \to p^{-1}} = (-1)^{r+4}p^{\binom{r+4}{2} - (r+9)s}\zeta^{\triangleleft}_{M_3 \times \mathbb{Z}^r, p}(s) ,$$

$$\zeta^{\leq}_{M_3 \times \mathbb{Z}^r, p}(s)\Big|_{p \to p^{-1}} = (-1)^{r+4}p^{\binom{r+4}{2} - (r+4)s}\zeta^{\leq}_{M_3 \times \mathbb{Z}^r, p}(s) .$$

The corresponding global zeta functions have abscissa of convergence $\alpha^{\triangleleft}_{M_3 \times \mathbb{Z}^r} = \alpha^{\leq}_{M_3 \times \mathbb{Z}^r} = r + 2$, *with* $\zeta^{\leq}_{M_3}(s)$ *having a quadruple pole at* $s = 2$.

The zeta functions counting ideals or all subrings in M_3 were first calculated by Taylor in [57]. The second author generalised the results to $M_3 \times \mathbb{Z}^r$ for $r \in \mathbb{N}$.

Theorem 2.27 ([64]).

$$\zeta^{\triangleleft}_{\mathcal{H} \times M_3, p}(s) = \zeta_{\mathbb{Z}^4, p}(s)\zeta_p(3s - 4)^2\zeta_p(4s - 4)\zeta_p(5s - 5)\zeta_p(6s - 5)\zeta_p(7s - 6)$$
$$\times \zeta_p(9s - 10)W^{\triangleleft}_{\mathcal{H} \times M_3}(p, p^{-s}) ,$$

where $W_{\mathcal{H} \times M_3}^{\triangleleft}(X, Y)$ is

$$1 - 2X^4Y^5 + X^5Y^5 - X^4Y^6 + X^4Y^7 - 2X^5Y^7 + X^8Y^9 - 2X^9Y^9 + 3X^9Y^{11}$$
$$- 2X^{10}Y^{11} + X^9Y^{12} + X^{10}Y^{13} + X^{13}Y^{14} + X^{14}Y^{15} - 2X^{13}Y^{16} + 3X^{14}Y^{16}$$
$$- 2X^{14}Y^{18} + X^{15}Y^{18} - 2X^{18}Y^{20} + X^{19}Y^{20} - X^{19}Y^{21} + X^{18}Y^{22}$$
$$- 2X^{19}Y^{22} + X^{23}Y^{27} .$$

This zeta function satisfies the functional equation

$$\zeta_{\mathcal{H} \times M_3, p}^{\triangleleft}(s)\big|_{p \to p^{-1}} = -p^{21-14s} \zeta_{\mathcal{H} \times M_3, p}^{\triangleleft}(s) .$$

The corresponding global zeta function has abscissa of convergence $\alpha_{\mathcal{H} \times M_3}^{\triangleleft} = 4$.

Theorem 2.28.

$$\zeta_{\mathcal{H}^2 \times M_3, p}^{\triangleleft}(s) = \zeta_{\mathbb{Z}^6, p}(s) \zeta_p(3s - 6)^3 \zeta_p(4s - 6) \zeta_p(5s - 7) \zeta_p(6s - 7) \zeta_p(7s - 8)$$
$$\times \zeta_p(8s - 8) \zeta_p(8s - 14) \zeta_p(9s - 9) \zeta_p(9s - 14) \zeta_p(10s - 15)$$
$$\times \zeta_p(11s - 16) \zeta_p(12s - 21) W_{\mathcal{H}^2 \times M_3}^{\triangleleft}(p, p^{-s})$$

for some polynomial $W_{\mathcal{H}^2 \times M_3}^{\triangleleft}(X, Y)$ of degrees 113 in X and 85 in Y. This zeta function satisfies the functional equation

$$\zeta_{\mathcal{H}^2 \times M_3, p}^{\triangleleft}(s)\big|_{p \to p^{-1}} = p^{45-19s} \zeta_{\mathcal{H}^2 \times M_3, p}^{\triangleleft}(s) .$$

The corresponding global zeta function has abscissa of convergence $\alpha_{\mathcal{H}^2 \times M_3}^{\triangleleft} = 6$.

Theorem 2.29.

$$\zeta_{M_3 \times M_3, p}^{\triangleleft}(s) = \zeta_{\mathbb{Z}^4, p}(s) \zeta_p(2s - 2) \zeta_p(3s - 4)^2 \zeta_p(4s - 4) \zeta_p(5s - 5) \zeta_p(6s - 5)$$
$$\times \zeta_p(7s - 5) \zeta_p(7s - 6) \zeta_p(8s - 6) \zeta_p(9s - 7) \zeta_p(9s - 10)$$
$$\times \zeta_p(10s - 10) \zeta_p(11s - 11) \zeta_p(12s - 12) \zeta_p(13s - 15)$$
$$\times W_{M_3 \times M_3}^{\triangleleft}(p, p^{-s})$$

for some polynomial $W_{M_3 \times M_3}^{\triangleleft}(X, Y)$ of degrees 84 in X and 95 in Y. This zeta function satisfies the functional equation

$$\zeta_{M_3 \times M_3, p}^{\triangleleft}(s)\big|_{p \to p^{-1}} = p^{28-18s} \zeta_{M_3 \times M_3, p}^{\triangleleft}(s) .$$

The corresponding global zeta function has abscissa of convergence $\alpha_{M_3 \times M_3}^{\triangleleft} = 4$.

Theorem 2.30. *Let the Lie ring $M_3 \times_{\mathbb{Z}} M_3$ have presentation*

$$\langle z_1, z_2, w_1, w_2, x_1, x_2, y : [z_1, w_1] = x_1, [z_2, w_2] = x_2, [z_1, x_1] = y, [z_2, x_2] = y \rangle \ .$$

Then

$$\zeta_{M_3 \times_{\mathbb{Z}} M_3, p}^{\triangleleft}(s) = \zeta_{\mathbb{Z}^4, p}(s) \zeta_p(3s-4)^2 \zeta_p(5s-5) \zeta_p(7s-4) \zeta_p(8s-5) \zeta_p(9s-6)$$
$$\times \zeta_p(12s-10) W_{M_3 \times_{\mathbb{Z}} M_3}^{\triangleleft}(p, p^{-s}) \ ,$$

where $W_{M_3 \times_{\mathbb{Z}} M_3}^{\triangleleft}(X, Y)$ is

$$1 - X^4 Y^5 - 2X^4 Y^8 + X^5 Y^8 + X^4 Y^9 - 2X^5 Y^9 + X^8 Y^{12} - 2X^9 Y^{12}$$
$$+ 3X^9 Y^{13} - 2X^{10} Y^{13} + X^{10} Y^{14} + X^9 Y^{17} + X^{14} Y^{17} + X^{13} Y^{20} - 2X^{13} Y^{21}$$
$$+ 3X^{14} Y^{21} - 2X^{14} Y^{22} + X^{15} Y^{22} - 2X^{18} Y^{25} + X^{19} Y^{25} + X^{18} Y^{26}$$
$$- 2X^{19} Y^{26} - X^{19} Y^{29} + X^{23} Y^{34} \ .$$

This zeta function satisfies the functional equation

$$\zeta_{M_3 \times_{\mathbb{Z}} M_3, p}^{\triangleleft}(s) \Big|_{p \to p^{-1}} = -p^{21-17s} \zeta_{M_3 \times_{\mathbb{Z}} M_3, p}^{\triangleleft}(s) \ .$$

The corresponding global zeta function has abscissa of convergence $\alpha_{M_3 \times_{\mathbb{Z}} M_3}^{\triangleleft} = 4$.

2.11 Lie Rings with Large Abelian Ideals

As we saw in Sect. 2.6, Voll has calculated $\zeta_{\mathcal{G}_n, p}^{\triangleleft}(s)$ and $\zeta_{\mathcal{G}_n, p}^{\leq}(s)$ for all $n \geq 2$. The Lie rings $\mathcal{G}_n$ have an abelian ideal of corank 1 (and thus of infinite index), and it is likely that this large ideal makes it easier to get a grasp on the structure of the lattices of ideals/subrings. Indeed the Lie rings M_n have this property too. In this section we consider some further Lie rings of nilpotency class 3 with this property.

Theorem 2.31 ([64]). *Let the Lie ring $L_{(3,3)}$ have presentation*

$$\langle z, w_1, w_2, x_1, x_2, y_1, y_2 : [z, w_1] = x_1, [z, w_2] = x_2, [z, x_1] = y_1, [z, x_2] = y_2 \rangle \ .$$

Then

$$\zeta_{L_{(3,3)}, p}^{\triangleleft}(s) = \zeta_{\mathbb{Z}^3, p}(s) \zeta_p(3s-4) \zeta_p(4s-5) \zeta_p(5s-6) \zeta_p(6s-7) \zeta_p(7s-6)$$
$$\times \zeta_p(8s-10) \zeta_p(9s-12) \zeta_p(11s-12) \zeta_p(4s-4)^{-1}$$
$$\times W_{L_{(3,3)}}^{\triangleleft}(p, p^{-s}) \ ,$$

where $W^{\triangleleft}_{L_{(3,3)}}(X,Y)$ *is*

$$1 + X^3Y^3 + 2X^4Y^4 - X^4Y^5 + X^6Y^5 + X^6Y^6 - X^6Y^7 + X^9Y^7 - X^6Y^8$$
$$+ 2X^8Y^8 - X^8Y^9 - X^{10}Y^9 - X^9Y^{10} + X^{12}Y^{10} - X^{10}Y^{11} - X^{12}Y^{11}$$
$$- X^{13}Y^{12} - X^{12}Y^{13} - X^{14}Y^{13} - 2X^{16}Y^{13} - 2X^{15}Y^{14} - X^{14}Y^{15} - X^{16}Y^{15}$$
$$- X^{18}Y^{15} + 2X^{16}Y^{16} - X^{18}Y^{16} - X^{19}Y^{16} - X^{18}Y^{17} - 2X^{20}Y^{17} + X^{18}Y^{18}$$
$$+ X^{20}Y^{18} - X^{21}Y^{18} + X^{19}Y^{19} - X^{20}Y^{19} - X^{22}Y^{19} + 2X^{20}Y^{20} + X^{22}Y^{20}$$
$$+ X^{21}Y^{21} + X^{22}Y^{21} - 2X^{24}Y^{21} + X^{22}Y^{22} + X^{24}Y^{22} + X^{26}Y^{22} + 2X^{25}Y^{23}$$
$$+ 2X^{24}Y^{24} + X^{26}Y^{24} + X^{28}Y^{24} + X^{27}Y^{25} + X^{28}Y^{26} + X^{30}Y^{26} - X^{28}Y^{27}$$
$$+ X^{31}Y^{27} + X^{30}Y^{28} + X^{32}Y^{28} - 2X^{32}Y^{29} + X^{34}Y^{29} - X^{31}Y^{30} + X^{34}Y^{30}$$
$$- X^{34}Y^{31} - X^{34}Y^{32} + X^{36}Y^{32} - 2X^{36}Y^{33} - X^{37}Y^{34} - X^{40}Y^{37} \ .$$

This zeta function satisfies the functional equation

$$\zeta^{\triangleleft}_{L_{(3,3)},p}(s)\Big|_{p \to p^{-1}} = -p^{21-15s}\zeta^{\triangleleft}_{L_{(3,3)},p}(s) \ .$$

The corresponding global zeta function has abscissa of convergence $\alpha^{\triangleleft}_{L_{(3,3)}} = 3.$

The second author also considered what happens when you delete generator y_2 from the presentation above:

Theorem 2.32 ([64]). *Let* $L_{(3,2)}$ *be given by the presentation*

$$\langle z, w_1, w_2, x_1, x_2, y : [z, w_1] = x_1, [z, w_2] = x_2, [z, x_1] = y \rangle \ .$$

Then

$$\zeta^{\triangleleft}_{L_{(3,2)},p}(s) = \zeta_{\mathbb{Z}^3,p}(s)\zeta_p(3s-4)\zeta_p(4s-4)\zeta_p(5s-5)\zeta_p(5s-6)\zeta_p(6s-6)$$
$$\times \zeta_p(9s-11)W^{\triangleleft}_{L_{(3,2)}}(p,p^{-s}) \ ,$$

where $W^{\triangleleft}_{L_{(3,2)}}(X,Y)$ *is*

$$1 + X^3Y^3 - X^4Y^5 - X^6Y^7 - X^7Y^7 + X^8Y^7 - X^8Y^8 - X^9Y^9 - X^{10}Y^9$$
$$+ X^{10}Y^{10} - X^{11}Y^{10} + X^{10}Y^{11} - X^{11}Y^{11} + X^{11}Y^{12} - X^{14}Y^{12} + X^{13}Y^{13}$$
$$- X^{14}Y^{13} + X^{14}Y^{14} + X^{15}Y^{14} + X^{17}Y^{16} + X^{18}Y^{17} + X^{20}Y^{18} - X^{21}Y^{21}$$
$$- X^{24}Y^{23} \ ,$$

and

$$\zeta^{\leq}_{L_{(3,2)},p}(s) = \zeta_{\mathbb{Z}^3,p}(s)\zeta_p(2s-4)\zeta_p(2s-5)^2\zeta_p(3s-7)\zeta_p(3s-8)\zeta_p(4s-10)$$
$$\times \zeta_p(5s-12)W^{\leq}_{L_{(3,2)}}(p,p^{-s}) \ ,$$

where $W_{L_{(3,2)}}^{\leq}(X, Y)$ *is*

$$1 + X^3Y^2 + X^4Y^2 - X^4Y^3 - X^5Y^3 + X^6Y^3 + X^7Y^3 - 2X^7Y^4 - 2X^8Y^4$$
$$+ X^9Y^5 - 2X^{10}Y^5 - 3X^{11}Y^5 + X^{11}Y^6 + X^{12}Y^6 - 2X^{13}Y^6 - 3X^{14}Y^6$$
$$+ X^{13}Y^7 + X^{14}Y^7 + 3X^{15}Y^7 - 2X^{16}Y^7 - X^{17}Y^7 + X^{16}Y^8 + X^{17}Y^8$$
$$+ 2X^{18}Y^8 + 2X^{18}Y^9 + 2X^{21}Y^9 + 2X^{21}Y^{10} + X^{22}Y^{10} + X^{23}Y^{10} - X^{22}Y^{11}$$
$$- 2X^{23}Y^{11} + 3X^{24}Y^{11} + X^{25}Y^{11} + X^{26}Y^{11} - 3X^{25}Y^{12} - 2X^{26}Y^{12}$$
$$+ X^{27}Y^{12} + X^{28}Y^{12} - 3X^{28}Y^{13} - 2X^{29}Y^{13} + X^{30}Y^{13} - 2X^{31}Y^{14}$$
$$- 2X^{32}Y^{14} + X^{32}Y^{15} + X^{33}Y^{15} - X^{34}Y^{15} - X^{35}Y^{15} + X^{35}Y^{16} + X^{36}Y^{16}$$
$$+ X^{39}Y^{18} \, .$$

The local zeta function counting all subrings satisfies the functional equation

$$\zeta_{L_{(3,2)},p}^{\leq}(s)\Big|_{p \to p^{-1}} = p^{15-6s}\zeta_{L_{(3,2)},p}^{\leq}(s) \, .$$

However, the local ideal zeta function satisfies no such functional equation. The corresponding global zeta functions have abscissa of convergence $\alpha_{L_{(3,2)}}^{\triangleleft} = \alpha_{L_{(3,2)}}^{\leq} = 3$, *with* $\zeta_{L_{(3,2)}}^{\leq}(s)$ *having a quadruple pole at* $s = 3$.

The zeta function counting ideals was the first calculated which satisfied no functional equation of the form (2.7).

A couple of Lie rings similar to $L_{(3,2)}$ were also considered. Their ideal zeta functions also satisfy no functional equation of the form seen numerous times before.

Theorem 2.33.

$$\zeta_{\mathcal{H} \times L_{(3,2)},p}^{\triangleleft}(s) = \zeta_{\mathbb{Z}^5,p}(s)\zeta_p(3s-5)\zeta_p(3s-6)\zeta_p(4s-6)\zeta_p(5s-7)\zeta_p(5s-10)$$
$$\times \zeta_p(6s-7)\zeta_p(6s-10)\zeta_p(7s-8)\zeta_p(7s-12)\zeta_p(8s-12)$$
$$\times \zeta_p(9s-14)\zeta_p(9s-17)\zeta_p(11s-19)\zeta_p(13s-20)$$
$$\times \zeta_p(13s-23)W_{\mathcal{H} \times L_{(3,2)}}^{\triangleleft}(p, p^{-s})$$

for some polynomial $W_{\mathcal{H} \times L_{(3,2)}}^{\triangleleft}(X, Y)$ *of degrees 150 in X and 97 in Y. This local zeta function satisfies no functional equation. The corresponding global zeta function has abscissa of convergence* $\alpha_{\mathcal{H} \times L_{(3,2)}}^{\triangleleft} = 5$.

Theorem 2.34. *Let the Lie ring* $L_{(3,2,2)}$ *have presentation*

$$\left\langle z, w_1, w_2, w_3, x_1, x_2, x_3, y : \begin{array}{l} [z, w_1] = x_1, [z, w_2] = x_2, \\ [z, w_3] = x_3, [z, x_1] = y \end{array} \right\rangle \, .$$

Then

$$\zeta_{L_{(3,2,2)},p}^{\triangleleft}(s) = \zeta_{\mathbb{Z}^4,p}(s)\zeta_p(2s-3)\zeta_p(3s-6)\zeta_p(5s-7)\zeta_p(5s-10)\zeta_p(6s-10)$$
$$\times \zeta_p(7s-12)\zeta_p(8s-12)\zeta_p(9s-17)\zeta_p(13s-23)$$
$$\times W_{L_{(3,2,2)}}^{\triangleleft}(p, p^{-s}) \, ,$$

where $W^{\triangleleft}_{L(3,2,2)}(X,Y)$ is given in Appendix A on p. 181. This local zeta function satisfies no functional equation. The corresponding global zeta function has abscissa of convergence $\alpha^{\triangleleft}_{L(3,2,2)} = 4$.

2.12 $F_{3,2}$

On p. 40 we considered the zeta functions of the free class-2 nilpotent Lie rings. The second author has added the zeta functions of the class-3, 2-generator nilpotent Lie ring.

Theorem 2.35 ([64]). *Let the Lie ring $F_{3,2}$ have presentation*

$$\langle x_1, x_2, y_1, z_1, z_2 : [x_1, x_2] = y_1, [x_1, y_1] = z_1, [x_2, y_1] = z_2 \rangle .$$

Then

$$\zeta^{\triangleleft}_{F_{3,2},p}(s) = \zeta_{\mathbb{Z}^2,p}(s)\zeta_p(3s-2)\zeta_p(4s-3)\zeta_p(5s-4)^2\zeta_p(7s-6)W^{\triangleleft}_{F_{3,2}}(p,p^{-s}) ,$$

where $W^{\triangleleft}_{F_{3,2}}(X,Y)$ is

$$1 + X^2Y^4 - X^2Y^5 - X^4Y^7 - X^6Y^9 - X^8Y^{11} + X^8Y^{12} + X^{10}Y^{16} ,$$

and

$$\zeta^{\leq}_{F_{3,2},p}(s) = \zeta_{\mathbb{Z}^2,p}(s)\zeta_p(2s-3)\zeta_p(2s-4)\zeta_p(3s-6)\zeta_p(4s-8)\zeta_p(5s-8)$$
$$\times \zeta_p(5s-9)W^{\leq}_{F_{3,2}}(p,p^{-s}) ,$$

where $W^{\leq}_{F_{3,2}}(X,Y)$ is

$$1 + X^2Y^2 + X^3Y^2 - X^3Y^3 + X^4Y^3 + 2X^5Y^3 - 2X^5Y^4 + 2X^7Y^4 - 2X^7Y^5$$
$$- 2X^8Y^5 - X^9Y^5 - X^{10}Y^6 - X^{11}Y^6 - X^{10}Y^7 - X^{13}Y^7 - 2X^{12}Y^8$$
$$- X^{13}Y^8 - X^{14}Y^8 - X^{15}Y^8 + X^{13}Y^9 - X^{16}Y^9 + X^{14}Y^{10} + X^{15}Y^{10}$$
$$+ X^{16}Y^{10} + 2X^{17}Y^{10} + X^{16}Y^{11} + X^{19}Y^{11} + X^{18}Y^{12} + X^{19}Y^{12} + X^{20}Y^{13}$$
$$+ 2X^{21}Y^{13} + 2X^{22}Y^{13} - 2X^{22}Y^{14} + 2X^{24}Y^{14} - 2X^{24}Y^{15} - X^{25}Y^{15}$$
$$+ X^{26}Y^{15} - X^{26}Y^{16} - X^{27}Y^{16} - X^{29}Y^{18} .$$

These zeta functions satisfy the functional equations

$$\zeta^{\triangleleft}_{F_{3,2},p}(s)\Big|_{p\to p^{-1}} = -p^{10-10s}\zeta^{\triangleleft}_{F_{3,2},p}(s) ,$$

$$\zeta^{\leq}_{F_{3,2},p}(s)\Big|_{p\to p^{-1}} = -p^{10-5s}\zeta^{\leq}_{F_{3,2},p}(s) .$$

The corresponding global zeta functions have abscissa of convergence $\alpha^{\triangleleft}_{F_{3,2}} = 2$, $\alpha^{\leq}_{F_{3,2}} = 5/2$.

Theorem 2.36 ([64]).

$$\zeta_{F_{3,2}\times\mathbb{Z},p}^{\lhd}(s) = \zeta_{\mathbb{Z}^3,p}(s)\zeta_p(3s-3)\zeta_p(4s-4)\zeta_p(5s-5)\zeta_p(5s-6)\zeta_p(7s-8)$$
$$\times W_{F_{3,2}\times\mathbb{Z}}^{\lhd}(p,p^{-s})\,,$$

where $W_{F_{3,2}\times\mathbb{Z}}^{\lhd}(X,Y)$ *is*

$$1 + X^3Y^4 - X^3Y^5 - X^6Y^7 - X^8Y^9 - X^{11}Y^{11} + X^{11}Y^{12} + X^{14}Y^{16}\,.$$

This zeta function satisfies the functional equation

$$\left.\zeta_{F_{3,2}\times\mathbb{Z},p}^{\lhd}(s)\right|_{p\to p^{-1}} = p^{15-11s}\zeta_{F_{3,2}\times\mathbb{Z},p}^{\lhd}(s)\,.$$

The corresponding global zeta function has abscissa of convergence $\alpha_{F_{3,2}\times\mathbb{Z}}^{\lhd}=3$.

2.13 The Maximal Class Lie Rings M_4 and Fil$_4$

We saw above that M_3 is in some sense the simplest Lie ring of nilpotency class 3. The Lie ring M_4 can be defined in a similar way, and in some sense it is the simplest of nilpotency class 4. The M_n family of Lie rings are *filiform*, in that the nilpotency class is maximal given the rank.

Theorem 2.37 ([57]). *Let the Lie ring M_4 have presentation*

$$\langle z, x_1, x_2, x_3, x_4 : [z,x_1]=x_2, [z,x_2]=x_3, [z,x_3]=x_4\rangle\,.$$

Then

$$\zeta_{M_4,p}^{\lhd}(s) = \zeta_{\mathbb{Z}^2,p}(s)\zeta_p(3s-2)\zeta_p(5s-2)\zeta_p(7s-4)\zeta_p(8s-5)\zeta_p(9s-6)$$
$$\times \zeta_p(11s-6)\zeta_p(12s-7)\zeta_p(6s-3)^{-1}W_{M_4}^{\lhd}(p,p^{-s})\,,$$

where $W_{M_4}^{\lhd}(X,Y)$ *is*

$$1 + X^2Y^4 - X^2Y^5 + X^3Y^5 - X^2Y^6 + 2X^3Y^6 - X^3Y^7 - X^5Y^9 + X^6Y^{10}$$
$$- 2X^5Y^{11} - X^7Y^{13} - X^8Y^{13} + X^7Y^{14} - X^8Y^{14} - X^8Y^{15} - X^9Y^{15}$$
$$+ X^9Y^{16} - X^9Y^{17} - X^{10}Y^{17} + 2X^9Y^{18} - X^{10}Y^{18} + X^{10}Y^{19} - 2X^{11}Y^{19}$$
$$+ X^{10}Y^{20} + X^{11}Y^{20} - X^{11}Y^{21} + X^{11}Y^{22} + X^{12}Y^{22} + X^{12}Y^{23} - X^{13}Y^{23}$$
$$+ X^{12}Y^{24} + X^{13}Y^{24} + 2X^{15}Y^{26} - X^{14}Y^{27} + X^{15}Y^{28} + X^{17}Y^{30} - 2X^{17}Y^{31}$$
$$+ X^{18}Y^{31} - X^{17}Y^{32} + X^{18}Y^{32} - X^{18}Y^{33} - X^{20}Y^{37}$$

and

$$\zeta_{M_4,p}^{\leq}(s) = \zeta_{\mathbb{Z}^2,p}(s)\zeta_p(2s-3)\zeta_p(2s-4)\zeta_p(3s-6)\zeta_p(4s-7)\zeta_p(4s-8)$$
$$\times \zeta_p(7s-12)W_{M_4}^{\leq}(p,p^{-s})\,,$$

where $W_{M_4}^{\leq}(X,Y)$ *is*

$$
\begin{aligned}
&1 + X^2Y^2 + X^3Y^2 - X^3Y^3 + X^4Y^3 + 2X^5Y^3 - 2X^5Y^4 + X^7Y^4 - 2X^7Y^5 \\
&- X^8Y^5 + X^9Y^5 - 2X^9Y^6 - 2X^{10}Y^6 - X^{11}Y^6 + X^{10}Y^7 - 2X^{12}Y^7 \\
&- X^{13}Y^7 + X^{13}Y^8 - X^{14}Y^8 - X^{16}Y^9 + X^{15}Y^{10} + X^{17}Y^{11} - X^{18}Y^{11} \\
&+ X^{18}Y^{12} + 2X^{19}Y^{12} - X^{21}Y^{12} + X^{20}Y^{13} + 2X^{21}Y^{13} + 2X^{22}Y^{13} \\
&- X^{22}Y^{14} + X^{23}Y^{14} + 2X^{24}Y^{14} - X^{24}Y^{15} + 2X^{26}Y^{15} - 2X^{26}Y^{16} \\
&- X^{27}Y^{16} + X^{28}Y^{16} - X^{28}Y^{17} - X^{29}Y^{17} - X^{31}Y^{19} \ .
\end{aligned}
$$

These zeta functions satisfy the functional equations

$$
\zeta_{M_4,p}^{\lhd}(s)\Big|_{p \to p^{-1}} = -p^{10-14s}\zeta_{M_4,p}^{\lhd}(s) \ ,
$$

$$
\zeta_{M_4,p}^{\leq}(s)\Big|_{p \to p^{-1}} = -p^{10-5s}\zeta_{M_4,p}^{\leq}(s) \ .
$$

The corresponding global zeta functions have abscissa of convergence $\alpha_{M_4}^{\lhd} = 2$, $\alpha_{M_4}^{\leq} = 5/2$.

Theorem 2.38 ([64]).

$$
\begin{aligned}
\zeta_{M_4 \times \mathbb{Z},p}^{\lhd}(s) = \ &\zeta_{\mathbb{Z}^3,p}(s)\zeta_p(3s-3)\zeta_p(5s-3)\zeta_p(7s-5)\zeta_p(8s-7)\zeta_p(9s-8) \\
&\times \zeta_p(11s-8)\zeta_p(12s-9)\zeta_p(6s-4)^{-1}W_{M_4 \times \mathbb{Z}}^{\lhd}(p,p^{-s}) \ ,
\end{aligned}
$$

where $W_{M_4 \times \mathbb{Z}}^{\lhd}(X,Y)$ *is*

$$
\begin{aligned}
&1 + X^3Y^4 - X^3Y^5 + X^4Y^5 - X^3Y^6 + 2X^4Y^6 - X^4Y^7 - X^7Y^9 + X^8Y^{10} \\
&- 2X^7Y^{11} - X^9Y^{13} - X^{11}Y^{13} + X^{10}Y^{14} - X^{11}Y^{14} - X^{11}Y^{15} - X^{12}Y^{15} \\
&+ X^{12}Y^{16} - X^{12}Y^{17} - X^{13}Y^{17} + 2X^{12}Y^{18} - X^{13}Y^{18} + X^{14}Y^{19} - 2X^{15}Y^{19} \\
&+ X^{14}Y^{20} + X^{15}Y^{20} - X^{15}Y^{21} + X^{15}Y^{22} + X^{16}Y^{22} + X^{16}Y^{23} - X^{17}Y^{23} \\
&+ X^{16}Y^{24} + X^{18}Y^{24} + 2X^{20}Y^{26} - X^{19}Y^{27} + X^{20}Y^{28} + X^{23}Y^{30} - 2X^{23}Y^{31} \\
&+ X^{24}Y^{31} - X^{23}Y^{32} + X^{24}Y^{32} - X^{24}Y^{33} - X^{27}Y^{37} \ .
\end{aligned}
$$

This zeta function satisfies the functional equation

$$
\zeta_{M_4 \times \mathbb{Z},p}^{\lhd}(s)\Big|_{p \to p^{-1}} = p^{15-15s}\zeta_{M_4 \times \mathbb{Z},p}^{\lhd}(s) \ .
$$

The corresponding global zeta function has abscissa of convergence $\alpha_{M_4 \times \mathbb{Z}}^{\lhd} = 3$.

M_4 is not the only filiform Lie ring of nilpotency class 4, up to isomorphism:

Theorem 2.39 ([64]). *Let the Lie ring* Fil_4 *have presentation*

$$
\langle z, x_1, x_2, x_3, x_4 : [z, x_1] = x_2, [z, x_2] = x_3, [z, x_3] = x_4, [x_1, x_2] = x_4 \rangle \ .
$$

Then

$$\zeta_{\mathrm{Fil}_4,p}^{\lhd}(s) = \zeta_{\mathbb{Z}^2,p}(s)\zeta_p(3s-2)\zeta_p(5s-2)\zeta_p(7s-4)\zeta_p(8s-5)\zeta_p(9s-6)$$
$$\times \zeta_p(10s-6)\zeta_p(12s-7)W_{\mathrm{Fil}_4}^{\lhd}(p,p^{-s})\,,$$

where $W_{\mathrm{Fil}_4}^{\lhd}(X,Y)$ *is*

$$1 + X^2Y^4 - X^2Y^5 + X^3Y^5 - X^2Y^6 + X^3Y^6 - X^3Y^7 - X^5Y^9 - X^5Y^{10}$$
$$- X^6Y^{11} - X^6Y^{12} + X^6Y^{13} - X^7Y^{13} - X^8Y^{13} - X^8Y^{14} + X^7Y^{15}$$
$$+ X^8Y^{15} - 2X^9Y^{15} + X^8Y^{17} + X^9Y^{17} - X^{10}Y^{17} + X^9Y^{19} + X^{10}Y^{19}$$
$$+ X^{11}Y^{20} + 2X^{11}Y^{21} - X^{11}Y^{22} + 2X^{12}Y^{22} + 2X^{13}Y^{23} - X^{13}Y^{24}$$
$$+ X^{14}Y^{24} - X^{13}Y^{25} + X^{14}Y^{25} + X^{15}Y^{25} - 2X^{14}Y^{27} + 2X^{15}Y^{27}$$
$$- 2X^{15}Y^{28} + X^{16}Y^{28} - X^{15}Y^{29} - X^{16}Y^{29} + X^{17}Y^{29} - 2X^{17}Y^{30} + X^{18}Y^{30}$$
$$- X^{18}Y^{31} - X^{18}Y^{32} - X^{18}Y^{33} - X^{20}Y^{35} + X^{20}Y^{36} - X^{21}Y^{36} + X^{20}Y^{37}$$
$$- X^{21}Y^{37} + X^{21}Y^{38} + X^{23}Y^{42}\,.$$

This local zeta function satisfies no functional equation. The corresponding global zeta function has abscissa of convergence $\alpha_{\mathrm{Fil}_4}^{\lhd} = 2$.

Despite repeated efforts, we have been unable to calculate $\zeta_{\mathrm{Fil}_4,p}^{\leq}(s)$. M_4 is the only Lie ring of nilpotency class 4 whose zeta function counting all subrings we have calculated.

Theorem 2.40 ([64]).

$$\zeta_{\mathrm{Fil}_4\times\mathbb{Z},p}^{\lhd}(s) = \zeta_{\mathbb{Z}^3,p}(s)\zeta_p(3s-3)\zeta_p(5s-3)\zeta_p(7s-5)\zeta_p(8s-7)\zeta_p(9s-8)$$
$$\times \zeta_p(10s-8)\zeta_p(12s-9)W_{\mathrm{Fil}_4\times\mathbb{Z}}^{\lhd}(p,p^{-s})\,,$$

where $W_{\mathrm{Fil}_4\times\mathbb{Z}}^{\lhd}(X,Y)$ *is*

$$1 + X^3Y^4 - X^3Y^5 + X^4Y^5 - X^3Y^6 + X^4Y^6 - X^4Y^7 - X^7Y^9 - X^7Y^{10}$$
$$- X^8Y^{11} - X^8Y^{12} + X^8Y^{13} - X^9Y^{13} - X^{11}Y^{13} - X^{11}Y^{14} + X^{10}Y^{15}$$
$$+ X^{11}Y^{15} - 2X^{12}Y^{15} + X^{11}Y^{17} + X^{12}Y^{17} - X^{13}Y^{17} + X^{12}Y^{19} + X^{14}Y^{19}$$
$$+ X^{15}Y^{20} + 2X^{15}Y^{21} - X^{15}Y^{22} + 2X^{16}Y^{22} + X^{17}Y^{23} + X^{18}Y^{23} - X^{18}Y^{24}$$
$$+ X^{19}Y^{24} - X^{18}Y^{25} + X^{19}Y^{25} + X^{20}Y^{25} - 2X^{19}Y^{27} + 2X^{20}Y^{27}$$
$$- 2X^{20}Y^{28} + X^{21}Y^{28} - X^{20}Y^{29} - X^{22}Y^{29} + X^{23}Y^{29} - 2X^{23}Y^{30} + X^{24}Y^{30}$$
$$- X^{24}Y^{31} - X^{24}Y^{32} - X^{24}Y^{33} - X^{27}Y^{35} + X^{27}Y^{36} - X^{28}Y^{36} + X^{27}Y^{37}$$
$$- X^{28}Y^{37} + X^{28}Y^{38} + X^{31}Y^{42}\,.$$

This zeta function satisfies no functional equation. The corresponding global zeta function has abscissa of convergence $\alpha_{\mathrm{Fil}_4\times\mathbb{Z}}^{\lhd} = 3$.

2.14 Nilpotent Lie Algebras of Dimension ≤ 6

A complete classification of the nilpotent Lie algebras over $\mathbb{R}$ of dimension ≤ 6 is given in [44].[2] We cannot hope to classify nilpotent Lie rings additively isomorphic to $\mathbb{Z}^d$ for some $d \leq 6$, but we can at least use a classification over $\mathbb{R}$ to produce Lie rings over $\mathbb{Z}$ which are guaranteed be non-isomorphic. For each Lie algebra, Magnin gives an $\mathbb{R}$-basis and a list of nonzero Lie brackets of the basis elements. The structure constants of each nilpotent Lie algebra L listed in [44] are (fortunately) all in $\mathbb{Z}$. Hence we can form Lie rings over $\mathbb{Z}$ (or $\mathbb{Z}_p$) by taking the $\mathbb{Z}$-span (or $\mathbb{Z}_p$-span) of the basis given.[3]

This approach has led to many new calculations of ideal zeta functions of Lie rings of rank 6, and some others arising from a Lie ring of rank 5:

Theorem 2.41 ([64]). *Let the Lie ring* $\mathfrak{g}_{5,3}$ *have presentation*

$$\langle x_1, x_2, x_3, x_4, x_5 : [x_1, x_2] = x_4, [x_1, x_4] = x_5, [x_2, x_3] = x_5 \rangle .$$

Then

$$\zeta^{\triangleleft}_{\mathfrak{g}_{5,3} \times \mathbb{Z}^r, p}(s) = \zeta_{\mathbb{Z}^{r+3}, p}(s)\zeta_p(3s - (r+3))\zeta_p(5s - (r+4)) ,$$

$$\zeta^{\leq}_{\mathfrak{g}_{5,3}, p}(s) = \zeta_{\mathbb{Z}^3, p}(s)\zeta_p(2s - 4)\zeta_p(3s - 4)\zeta_p(3s - 6)\zeta_p(6s - 11)\zeta_p(6s - 12)$$
$$\times W^{\leq}_{\mathfrak{g}_{5,3}}(p, p^{-s}) ,$$

where $W^{\leq}_{\mathfrak{g}_{5,3}}(X, Y)$ *is*

$$1 + X^3Y^2 - X^4Y^3 + X^5Y^3 - X^5Y^4 + X^7Y^4 + X^8Y^4 - 2X^7Y^5 - 2X^8Y^5$$
$$- X^9Y^5 + X^8Y^6 + X^9Y^6 + X^{10}Y^6 - X^{10}Y^7 - 2X^{11}Y^7 - 2X^{12}Y^7 + X^{11}Y^8$$
$$+ X^{12}Y^8 - X^{14}Y^8 - X^{15}Y^8 + X^{15}Y^{10} + X^{16}Y^{10} - X^{18}Y^{10} - X^{19}Y^{10}$$
$$+ 2X^{18}Y^{11} + 2X^{19}Y^{11} + X^{20}Y^{11} - X^{20}Y^{12} - X^{21}Y^{12} - X^{22}Y^{12} + X^{21}Y^{13}$$
$$+ 2X^{22}Y^{13} + 2X^{23}Y^{13} - X^{22}Y^{14} - X^{23}Y^{14} + X^{25}Y^{14} - X^{25}Y^{15} + X^{26}Y^{15}$$
$$- X^{27}Y^{16} - X^{30}Y^{18} .$$

These zeta functions satisfy the functional equations

$$\zeta^{\triangleleft}_{\mathfrak{g}_{5,3} \times \mathbb{Z}^r, p}(s)\Big|_{p \to p^{-1}} = (-1)^{r+5}p^{\binom{r+5}{2} - (r+11)s}\zeta^{\triangleleft}_{\mathfrak{g}_{5,3} \times \mathbb{Z}^r, p}(s) ,$$

$$\zeta^{\leq}_{\mathfrak{g}_{5,3}, p}(s)\Big|_{p \to p^{-1}} = -p^{10-5s}\zeta^{\leq}_{\mathfrak{g}_{5,3}, p}(s) .$$

The corresponding global zeta functions have abscissa of convergence $\alpha^{\triangleleft}_{\mathfrak{g}_{5,3}} = \alpha^{\leq}_{\mathfrak{g}_{5,3}} = 3.$

---

[2] The classification was first given in [46], but we refer to [44] as this article is likely to be more accessible.

[3] We have permuted some of the bases of the Lie algebras from [44]; the bases we give are those that make the calculations of the zeta functions easiest.

Theorem 2.42.

$$\zeta_{\mathcal{H}\times\mathfrak{g}_{5,3},p}^{\triangleleft}(s) = \zeta_{\mathbb{Z}^5,p}(s)\zeta_p(3s-5)^2\zeta_p(5s-6)^2\zeta_p(7s-7)\zeta_p(5s-5)^{-1}$$
$$\times \zeta_p(7s-6)^{-1} .$$

This zeta function satisfies the functional equation

$$\zeta_{\mathcal{H}\times\mathfrak{g}_{5,3},p}^{\triangleleft}(s)\bigg|_{p\to p^{-1}} = p^{28-16s}\zeta_{\mathcal{H}\times\mathfrak{g}_{5,3},p}^{\triangleleft}(s) .$$

The corresponding global zeta function has abscissa of convergence $\alpha_{\mathcal{H}\times\mathfrak{g}_{5,3}}^{\triangleleft} = 5$.

Theorem 2.43.

$$\zeta_{\mathcal{G}_3\times\mathfrak{g}_{5,3},p}^{\triangleleft}(s) = \zeta_{\mathbb{Z}^6,p}(s)\zeta_p(3s-6)\zeta_p(3s-7)\zeta_p(5s-7)\zeta_p(5s-8)\zeta_p(5s-12)$$
$$\times \zeta_p(7s-9)\zeta_p(7s-14)\zeta_p(9s-15)\zeta_p(11s-16)$$
$$\times W_{\mathcal{G}_3\times\mathfrak{g}_{5,3}}^{\triangleleft}(p,p^{-s}) ,$$

where $W_{\mathcal{G}_3\times\mathfrak{g}_{5,3}}^{\triangleleft}(X,Y)$ *is given in Appendix A on p. 182. This zeta function satisfies the functional equation*

$$\zeta_{\mathcal{G}_3\times\mathfrak{g}_{5,3},p}^{\triangleleft}(s)\bigg|_{p\to p^{-1}} = p^{45-19s}\zeta_{\mathcal{G}_3\times\mathfrak{g}_{5,3},p}^{\triangleleft}(s) .$$

The corresponding global function has abscissa of convergence $\alpha_{\mathcal{G}_3\times\mathfrak{g}_{5,3}}^{\triangleleft} = 6$.

We write $\mathfrak{g}_{6,n}$ for a Lie ring whose presentation is taken from that of the nth Lie algebra in the list in [44]. We have already seen several examples of rank 6, $\mathfrak{g}_{6,1} = L_{(3,2)}$, $\mathfrak{g}_{6,3} = F_{2,3}$, $\mathfrak{g}_{6,4} = F_{2,3}/\langle z\rangle \cdot \mathbb{Z}$ and $\mathfrak{g}_{6,5} = U_3(R_2)$ where R_2 is the ring of integers of a quadratic number field. $\mathfrak{g}_{6,2} = M_5$, whose local zeta functions we have been unable to calculate.

Theorem 2.44 ([64]). *Let the Lie ring* $\mathfrak{g}_{6,6}$ *have presentation*

$$\langle x_1,\ldots,x_6 : [x_1,x_2]=x_4, [x_1,x_3]=x_5, [x_1,x_4]=x_6, [x_2,x_3]=x_6\rangle .$$

Then

$$\zeta_{\mathfrak{g}_{6,6},p}^{\triangleleft}(s) = \zeta_{\mathbb{Z}^3,p}(s)\zeta_p(3s-4)\zeta_p(5s-5)\zeta_p(5s-6)\zeta_p(6s-6)\zeta_p(7s-8)$$
$$\times \zeta_p(9s-11)W_{\mathfrak{g}_{6,6}}^{\triangleleft}(p,p^{-s}) ,$$

where $W_{\mathfrak{g}_{6,6}}^{\triangleleft}(X,Y)$ *is*

$$1 + X^3Y^3 - X^6Y^7 - X^8Y^8 - X^9Y^9 - 2X^{11}Y^{10} - X^{14}Y^{12} + X^{14}Y^{14}$$
$$- X^{15}Y^{14} + X^{15}Y^{15} + X^{17}Y^{16} + X^{17}Y^{17} + X^{19}Y^{17} + X^{20}Y^{19} + X^{21}Y^{19}$$
$$- X^{21}Y^{20} + X^{22}Y^{20} - X^{25}Y^{24} - X^{28}Y^{26} .$$

This local zeta function satisfies no functional equation. The corresponding global zeta function has abscissa of convergence $\alpha_{\mathfrak{g}_{6,6}}^{\triangleleft} = 3$.

Theorem 2.45 ([64]). *Let the Lie ring $\mathfrak{g}_{6,7}$ have presentation*

$$\langle x_1, \ldots, x_6 : [x_1, x_3] = x_4, [x_1, x_4] = x_5, [x_2, x_3] = x_6 \rangle .$$

Then

$$\zeta^{\triangleleft}_{\mathfrak{g}_{6,7},p}(s) = \zeta_{\mathbb{Z}^3,p}(s)\zeta_p(3s-4)\zeta_p(4s-3)\zeta_p(5s-5)\zeta_p(5s-6)\zeta_p(6s-6)$$
$$\times \zeta_p(7s-7)W^{\triangleleft}_{\mathfrak{g}_{6,7}}(p, p^{-s}) ,$$

where $W^{\triangleleft}_{\mathfrak{g}_{6,7}}(X, Y)$ is

$$1 + X^3Y^3 - X^3Y^5 - 2X^6Y^7 - X^7Y^8 - X^9Y^9 - X^{10}Y^{10} + X^9Y^{11} - X^{10}Y^{11}$$
$$+ 2X^{10}Y^{12} + X^{12}Y^{14} + X^{13}Y^{14} + X^{13}Y^{15} + X^{16}Y^{16} - X^{16}Y^{19} - X^{19}Y^{21} .$$

This local zeta function satisfies no functional equation. The corresponding global zeta function has abscissa of convergence $\alpha^{\triangleleft}_{\mathfrak{g}_{6,7}} = 3$.

Theorem 2.46 ([64]). *Let the Lie ring $\mathfrak{g}_{6,8}$ have presentation*

$$\langle x_1, \ldots, x_6 : [x_1, x_2] = x_3 + x_4, [x_1, x_3] = x_5, [x_2, x_4] = x_6 \rangle .$$

Then

$$\zeta^{\triangleleft}_{\mathfrak{g}_{6,8},p}(s) = \zeta_{\mathbb{Z}^3,p}(s)\zeta_p(3s-3)\zeta_p(4s-3)\zeta_p(5s-5)\zeta_p(6s-6)\zeta_p(7s-7)$$
$$\times \zeta_p(8s-8)(1+p^{1-s})W^{\triangleleft}_{\mathfrak{g}_{6,8}}(p, p^{-s}) ,$$

where $W^{\triangleleft}_{\mathfrak{g}_{6,8}}(X, Y)$ is

$$1 - XY + X^2Y^2 - X^3Y^3 + X^3Y^4 + X^4Y^4 - 2X^3Y^5 - X^5Y^5 + 2X^4Y^6$$
$$+ X^6Y^6 - 2X^5Y^7 - 2X^6Y^7 + 3X^6Y^8 - 4X^7Y^9 + 4X^8Y^{10} - 4X^9Y^{11}$$
$$- X^{10}Y^{11} + X^9Y^{12} + 4X^{10}Y^{12} - 4X^{11}Y^{13} + 4X^{12}Y^{14} - 3X^{13}Y^{15}$$
$$+ 2X^{13}Y^{16} + 2X^{14}Y^{16} - X^{13}Y^{17} - 2X^{15}Y^{17} + X^{14}Y^{18} + 2X^{16}Y^{18}$$
$$- X^{15}Y^{19} - X^{16}Y^{19} + X^{16}Y^{20} - X^{17}Y^{21} + X^{18}Y^{22} - X^{19}Y^{23} .$$

This zeta function satisfies the functional equation

$$\zeta^{\triangleleft}_{\mathfrak{g}_{6,8},p}(s)\Big|_{p \to p^{-1}} = p^{15-12s}\zeta^{\triangleleft}_{\mathfrak{g}_{6,8},p}(s) .$$

The corresponding global zeta function has abscissa of convergence $\alpha^{\triangleleft}_{\mathfrak{g}_{6,8}} = 3$.

Theorem 2.47 ([64]). *Let the Lie ring $\mathfrak{g}_{6,9}$ have presentation*

$$\langle x_1, \ldots, x_6 : [x_1, x_2] = x_4, [x_1, x_4] = x_5, [x_1, x_3] = x_6, [x_2, x_4] = x_6 \rangle .$$

Then

$$\zeta^{\triangleleft}_{\mathfrak{g}_{6,9},p}(s) = \zeta_{\mathbb{Z}^3,p}(s)\zeta_p(5s - 5)\zeta_p(6s - 6)\zeta_p(8s - 7)\zeta_p(8s - 8)\zeta_p(14s - 15)$$
$$\times\ W^{\triangleleft}_{\mathfrak{g}_{6,9}}(p, p^{-s})\,,$$

where $W^{\triangleleft}_{\mathfrak{g}_{6,9}}(X, Y)$ *is*

$$1 + X^3Y^3 + X^3Y^4 - X^3Y^5 + X^4Y^5 + X^6Y^6 + X^7Y^7 - X^6Y^8 - X^7Y^9$$
$$+ X^9Y^9 + X^{10}Y^{10} - X^9Y^{11} - X^{10}Y^{11} + X^{11}Y^{11} - X^{10}Y^{12} - X^{11}Y^{12}$$
$$+ X^{12}Y^{12} - X^{11}Y^{13} + X^{13}Y^{13} - X^{12}Y^{14} - X^{13}Y^{14} - X^{13}Y^{15} + X^{13}Y^{16}$$
$$- X^{14}Y^{16} - X^{15}Y^{16} + X^{16}Y^{16} - X^{16}Y^{17} - X^{16}Y^{18} - X^{17}Y^{18} + X^{16}Y^{19}$$
$$- X^{18}Y^{19} + X^{17}Y^{20} - X^{18}Y^{20} - X^{19}Y^{20} + X^{18}Y^{21} - X^{19}Y^{21} - X^{20}Y^{21}$$
$$+ X^{19}Y^{22} + X^{20}Y^{23} - X^{22}Y^{23} - X^{23}Y^{24} + X^{22}Y^{25} + X^{23}Y^{26} + X^{25}Y^{27}$$
$$- X^{26}Y^{27} + X^{26}Y^{28} + X^{26}Y^{29} + X^{29}Y^{32}\,.$$

This zeta function satisfies the functional equation

$$\zeta^{\triangleleft}_{\mathfrak{g}_{6,9},p}(s)\Big|_{p \to p^{-1}} = p^{15-12s}\zeta^{\triangleleft}_{\mathfrak{g}_{6,9},p}(s)\,.$$

The corresponding global zeta function has abscissa of convergence $\alpha^{\triangleleft}_{\mathfrak{g}_{6,9}} = 3$.

Theorem 2.48 ([64]). *Let* $\gamma \in \mathbb{Z} \setminus \{0, 1\}$ *be a squarefree integer. Let the Lie ring* $\mathfrak{g}_{6,10}(\gamma)$ *have presentation*

$$\left\langle x_1, \ldots, x_6 : \begin{array}{l} [x_1, x_2] = x_4, [x_1, x_4] = x_6, [x_1, x_3] = x_5, \\ [x_2, x_3] = x_6, [x_2, x_4] = \alpha x_5 + \beta x_6 \end{array} \right\rangle,$$

where

$$\alpha x_5 + \beta x_6 = \begin{cases} \gamma x_5 & \text{if } \gamma \equiv 2, 3 \pmod 4, \\ \frac{1}{4}(\gamma - 1)x_5 + x_6 & \text{if } \gamma \equiv 1 \pmod 4. \end{cases}$$

Then, if p *is inert in* $\mathbb{Q}(\sqrt{\gamma})$,

$$\zeta^{\triangleleft}_{\mathfrak{g}_{6,10}(\gamma),p}(s) = \zeta_{\mathbb{Z}^3,p}(s)\zeta_p(3s - 3)\zeta_p(5s - 4)\zeta_p(5s - 5)\zeta_p(6s - 6)$$
$$\times\ \zeta_p(8s - 8)\zeta_p(8s - 6)^{-1}\zeta_p(10s - 8)^{-1}\,.$$

If p *splits in* $\mathbb{Q}(\sqrt{\gamma})$ *and either*

- $\gamma \equiv 1 \pmod 4$ *and* $p \nmid \frac{1}{4}(\gamma - 1)$, *or*
- $\gamma \not\equiv 1 \pmod 4$,

then

$$\zeta^{\triangleleft}_{\mathfrak{g}_{6,10}(\gamma),p}(s) = \zeta_{\mathbb{Z}^3,p}(s)\zeta_p(3s - 3)\zeta_p(4s - 3)\zeta_p(5s - 5)\zeta_p(6s - 6)\zeta_p(7s - 7)$$
$$\times\ \zeta_p(8s - 8)(1 + p^{1-s})W^{\triangleleft}_{\mathfrak{g}_{6,8}}(p, p^{-s})\,,$$

where $W^{\lhd}_{\mathfrak{g}_{6,8}}(X,Y)$ is given above on p. 57. For all but finitely many primes, the local zeta function satisfies the functional equation

$$\zeta^{\lhd}_{\mathfrak{g}_{6,10}(\gamma),p}(s)\Big|_{p\to p^{-1}} = p^{15-12s}\zeta^{\lhd}_{\mathfrak{g}_{6,10}(\gamma),p}(s) .$$

Theorem 2.49 ([64]). *Let the Lie ring $\mathfrak{g}_{6,12}$ have presentation*

$$\langle x_1,\ldots,x_6 : [x_1,x_3] = x_5, [x_1,x_5] = x_6, [x_2,x_4] = x_6 \rangle .$$

Then

$$\zeta^{\lhd}_{\mathfrak{g}_{6,12},p}(s) = \zeta_{\mathbb{Z}^4,p}(s)\zeta_p(3s-4)\zeta_p(6s-4)\zeta_p(7s-5)\zeta_p(7s-4)^{-1} ,$$

$$\zeta^{\leq}_{\mathfrak{g}_{6,12},p}(s) = \zeta_{\mathbb{Z}^4,p}(s)\zeta_p(2s-5)\zeta_p(3s-5)\zeta_p(3s-6)\zeta_p(4s-8)\zeta_p(4s-9)$$
$$\times \zeta_p(5s-12)\zeta_p(6s-12)\zeta_p(6s-13)\zeta_p(7s-16)\zeta_p(s-2)^{-1}$$
$$\times W^{\leq}_{\mathfrak{g}_{6,12}}(p,p^{-s}) ,$$

where $W^{\leq}_{\mathfrak{g}_{6,12}}(X,Y)$ is given in Appendix A on p. 183. These zeta functions satisfy the functional equations

$$\zeta^{\lhd}_{\mathfrak{g}_{6,12},p}(s)\Big|_{p\to p^{-1}} = p^{15-13s}\zeta^{\lhd}_{\mathfrak{g}_{6,12},p}(s) ,$$

$$\zeta^{\leq}_{\mathfrak{g}_{6,12},p}(s)\Big|_{p\to p^{-1}} = p^{15-6s}\zeta^{\leq}_{\mathfrak{g}_{6,12},p}(s) .$$

The corresponding global zeta functions have abscissa of convergence $\alpha^{\lhd}_{\mathfrak{g}_{6,12}} = \alpha^{\leq}_{\mathfrak{g}_{6,12}} = 4$.

It can easily be seen that $\mathfrak{g}_{6,12}$ is the direct product with central amalgamation of $\mathcal{H}$ with M_3.

Theorem 2.50.

$$\zeta^{\lhd}_{\mathcal{H}\times\mathfrak{g}_{6,12},p}(s) = \zeta_{\mathbb{Z}^6,p}(s)\zeta_p(3s-6)^2\zeta_p(5s-7)\zeta_p(6s-6)\zeta_p(7s-7)\zeta_p(8s-7)$$
$$\times \zeta_p(9s-8)\zeta_p(11s-14)W^{\lhd}_{\mathcal{H}\times\mathfrak{g}_{6,12}}(p,p^{-s}) ,$$

where $W^{\lhd}_{\mathcal{H}\times\mathfrak{g}_{6,12}}(X,Y)$ is

$$1 - X^6Y^5 - X^6Y^7 - X^6Y^8 + X^6Y^9 - 2X^7Y^9 + X^{12}Y^{11} - 2X^{13}Y^{11}$$
$$+ 2X^{13}Y^{12} - X^{14}Y^{12} + 2X^{13}Y^{13} - X^{14}Y^{13} + X^{14}Y^{14} + 2X^{13}Y^{15}$$
$$- X^{14}Y^{15} + X^{14}Y^{16} + X^{20}Y^{16} + X^{14}Y^{17} + X^{20}Y^{18} + X^{20}Y^{19} - 2X^{19}Y^{20}$$
$$+ 2X^{21}Y^{20} - X^{20}Y^{21} - X^{20}Y^{22} - X^{26}Y^{23} - X^{20}Y^{24} - X^{26}Y^{24} + X^{26}Y^{25}$$
$$- 2X^{27}Y^{25} - X^{26}Y^{26} + X^{26}Y^{27} - 2X^{27}Y^{27} + X^{26}Y^{28} - 2X^{27}Y^{28}$$
$$+ 2X^{27}Y^{29} - X^{28}Y^{29} + 2X^{33}Y^{31} - X^{34}Y^{31} + X^{34}Y^{32} + X^{34}Y^{33} + X^{34}Y^{35}$$
$$- X^{40}Y^{40} .$$

This zeta function satisfies the functional equation

$$\zeta^{\triangleleft}_{\mathcal{H} \times \mathfrak{g}_{6,12},p}(s)\Big|_{p \to p^{-1}} = -p^{36-18s}\zeta^{\triangleleft}_{\mathcal{H} \times \mathfrak{g}_{6,12},p}(s) \ .$$

The corresponding global zeta function has abscissa of convergence $\alpha^{\triangleleft}_{\mathcal{H} \times \mathfrak{g}_{6,12}} = 6$.

Theorem 2.51 ([64]). *Let the Lie ring $\mathfrak{g}_{6,13}$ have presentation*

$$\langle x_1, \ldots, x_6 : [x_1, x_2] = x_5, [x_1, x_3] = x_4, [x_1, x_4] = x_6, [x_2, x_5] = x_6 \rangle \ .$$

Then

$$\zeta^{\triangleleft}_{\mathfrak{g}_{6,13},p}(s) = \zeta_{\mathbb{Z}^3,p}(s)\zeta_p(3s-4)\zeta_p(5s-6)\zeta_p(6s-4)\zeta_p(7s-5)\zeta_p(9s-8)$$
$$\times W^{\triangleleft}_{\mathfrak{g}_{6,13}}(p, p^{-s}) \ ,$$

where $W^{\triangleleft}_{\mathfrak{g}_{6,13}}(X, Y)$ is

$$1 + X^3Y^3 - X^4Y^7 - X^7Y^9 - X^8Y^{10} - X^{11}Y^{12} + X^{12}Y^{16} + X^{15}Y^{19} \ .$$

This zeta function satisfies the functional equation

$$\zeta^{\triangleleft}_{\mathfrak{g}_{6,13},p}(s)\Big|_{p \to p^{-1}} = p^{15-14s}\zeta^{\triangleleft}_{\mathfrak{g}_{6,13},p}(s) \ .$$

The corresponding global zeta function has abscissa of convergence $\alpha^{\triangleleft}_{\mathfrak{g}_{6,13}} = 3$.

Theorem 2.52 ([64]). *Let $\gamma \in \mathbb{Z}$ be a nonzero integer, and let $\mathfrak{g}_{6,14}(\gamma)$ have presentation*

$$\langle x_1, \ldots, x_6 : [x_1, x_3] = x_4, [x_1, x_4] = x_6, [x_2, x_3] = x_5, [x_2, x_5] = \gamma x_6 \rangle \ .$$

Then, for all primes p not dividing γ,

$$\zeta^{\triangleleft}_{\mathfrak{g}_{6,14}(\gamma),p}(s) = \zeta_{\mathbb{Z}^3,p}(s)\zeta_p(3s-3)\zeta_p(3s-4)\zeta_p(5s-6)\zeta_p(6s-3)\zeta_p(7s-5)$$
$$\times \zeta_p(6s-6)^{-1}\zeta_p(7s-3)^{-1} \ .$$

If $p \nmid \gamma$, the local zeta function satisfies the functional equation

$$\zeta^{\triangleleft}_{\mathfrak{g}_{6,14}(\gamma),p}(s)\Big|_{p \to p^{-1}} = p^{15-14s}\zeta^{\triangleleft}_{\mathfrak{g}_{6,14}(\gamma),p}(s) \ .$$

For $\gamma = \pm 1$, the corresponding global zeta function has abscissa of convergence $\alpha^{\triangleleft}_{\mathfrak{g}_{6,14}(\pm 1)} = 3$.

The following proposition has a routine proof which we do not repeat.

Proposition 2.53. *For $\gamma_1, \gamma_2 \neq 0$, let $\mathfrak{g}_{6,14}(\gamma_1)$ and $\mathfrak{g}_{6,14}(\gamma_2)$ be defined over any integral domain or field R. Then $\mathfrak{g}_{6,14}(\gamma_1) \cong \mathfrak{g}_{6,14}(\gamma_2)$ iff $\gamma_1 = u^2\gamma_2$ for some $u \in R^*$.*

It can also be shown that the local zeta functions depend only on the power of p dividing γ. We therefore have the following

Corollary 2.54. *Let $\gamma \in \mathbb{Z}$ be a nonzero integer. Then $\mathfrak{g}_{6,14}(\gamma) \not\cong \mathfrak{g}_{6,14}(-\gamma)$ but $\zeta^{\triangleleft}_{\mathfrak{g}_{6,14}(\gamma)}(s) = \zeta^{\triangleleft}_{\mathfrak{g}_{6,14}(-\gamma)}(s)$.*

The classification of six-dimensional Lie algebras has also given rise to some new calculations in nilpotency class 4. In particular, the second author found the following:

Theorem 2.55 ([64]). *Define the two Lie rings $\mathfrak{g}_{6,15}$ and $\mathfrak{g}_{6,17}$ by the presentations*

$$\mathfrak{g}_{6,15} = \left\langle x_1, x_2, x_3, x_4, x_5, x_6 : \begin{array}{l} [x_1, x_2] = x_3 + x_4, [x_1, x_4] = x_5, \\ [x_1, x_5] = x_6, [x_2, x_3] = x_6 \end{array} \right\rangle,$$

$$\mathfrak{g}_{6,17} = \left\langle x_1, x_2, x_3, x_4, x_5, x_6 : \begin{array}{l} [x_1, x_2] = x_4, [x_1, x_4] = x_5, \\ [x_1, x_5] = x_6, [x_2, x_3] = x_6 \end{array} \right\rangle.$$

Then

$$\begin{aligned} \zeta^{\triangleleft}_{\mathfrak{g}_{6,15},p}(s) = \zeta^{\triangleleft}_{\mathfrak{g}_{6,17},p}(s) &= \zeta_{\mathbb{Z}^3,p}(s)\zeta_p(3s-3)\zeta_p(4s-3)\zeta_p(6s-4)\zeta_p(7s-5) \\ &\times \zeta_p(9s-8) W^{\triangleleft}_{\mathfrak{g}_{6,15}}(p, p^{-s}), \end{aligned} \tag{2.10}$$

where $W^{\triangleleft}_{\mathfrak{g}_{6,15}}(X, Y)$ is

$$1 - X^3 Y^5 + X^4 Y^5 - X^4 Y^7 - X^7 Y^9 + X^7 Y^{11} - X^8 Y^{11} + X^{11} Y^{16}.$$

This zeta function satisfies the functional equation

$$\zeta^{\triangleleft}_{\mathfrak{g}_{6,15},p}(s)\Big|_{p \to p^{-1}} = p^{15-16s} \zeta^{\triangleleft}_{\mathfrak{g}_{6,15},p}(s).$$

The corresponding global zeta function has abscissa of convergence $\alpha^{\triangleleft}_{\mathfrak{g}_{6,15}} = 3$.

It follows from the classification [44] that $\mathfrak{g}_{6,15} \not\cong \mathfrak{g}_{6,17}$, but an appeal to a classification is not an enlightening proof. To be sure, we verify

Proposition 2.56. *$\mathfrak{g}_{6,15}$ and $\mathfrak{g}_{6,17}$ are not isomorphic.*

Proof. The rank of the centraliser of the derived subring is invariant under isomorphism. Firstly, $\mathfrak{g}'_{6,15} = \langle y_3 + y_4, y_5, y_6 \rangle$, which has centraliser $\langle y_3, y_4, y_5, y_6 \rangle$. Secondly, $\mathfrak{g}'_{6,17} = \langle x_4, x_5, x_6 \rangle$, which is centralised by $\langle x_2, x_3, x_4, x_5, x_6 \rangle$. Thus $\mathfrak{g}_{6,15} \not\cong \mathfrak{g}_{6,17}$. $\square$

The only other calculation at nilpotency class 4 this classification leads to is the following:

Theorem 2.57 ([64]). *Let the Lie ring* $\mathfrak{g}_{6,16}$ *have presentation*

$$\left\langle x_1, x_2, x_3, x_4, x_5, x_6 : \begin{array}{c} [x_1, x_3] = x_4, [x_1, x_4] = x_5, [x_1, x_5] = x_6, \\ [x_2, x_3] = x_5, [x_2, x_4] = x_6 \end{array} \right\rangle .$$

Then

$$\zeta_{\mathfrak{g}_{6,16},p}^{\triangleleft}(s) = \zeta_{\mathbb{Z}^3,p}(s)\zeta_p(3s-3)\zeta_p(5s-4)\zeta_p(6s-3)\zeta_p(7s-5)\zeta_p(7s-3)^{-1} .$$

This zeta function satisfies the functional equation

$$\zeta_{\mathfrak{g}_{6,16},p}^{\triangleleft}(s)\Big|_{p \to p^{-1}} = p^{15-17s}\zeta_{\mathfrak{g}_{6,16},p}^{\triangleleft}(s) .$$

The corresponding global zeta function has abscissa of convergence $\alpha_{\mathfrak{g}_{6,16}}^{\triangleleft} = 3$.

2.15 Nilpotent Lie Algebras of Dimension 7

The Lie algebras of dimension 7 over algebraically closed fields and $\mathbb{R}$ were first classified successfully by Gong [26]. Once again, the structure constants of each Lie algebra are all rational integers. This includes the six one-parameter families, providing we restrict the parameter to $\mathbb{Z}$. Hence we can also use this classification to obtain presentations of $\mathbb{Z}$-Lie rings of rank 7.

We write $\mathfrak{g}_{\text{name}}$ for the $\mathbb{Z}$-Lie ring corresponding to the Lie algebra with the label (name) in [26]. For example, $\mathfrak{g}_{1357\text{F}}$ corresponds to (1357F) in [26]. The digits are the dimensions of the terms in the upper-central series, and the suffix letter (when shown) distinguishes non-isomorphic Lie algebras with the same upper-central series dimensions. We have encountered some of these Lie rings before, in particular $\mathfrak{g}_{17} \cong G(3,0)$, $\mathfrak{g}_{37\text{A}} \cong \mathcal{G}_4$, $\mathfrak{g}_{37\text{B}} \cong T_4$, $\mathfrak{g}_{137\text{A}} \cong M_3 \times_{\mathbb{Z}} M_3$ and $\mathfrak{g}_{247\text{A}} \cong L_{(3,3)}$. Furthermore, some of them arise as direct products with central amalgamation: $\mathfrak{g}_{157}$, $\mathfrak{g}_{257\text{K}}$, $\mathfrak{g}_{1457\text{A}}$ and $\mathfrak{g}_{1457\text{B}}$ are the direct products with central amalgamation of $\mathcal{H}$ with $\mathfrak{g}_{5,3}$, $F_{3,2}$, M_4 and Fil_4 respectively.

We saw above that $\mathfrak{g}_{6,15}$ and $\mathfrak{g}_{6,17}$ are non-isomorphic yet their ideal zeta functions are equal. Amongst those calculations in rank 7 we have so far completed, there are no less than seven pairs of normally isospectral Lie rings. We do not provide proof that the Lie rings are non-isomorphic, instead referring the curious reader to [26].

Theorem 2.58. *Let the Lie ring* $\mathfrak{g}_{27\text{A}}$ *have presentation*

$$\langle x_1, x_2, x_3, x_4, x_5, x_6, x_7 : [x_1, x_2] = x_6, [x_1, x_4] = x_7, [x_3, x_5] = x_7 \rangle .$$

Then

$$\zeta_{\mathfrak{g}_{27\text{A}},p}^{\triangleleft}(s) = \zeta_{\mathbb{Z}^5,p}(s)\zeta_p(3s-5)\zeta_p(5s-6)\zeta_p(7s-10)\zeta_p(8s-10)^{-1} .$$

This zeta function satisfies the functional equation

$$\zeta_{\mathfrak{g}_{27\text{A}},p}^{\triangleleft}(s)\Big|_{p \to p^{-1}} = -p^{21-12s}\zeta_{\mathfrak{g}_{27\text{A}},p}^{\triangleleft}(s) .$$

The corresponding global zeta function has abscissa of convergence $\alpha_{\mathfrak{g}_{27\text{A}}}^{\triangleleft} = 5$.

Theorem 2.59. *Let the Lie ring* $\mathfrak{g}_{27B}$ *have presentation*

$$\langle x_1, \ldots, x_7 : [x_1, x_2] = x_6, [x_1, x_5] = x_7, [x_2, x_3] = x_7, [x_3, x_4] = x_6 \rangle .$$

Then

$$\zeta^{\triangleleft}_{\mathfrak{g}_{27B}, p}(s) = \zeta_{\mathbb{Z}^5, p}(s) \zeta_p(5s - 5) \zeta_p(5s - 6) \zeta_p(7s - 10) \zeta_p(10s - 10)^{-1} .$$

This zeta function satisfies the functional equation

$$\zeta^{\triangleleft}_{\mathfrak{g}_{27B}, p}(s) \Big|_{p \to p^{-1}} = -p^{21 - 12s} \zeta^{\triangleleft}_{\mathfrak{g}_{27B}, p}(s) .$$

The corresponding global zeta function has abscissa of convergence $\alpha^{\triangleleft}_{\mathfrak{g}_{27B}} = 5$.

Theorem 2.60. *Let the Lie ring* $\mathfrak{g}_{37C}$ *have presentation*

$$\langle x_1, \ldots, x_7 : [x_1, x_2] = x_5, [x_2, x_3] = x_6, [x_2, x_4] = x_7, [x_3, x_4] = x_5 \rangle .$$

Then $\zeta^{\triangleleft}_{\mathfrak{g}_{37C}, p}(s) = \zeta^{\triangleleft}_{T_4, p}(s)$ *(p. 45).*

Theorem 2.61. *Let the Lie ring* $\mathfrak{g}_{37D}$ *have presentation*

$$\langle x_1, \ldots, x_7 : [x_1, x_2] = x_5, [x_1, x_3] = x_7, [x_2, x_4] = x_7, [x_3, x_4] = x_6 \rangle .$$

Then

$$\zeta^{\triangleleft}_{\mathfrak{g}_{37D}, p}(s) = \zeta_{\mathbb{Z}^4, p}(s) \zeta_p(3s - 5) \zeta_p(5s - 6) \zeta_p(6s - 10) \zeta_p(7s - 12) W^{\triangleleft}_{\mathfrak{g}_{37D}}(p, p^{-s}) ,$$

where $W^{\triangleleft}_{\mathfrak{g}_{37D}}(X, Y)$ *is*

$$1 + X^4 Y^3 + X^8 Y^6 + X^9 Y^6 - X^9 Y^8 - X^{10} Y^8 - X^{14} Y^{11} - X^{18} Y^{14} .$$

This zeta function satisfies the functional equation

$$\zeta^{\triangleleft}_{\mathfrak{g}_{37D}, p}(s) \Big|_{p \to p^{-1}} = -p^{21 - 11s} \zeta^{\triangleleft}_{\mathfrak{g}_{37D}, p}(s) .$$

The corresponding global zeta function has abscissa of convergence $\alpha^{\triangleleft}_{\mathfrak{g}_{37D}} = 4$.

Theorem 2.62. *Let the Lie ring* $\mathfrak{g}_{137B}$ *have presentation*

$$\left\langle x_1, x_2, x_3, x_4, x_5, x_6, x_7 : \begin{array}{l} [x_1, x_2] = x_5, [x_1, x_5] = x_7, [x_2, x_4] = x_7, \\ [x_3, x_4] = x_6, [x_3, x_6] = x_7 \end{array} \right\rangle .$$

Then $\zeta^{\triangleleft}_{\mathfrak{g}_{137B}, p}(s) = \zeta^{\triangleleft}_{M_3 \times_{\mathbb{Z}} M_3, p}(s)$ *(p. 48).*

Theorem 2.63. *Let the Lie rings* $\mathfrak{g}_{137C}$ *and* $\mathfrak{g}_{137D}$ *have presentations*

$$\mathfrak{g}_{137C} = \left\langle x_1, \ldots, x_7 : \begin{array}{l} [x_1, x_2] = x_5, [x_1, x_4] = x_6, [x_1, x_6] = x_7, \\ [x_2, x_3] = x_6, [x_3, x_5] = -x_7 \end{array} \right\rangle ,$$

$$\mathfrak{g}_{137D} = \left\langle x_1, \ldots, x_7 : \begin{array}{l} [x_1, x_2] = x_5, [x_1, x_4] = x_6, [x_1, x_6] = x_7, \\ [x_2, x_3] = x_6, [x_2, x_4] = x_7, [x_3, x_5] = -x_7 \end{array} \right\rangle .$$

Then

$$\zeta^{\triangleleft}_{\mathfrak{g}_{137C},p}(s) = \zeta^{\triangleleft}_{\mathfrak{g}_{137D},p}(s) = \zeta_{\mathbb{Z}^4,p}(s)\zeta_p(3s-4)\zeta_p(5s-5)\zeta_p(6s-9)\zeta_p(7s-4)$$
$$\times\ \zeta_p(9s-6)\zeta_p(11s-10)\zeta_p(12s-10)$$
$$\times\ \zeta_p(16s-11)W^{\triangleleft}_{\mathfrak{g}_{137C}}(p,p^{-s})\,,$$

where $W^{\triangleleft}_{\mathfrak{g}_{137C}}(X,Y)$ *is*

$$1 - X^4Y^8 + X^5Y^8 - X^9Y^8 - X^5Y^9 - X^9Y^{11} - X^{10}Y^{12} + X^9Y^{13} - X^{10}Y^{13}$$
$$+ X^{13}Y^{15} - X^{14}Y^{15} - X^{10}Y^{16} + X^{14}Y^{16} - X^{15}Y^{16} + X^{10}Y^{17} - X^{11}Y^{17}$$
$$+ X^{15}Y^{17} + X^{14}Y^{19} - X^{15}Y^{19} + X^{19}Y^{19} + X^{15}Y^{20} + X^{19}Y^{20} + X^{14}Y^{21}$$
$$+ X^{15}Y^{21} - X^{16}Y^{21} - X^{15}Y^{22} + X^{16}Y^{22} + X^{18}Y^{23} + X^{19}Y^{23} - X^{20}Y^{23}$$
$$- X^{18}Y^{24} - X^{19}Y^{24} + 3X^{20}Y^{24} + X^{15}Y^{25} - X^{23}Y^{26} + X^{24}Y^{26} + X^{19}Y^{27}$$
$$- X^{19}Y^{28} + X^{20}Y^{28} + X^{21}Y^{28} - X^{23}Y^{28} - X^{24}Y^{28} + X^{25}Y^{28} - X^{25}Y^{29}$$
$$- X^{20}Y^{30} + X^{21}Y^{30} - X^{29}Y^{31} - 3X^{24}Y^{32} + X^{25}Y^{32} + X^{26}Y^{32} + X^{24}Y^{33}$$
$$- X^{25}Y^{33} - X^{26}Y^{33} - X^{28}Y^{34} + X^{29}Y^{34} + X^{28}Y^{35} - X^{29}Y^{35} - X^{30}Y^{35}$$
$$- X^{25}Y^{36} - X^{29}Y^{36} - X^{25}Y^{37} + X^{29}Y^{37} - X^{30}Y^{37} - X^{29}Y^{39} + X^{33}Y^{39}$$
$$- X^{34}Y^{39} + X^{29}Y^{40} - X^{30}Y^{40} + X^{34}Y^{40} + X^{30}Y^{41} - X^{31}Y^{41} + X^{34}Y^{43}$$
$$- X^{35}Y^{43} + X^{34}Y^{44} + X^{35}Y^{45} + X^{39}Y^{47} + X^{35}Y^{48} - X^{39}Y^{48} + X^{40}Y^{48}$$
$$- X^{44}Y^{56}\,.$$

This zeta function satisfies the functional equation

$$\zeta^{\triangleleft}_{\mathfrak{g}_{137C},p}(s)\Big|_{p\to p^{-1}} = -p^{21-17s}\zeta^{\triangleleft}_{\mathfrak{g}_{137C},p}(s)\,.$$

The corresponding global zeta function has abscissa of convergence $\alpha^{\triangleleft}_{\mathfrak{g}_{137C}} = 4.$

Theorem 2.64. *Let the Lie rings* $\mathfrak{g}_{147A}$ *and* $\mathfrak{g}_{147B}$ *have presentations*

$$\mathfrak{g}_{147A} = \left\langle x_1,\dots,x_7 : \begin{array}{l}[x_1,x_2]=x_4,[x_1,x_3]=x_5,[x_1,x_6]=x_7,\\ [x_2,x_5]=x_7,[x_3,x_4]=x_7\end{array}\right\rangle,$$

$$\mathfrak{g}_{147B} = \left\langle x_1,\dots,x_7 : \begin{array}{l}[x_1,x_2]=x_4,[x_1,x_3]=x_5,[x_1,x_4]=x_7,\\ [x_2,x_6]=x_7,[x_3,x_5]=x_7\end{array}\right\rangle.$$

Then

$$\zeta^{\triangleleft}_{\mathfrak{g}_{147A},p}(s) = \zeta^{\triangleleft}_{\mathfrak{g}_{147B},p}(s) = \zeta_{\mathbb{Z}^4,p}(s)\zeta_p(3s-4)\zeta_p(3s-5)\zeta_p(5s-8)\zeta_p(7s-6)$$
$$\times\ \zeta_p(6s-8)^{-1}\,.$$

This zeta function satisfies the functional equation

$$\zeta^{\triangleleft}_{\mathfrak{g}_{147A},p}(s)\Big|_{p\to p^{-1}} = -p^{21-16s}\zeta^{\triangleleft}_{\mathfrak{g}_{147A},p}(s)\,.$$

The corresponding global zeta function has abscissa of convergence $\alpha^{\triangleleft}_{\mathfrak{g}_{147A}} = 4.$

Theorem 2.65. *Let $\mathfrak{g}_{157}$ have presentation*

$$\langle x_1, \ldots, x_7 : [x_1, x_2] = x_3, [x_1, x_3] = x_7, [x_2, x_4] = x_7, [x_5, x_6] = x_7 \rangle .$$

Then

$$\zeta^{\triangleleft}_{\mathfrak{g}_{157},p}(s) = \zeta_{\mathbb{Z}^5,p}(s)\zeta_p(3s - 5)\zeta_p(7s - 6) .$$

This zeta function satisfies the functional equation

$$\zeta^{\triangleleft}_{\mathfrak{g}_{157A},p}(s)\Big|_{p \to p^{-1}} = -p^{21-15s}\zeta^{\triangleleft}_{\mathfrak{g}_{157A},p}(s) .$$

The corresponding global zeta function has abscissa of convergence $\alpha^{\triangleleft}_{\mathfrak{g}_{157}} = 5$.

Theorem 2.66. *Let the Lie ring $\mathfrak{g}_{247B}$ have presentation*

$$\langle x_1, \ldots, x_7 : [x_1, x_2] = x_4, [x_1, x_3] = x_5, [x_1, x_4] = x_6, [x_3, x_5] = x_7 \rangle .$$

Then

$$\begin{aligned}
\zeta^{\triangleleft}_{\mathfrak{g}_{247B},p}(s) = {} & \zeta_{\mathbb{Z}^3,p}(s)\zeta_p(3s - 4)\zeta_p(4s - 3)\zeta_p(5s - 5)\zeta_p(5s - 6)\zeta_p(6s - 5) \\
& \times \zeta_p(6s - 6)\zeta_p(7s - 6)\zeta_p(7s - 7)\zeta_p(8s - 7)\zeta_p(8s - 8) \\
& \times \zeta_p(9s - 10)\zeta_p(9s - 11)\zeta_p(10s - 9)\zeta_p(10s - 11)\zeta_p(11s - 10) \\
& \times \zeta_p(11s - 12)\zeta_p(12s - 12)\zeta_p(13s - 13)\zeta_p(s - 1)^{-2} \\
& \times \zeta_p(2s - 2)^{-1}W^{\triangleleft}_{\mathfrak{g}_{247B}}(p, p^{-s})
\end{aligned}$$

for some polynomial $W^{\triangleleft}_{\mathfrak{g}_{247B}}(X, Y)$ of degrees 123 in X and 128 in Y. This zeta function satisfies the functional equation

$$\zeta^{\triangleleft}_{\mathfrak{g}_{247B},p}(s)\Big|_{p \to p^{-1}} = -p^{21-15s}\zeta^{\triangleleft}_{\mathfrak{g}_{247B},p}(s) .$$

The corresponding global zeta function has abscissa of convergence $\alpha^{\triangleleft}_{\mathfrak{g}_{247B}} = 3$.

Theorem 2.67. *Let the Lie rings $\mathfrak{g}_{257A}$ and $\mathfrak{g}_{257C}$ have presentations*

$$\mathfrak{g}_{257A} = \langle x_1, \ldots, x_7 : [x_1, x_2] = x_3, [x_1, x_3] = x_6, [x_1, x_5] = x_7, [x_2, x_4] = x_6 \rangle ,$$
$$\mathfrak{g}_{257C} = \langle x_1, \ldots, x_7 : [x_1, x_2] = x_3, [x_1, x_3] = x_6, [x_2, x_4] = x_6, [x_2, x_5] = x_7 \rangle .$$

Then

$$\begin{aligned}
\zeta^{\triangleleft}_{\mathfrak{g}_{257A},p}(s) = \zeta^{\triangleleft}_{\mathfrak{g}_{257C},p}(s) = {} & \zeta_{\mathbb{Z}^4,p}(s)\zeta_p(3s - 5)\zeta_p(5s - 6)\zeta_p(5s - 8)\zeta_p(7s - 9) \\
& \times W^{\triangleleft}_{\mathfrak{g}_{257A}}(p, p^{-s}) ,
\end{aligned}$$

where

$$W^{\triangleleft}_{\mathfrak{g}_{257A}}(X, Y) = 1 + X^4Y^3 - X^9Y^8 - X^{13}Y^{10} .$$

This zeta function satisfies no functional equation. The corresponding global zeta function has abscissa of convergence $\alpha^{\triangleleft}_{\mathfrak{g}_{257A}} = 4$.

Theorem 2.68. *Let the Lie ring* $\mathfrak{g}_{257B}$ *have presentation*

$$\langle x_1, \ldots, x_7 : [x_1, x_2] = x_3, [x_1, x_3] = x_6, [x_1, x_4] = x_7, [x_2, x_5] = x_7 \rangle \ .$$

Then

$$\zeta_{\mathfrak{g}_{257B}, p}^{\triangleleft}(s) = \zeta_{\mathbb{Z}^4, p}(s)\zeta_p(3s-4)\zeta_p(4s-4)\zeta_p(5s-6)\zeta_p(6s-9)\zeta_p(7s-9)$$
$$\times \zeta_p(8s-10)\zeta_p(12s-15)W_{\mathfrak{g}_{257B}}^{\triangleleft}(p, p^{-s}) \ ,$$

where $W_{\mathfrak{g}_{257B}}^{\triangleleft}(X, Y)$ *is*

$$1 - X^4Y^5 + X^5Y^5 - 2X^9Y^8 - X^9Y^9 - X^{13}Y^{10} + X^{13}Y^{11} - X^{14}Y^{11}$$
$$+ 2X^{13}Y^{12} - 2X^{14}Y^{12} + X^{14}Y^{13} - X^{15}Y^{13} + 2X^{18}Y^{15} - X^{19}Y^{15}$$
$$+ X^{18}Y^{16} + 2X^{19}Y^{17} - X^{20}Y^{17} + X^{23}Y^{18} - X^{22}Y^{19} + X^{23}Y^{19} - X^{23}Y^{20}$$
$$+ 2X^{24}Y^{20} + X^{24}Y^{21} + X^{28}Y^{22} - X^{27}Y^{23} - X^{28}Y^{23} + X^{29}Y^{23} - 2X^{28}Y^{24}$$
$$+ X^{29}Y^{24} - X^{33}Y^{27} - X^{33}Y^{28} - X^{33}Y^{29} - X^{38}Y^{30} + X^{37}Y^{32} + X^{42}Y^{35} \ .$$

This zeta function satisfies no functional equation. The corresponding global zeta function has abscissa of convergence $\alpha_{\mathfrak{g}_{257B}}^{\triangleleft} = 4$.

Theorem 2.69. *Let* $\mathfrak{g}_{257K}$ *have presentation*

$$\langle x_1, \ldots, x_7 : [x_1, x_2] = x_5, [x_1, x_5] = x_6, [x_2, x_5] = x_7, [x_3, x_4] = x_7 \rangle \ .$$

Then

$$\zeta_{\mathfrak{g}_{257K}, p}^{\triangleleft}(s) = \zeta_{\mathbb{Z}^4, p}(s)\zeta_p(3s-4)\zeta_p(4s-4)\zeta_p(5s-5)\zeta_p(6s-5)\zeta_p(7s-6)$$
$$\times \zeta_p(7s-8)\zeta_p(9s-10)W_{\mathfrak{g}_{257K}}^{\triangleleft}(p, p^{-s}) \ ,$$

where $W_{\mathfrak{g}_{257K}}^{\triangleleft}(X, Y)$ *is*

$$1 - X^4Y^5 - X^5Y^7 - X^8Y^9 - X^8Y^{10} + X^8Y^{11} - X^{10}Y^{11} + X^9Y^{12}$$
$$+ X^{12}Y^{13} - X^{13}Y^{13} + X^{13}Y^{14} + 2X^{13}Y^{15} - X^{14}Y^{15} - X^{13}Y^{16} + 2X^{14}Y^{16}$$
$$+ X^{14}Y^{17} - X^{14}Y^{18} + X^{15}Y^{18} + X^{18}Y^{19} - X^{17}Y^{20} + X^{19}Y^{20} - X^{19}Y^{21}$$
$$- X^{19}Y^{22} - X^{22}Y^{24} - X^{23}Y^{26} + X^{27}Y^{31} \ .$$

This zeta function satisfies the functional equation

$$\zeta_{\mathfrak{g}_{257K}, p}^{\triangleleft}(s)\big|_{p \to p^{-1}} = -p^{21-14s}\zeta_{\mathfrak{g}_{257K}, p}^{\triangleleft}(s) \ .$$

The corresponding global zeta function has abscissa of convergence $\alpha_{\mathfrak{g}_{257K}}^{\triangleleft} = 4$.

Theorem 2.70. *Let the Lie ring* $\mathfrak{g}_{1357A}$ *have presentation*

$$\left\langle x_1, \ldots, x_7 : \begin{array}{l} [x_1, x_2] = x_4, [x_1, x_4] = x_5, [x_1, x_5] = x_7, \\ [x_2, x_3] = x_5, [x_2, x_6] = x_7, [x_3, x_4] = -x_7 \end{array} \right\rangle \ .$$

Then

$$\zeta^{\triangleleft}_{\mathfrak{g}1357A,p}(s) = \zeta_{\mathbb{Z}^4,p}(s)\zeta_p(3s - 4)\zeta_p(5s - 5)\zeta_p(7s - 6) .$$

This zeta function satisfies the functional equation

$$\zeta^{\triangleleft}_{\mathfrak{g}1357A,p}(s)\Big|_{p\to p^{-1}} = -p^{21-19s}\zeta^{\triangleleft}_{\mathfrak{g}1357A,p}(s) .$$

The corresponding global zeta function has abscissa of convergence $\alpha^{\triangleleft}_{\mathfrak{g}1357A} = 4.$

Theorem 2.71. *Let the Lie rings* $\mathfrak{g}1357B$ *and* $\mathfrak{g}1357C$ *have presentations*

$$\mathfrak{g}1357B = \left\langle x_1,\ldots,x_7 : \begin{array}{l} [x_1,x_2] = x_4, [x_1,x_4] = x_5, [x_1,x_5] = x_7 \\ [x_2,x_3] = x_5, [x_3,x_4] = -x_7, [x_3,x_6] = x_7 \end{array} \right\rangle ,$$

$$\mathfrak{g}1357C = \left\langle x_1,\ldots,x_7 : \begin{array}{l} [x_1,x_2] = x_4, [x_1,x_4] = x_5, [x_1,x_5] = x_7, \\ [x_2,x_3] = x_5, [x_2,x_4] = x_7, \\ [x_3,x_4] = -x_7, [x_3,x_6] = x_7 \end{array} \right\rangle .$$

Then

$$\zeta^{\triangleleft}_{\mathfrak{g}1357B,p}(s) = \zeta^{\triangleleft}_{\mathfrak{g}1357C,p}(s) = \zeta_{\mathbb{Z}^4,p}(s)\zeta_p(3s - 4)\zeta_p(5s - 5)\zeta_p(7s - 4)\zeta_p(9s - 6)$$
$$\times \zeta_p(11s - 10)\zeta_p(16s - 11)W^{\triangleleft}_{\mathfrak{g}1357B}(p,p^{-s}) ,$$

where $W^{\triangleleft}_{\mathfrak{g}1357B}(X,Y)$ *is*

$$1 - X^4Y^8 + X^5Y^8 - X^5Y^9 - X^9Y^{11} + X^9Y^{12} - X^{10}Y^{12} - X^{10}Y^{16}$$
$$+ X^{10}Y^{17} - X^{11}Y^{17} + X^{14}Y^{19} - X^{15}Y^{19} + X^{15}Y^{20} + X^{15}Y^{25} + X^{19}Y^{27}$$
$$- X^{19}Y^{28} + X^{21}Y^{28} - X^{25}Y^{36} .$$

This zeta function satisfies no functional equation. The corresponding global zeta function has abscissa of convergence $\alpha^{\triangleleft}_{\mathfrak{g}1357B} = 4.$

Theorem 2.72. *Let the Lie rings* $\mathfrak{g}1357G$ *and* $\mathfrak{g}1357H$ *have presentations*

$$\mathfrak{g}1357G = \left\langle x_1,\ldots,x_7 : \begin{array}{l} [x_1,x_2] = x_3, [x_1,x_4] = x_6, [x_1,x_6] = x_7 \\ [x_2,x_3] = x_5, [x_2,x_5] = x_7 \end{array} \right\rangle ,$$

$$\mathfrak{g}1357H = \left\langle x_1,\ldots,x_7 : \begin{array}{l} [x_1,x_2] = x_3, [x_1,x_4] = x_6, [x_1,x_6] = x_7, \\ [x_2,x_3] = x_5, [x_2,x_5] = x_7, [x_2,x_6] = x_7, \\ [x_3,x_4] = -x_7 \end{array} \right\rangle .$$

Then

$$\zeta^{\triangleleft}_{\mathfrak{g}1357G,p}(s) = \zeta^{\triangleleft}_{\mathfrak{g}1357H,p}(s) = \zeta_{\mathbb{Z}^3,p}(s)\zeta_p(3s - 4)\zeta_p(4s - 3)\zeta_p(5s - 5)\zeta_p(5s - 6)$$
$$\times \zeta_p(6s - 6)\zeta_p(7s - 4)\zeta_p(7s - 7)\zeta_p(8s - 5)$$
$$\times \zeta_p(9s - 6)\zeta_p(10s - 9)\zeta_p(11s - 8)\zeta_p(12s - 10)$$
$$\times \zeta_p(12s - 11)W^{\triangleleft}_{\mathfrak{g}1357G}(p,p^{-s})$$

where $W^{\triangleleft}_{\mathfrak{g}1357G}(X,Y)$ *is given in Appendix A on p. 184. This zeta function satisfies no functional equation. The corresponding global zeta function has abscissa of convergence* $\alpha^{\triangleleft}_{\mathfrak{g}1357G} = 3.$

Theorem 2.73. *Let* $\mathfrak{g}_{1457A}$ *have the presentation*

$$\langle x_1, \ldots, x_7 : [x_1, x_2] = x_5, [x_1, x_5] = x_6, [x_1, x_6] = x_7, [x_3, x_4] = x_7 \rangle .$$

Then

$$\begin{aligned}
\zeta^{\triangleleft}_{\mathfrak{g}_{1457A}, p}(s) = {}& \zeta_{\mathbb{Z}^4, p}(s) \zeta_p(3s - 4) \zeta_p(4s - 4) \zeta_p(5s - 5) \zeta_p(7s - 4) \zeta_p(9s - 6) \\
& \times \zeta_p(10s - 9) \zeta_p(11s - 10) \zeta_p(12s - 10) \zeta_p(15s - 10) \\
& \times \zeta_p(16s - 11) W^{\triangleleft}_{\mathfrak{g}_{1457A}}(p, p^{-s}) ,
\end{aligned}$$

where $W^{\triangleleft}_{\mathfrak{g}_{1457A}}(X, Y)$ *is given in Appendix A on p. 186. This zeta function satisfies the functional equation*

$$\zeta^{\triangleleft}_{\mathfrak{g}_{1457A}, p}(s)\Big|_{p \to p^{-1}} = -p^{21-18s} \zeta^{\triangleleft}_{\mathfrak{g}_{1457A}, p}(s) .$$

The corresponding global zeta function has abscissa of convergence $\alpha^{\triangleleft}_{\mathfrak{g}_{1457A}} = 4$.

Theorem 2.74. *Let* $\mathfrak{g}_{1457B}$ *have presentation*

$$\left\langle x_1, \ldots, x_7 : \begin{array}{c} [x_1, x_2] = x_5, [x_1, x_5] = x_6, [x_1, x_6] = x_7, \\ [x_2, x_5] = x_7, [x_3, x_4] = x_7 \end{array} \right\rangle .$$

Then

$$\begin{aligned}
\zeta^{\triangleleft}_{\mathfrak{g}_{1457B}, p}(s) = {}& \zeta_{\mathbb{Z}^4, p}(s) \zeta_p(3s - 4) \zeta_p(4s - 4) \zeta_p(5s - 5) \zeta_p(7s - 4) \zeta_p(9s - 6) \\
& \times \zeta_p(10s - 9) \zeta_p(11s - 10) \zeta_p(12s - 10) \zeta_p(16s - 11) \\
& \times W^{\triangleleft}_{\mathfrak{g}_{1457B}}(p, p^{-s}) ,
\end{aligned}$$

where $W^{\triangleleft}_{\mathfrak{g}_{1457B}}(X, Y)$ *is given in Appendix A on p. 187. This zeta function satisfies no functional equation. The corresponding global zeta function has abscissa of convergence* $\alpha^{\triangleleft}_{\mathfrak{g}_{1457B}} = 4$.

3

Soluble Lie Rings

3.1 Introduction

In this chapter, we present some calculations of zeta functions of soluble (but non-nilpotent) Lie rings over $\mathbb{Z}$. Since these Lie rings are not nilpotent, the Mal'cev correspondence cannot be used, and so there is no corresponding $\mathfrak{T}$-group whose local zeta functions we are also calculating. We prove that the zeta functions we consider behave in a similar fashion to those of nilpotent Lie rings. What is remarkable is that the uniform behaviour is 'stronger' than that seen with the nilpotent Lie rings.

Theorem 3.1. *For $n \in \mathbb{N}_{>0}$, let $\mathfrak{tr}_n(\mathbb{Z})$ denote the set of upper-triangular $n \times n$ matrices, with the Lie bracket given by the familiar commutator $[x, y] = xy - yx$. For each $n \in \mathbb{N}_{>0}$ there exists a univariate rational function $R_n(Y)$, with $R_n(0) = 1$, such that*

$$\zeta^{\triangleleft}_{\mathfrak{tr}_n(\mathbb{Z}),p}(s) = \zeta^{\triangleleft}_{\mathbb{Z}^n,p}(s) R_n(p^{-s})$$

for all primes p. Furthermore,

$$\zeta^{\triangleleft}_{\mathfrak{tr}_n(\mathbb{Z}),p}(s)|_{p \to p^{-1}} = (-1)^{\frac{1}{2}n(n+1)} p^{\binom{n}{2} - \frac{1}{6}(2n^3 + 3n^2 - 5n + 6)s} \zeta^{\triangleleft}_{\mathfrak{tr}_n(\mathbb{Z}),p}(s) \qquad (3.1)$$

for all primes p.

We note in passing the following corollary of Theorem 3.1.

Corollary 3.2. *The abscissa of convergence of $\zeta^{\triangleleft}_{\mathfrak{tr}_n(\mathbb{Z})}(s)$ is $\alpha^{\triangleleft}_{\mathfrak{tr}_n(\mathbb{Z})} = n$, with a simple pole at $s = n$.*

Proof. We have that

$$\zeta^{\triangleleft}_{\mathfrak{tr}_n(\mathbb{Z})}(s) = \zeta^{\triangleleft}_{\mathbb{Z}^n}(s) \prod_p R_n(p^{-s}) .$$

It is well-known that $\zeta^{\triangleleft}_{\mathbb{Z}^n}(s)$ has abscissa of convergence n with a simple pole at $s = n$. $\mathfrak{tr}_1(\mathbb{Z}) \cong \mathbb{Z}$, so the result is clear for $n = 1$, and for $n \geq 2$, $\prod_p R_n(p^{-s})$ converges for $\Re(s) > 1$. $\qquad \square$

The proof of Theorem 3.1 is combinatorial. The following result, due to Stanley [55, Proposition 7.1], plays a crucial part in the proof:

Theorem 3.3. *Let* $\mathbf{E}$ *be a system of homogeneous linear equations in* k *variables* $\mathbf{a} = (a_1, \ldots, a_k)$ *with coefficients in* $\mathbb{Z}$. *Let* $\mathbf{S_E}$ *be the solution set of* $\mathbf{E}$ *over* $\mathbb{N}$ *and* $\bar{\mathbf{S}}_{\mathbf{E}}$ *the solution set over* $\mathbb{N}_{>0}$. *Let* $\mathbf{X} = (X_1, \ldots, X_k)$ *be* k *commuting indeterminates and use the notation* $\mathbf{X^a} = X_1^{a_1} \ldots X_k^{a_k}$ *and* $1/\mathbf{X} = (1/X_1, \ldots, 1/X_k)$. *Define the generating functions*

$$F(\mathbf{E}; \mathbf{X}) = \sum_{\mathbf{a} \in \mathbf{S_E}} \mathbf{X^a} , \qquad \bar{F}(\mathbf{E}; \mathbf{X}) = \sum_{\mathbf{a} \in \bar{\mathbf{S}}_{\mathbf{E}}} \mathbf{X^a} .$$

Then F *and* $\bar{F}$ *are rational functions in* $\mathbf{X}$. *Furthermore, if* $\bar{\mathbf{S}}_{\mathbf{E}} \neq \varnothing$, *then*

$$F(\mathbf{E}; 1/\mathbf{X}) = (-1)^{\kappa} \bar{F}(\mathbf{E}; \mathbf{X}) , \qquad (3.2)$$

where $\kappa = \kappa(\mathbf{E})$ *is the corank of* $\mathbf{E}$.

Stanley's theorem applies to systems of linear equations, and it can easily be generalised to linear inequalities:

Corollary 3.4. *Let* $\mathbf{I}$ *be a system of* $k - r$ *homogeneous linear inequalities in* r *variables* $\mathbf{b} = (b_1, \ldots, b_r)$ *with coefficients in* $\mathbb{Z}$. *Let* $\mathbf{S_I}$ *be the solution set of* $\mathbf{I}$ *over* $\mathbb{N}$, *and let* $\mathbf{Y} = (Y_1, \ldots, Y_r)$ *be* r *commuting indeterminates. There exists a system of linear equations* $\mathbf{E}$ *of corank* r *such that we may write*

$$F(\mathbf{I}; \mathbf{Y}) := \sum_{\mathbf{b} \in \mathbf{S_I}} \mathbf{Y^b} = F(\mathbf{E}; \mathbf{X})$$

for suitable $\mathbf{X} = (X_1, \ldots, X_k)$ *depending on* $\mathbf{Y}$. *In particular,* $F(\mathbf{I}; \mathbf{Y})$ *is a rational function in* $\mathbf{Y}$.

Proof. The system of equations $\mathbf{E}$ is obtained from $\mathbf{I}$ by adding a distinct slack variable to the inferior side of each inequality. Clearly, there is a bijective correspondence between $\mathbf{S_I}$ and $\mathbf{S_E}$. Since each equation in $\mathbf{E}$ has a unique slack variable, the corank $\kappa(\mathbf{E}) = r$. Hence we may consider the generating function of $\mathbf{S_E}$, $F(\mathbf{E}; \mathbf{X})$, where $\mathbf{X} = (X_1, \ldots, X_k)$ is a vector of commuting indeterminates. To obtain $F(\mathbf{I}; \mathbf{Y})$ from $F(\mathbf{E}; \mathbf{X})$, we set $X_i = Y_i$ for $1 \leq i \leq r$ and $X_i = 1$ for $r + 1 \leq i \leq k$.

However, we must check that setting $X_i = 1$ for $r + 1 \leq i \leq k$ gives us a well-defined rational function. $F(\mathbf{E}; (0, \ldots, 0, X_{r+1}, \ldots, X_k))$ is the generating function of all solutions to $\mathbf{E}$ with $a_1 = \cdots = a_r = 0$. Clearly this counts only the trivial solution $\mathbf{0}$, i.e. $F(\mathbf{E}; (0, \ldots, 0, X_{r+1}, \ldots, X_k)) = 1$. Hence the denominator of $F(\mathbf{E}; \mathbf{X})$ can have no factors of the form $(1 - \mathbf{X^c})$ with $c_i = 0$ for all $1 \leq i \leq r$, so by setting $X_i = Y_i$ for $1 \leq i \leq r$ and $X_i = 1$ for $r + 1 \leq i \leq k$, we obtain a well-defined rational function in $\mathbf{Y}$. $\square$

Corollary 3.5. *Assume the notation of Theorem 3.3. Suppose in addition that $\mathbf{1} = (1, 1, \ldots, 1) \in \bar{\mathbf{S}}_{\mathbf{E}}$. Then*

$$F(\mathbf{E}; 1/\mathbf{X}) = (-1)^{\kappa} \mathbf{X}^{\mathbf{1}} F(\mathbf{E}; \mathbf{X}) . \tag{3.3}$$

Proof. If $\mathbf{a} \in \mathbb{N}_{>0}^k$, then $\mathbf{a} \in \bar{\mathbf{S}}_{\mathbf{E}}$ if and only if $\mathbf{a} - \mathbf{1} \in \mathbf{S}_{\mathbf{E}}$. Hence

$$\bar{F}(\mathbf{E}; \mathbf{X}) = \sum_{\mathbf{a} \in \bar{\mathbf{S}}_{\mathbf{E}}} \mathbf{X}^{\mathbf{a}} = \sum_{\mathbf{a} - \mathbf{1} \in \mathbf{S}_{\mathbf{E}}} \mathbf{X}^{\mathbf{a}} = \sum_{\mathbf{a} \in \mathbf{S}_{\mathbf{E}}} \mathbf{X}^{\mathbf{a}+1} = \mathbf{X}^{\mathbf{1}} F(\mathbf{E}; \mathbf{X}) .$$

Clearly $\bar{\mathbf{S}}_{\mathbf{E}} \neq \varnothing$, so Theorem 3.3 implies the result. $\qquad\square$

3.2 Proof of Theorem 3.1

We prove Theorem 3.1 by representing $\zeta_{\mathrm{tr}_n(\mathbb{Z}),p}^{\triangleleft}(s)$ as a generating function of the form $F(\mathbf{I}; \mathbf{Y})$ for the ideals of $\mathrm{tr}_n(\mathbb{Z})$ of p-power index in $\mathrm{tr}_n(\mathbb{Z})$. The proof is broken up into a number of stages.

3.2.1 Choosing a Basis for $\mathrm{tr}_n(\mathbb{Z})$

The most obvious basis for $\mathrm{tr}_n(\mathbb{Z})$ is the $N := \frac{1}{2}n(n+1)$ elementary $n \times n$ matrices whose nonzero entries are on or above the leading diagonal. However, this basis is unsuitable for our purposes. We therefore present an alternative choice of basis.

Let $E_{j,k}$ denote the elementary matrix with a 1 in the (j, k) entry and zeros elsewhere. Let f be any bijection $f : \{ (j, k) : 1 \leq j \leq k \leq n \} \to \{1, \ldots, N\}$ such that for all $1 \leq j_1 \leq k_1 \leq n$, $1 \leq j_2 \leq k_2 \leq n$ with $k_2 - j_2 > k_1 - j_1$, $f(j_2, k_2) > f(j_1, k_1)$. We choose the basis for $\mathrm{tr}_n(\mathbb{Z})$ to be $(\mathbf{e}_1, \ldots, \mathbf{e}_N)$, where

$$\mathbf{e}_i = \begin{cases} \sum_{j=i}^{n} E_{j,j} & \text{if } 1 \leq i \leq n, \\ E_{j,k} & \text{if } n+1 \leq i \leq N, \text{ where } (j, k) = f^{-1}(i) . \end{cases} \tag{3.4}$$

Intuitively, the n diagonal basis elements are followed by the $n - 1$ elementary matrices whose nonzero entry is on the first superdiagonal, and then the $n - 2$ elementary matrices with nonzero entry on the second superdiagonal, and so on. This basis has the property that if $1 \leq j_1 \leq j_2 < k_2 \leq k_2 \leq n$ and $(j_1, k_1) \neq (j_2, k_2)$, then E_{j_2,k_2} precedes E_{j_1,k_1}. Also, $\mathbf{e}_1$ is the identity matrix.

The following lemma provides justification for our choice of diagonal basis elements:

Lemma 3.6. *Let $1 \leq j < k \leq n$, $1 \leq i \leq n$. Then*

$$[E_{j,k}, \mathbf{e}_i] = \begin{cases} E_{j,k} & \text{if } j < i \leq k , \\ 0 & \text{otherwise.} \end{cases}$$

Proof. We split into three cases:

1. $i \leq j < k$: $[E_{j,k}, \mathbf{e}_i] = [E_{j,k}, E_{j,j}] + [E_{j,k}, E_{k,k}] = 0$,
2. $j < i \leq k$: $[E_{j,k}, \mathbf{e}_i] = [E_{j,k}, E_{k,k}] = E_{j,k}$,
3. $j < k < i$: $[E_{j,k}, \mathbf{e}_i] = 0$.

□

Corollary 3.7. *Suppose* $1 \leq j < k \leq n$, $1 \leq j_1 < k_1 \leq n$. *Then*

$$[E_{j_1,k_1}, \mathbf{e}_{j+1}, \mathbf{e}_k] = \begin{cases} E_{j',k'} & \text{if } 1 \leq j_1 \leq j < k \leq k_1 \leq n, \\ 0 & \text{otherwise}. \end{cases}$$

Proof. Immediate from Lemma 3.6.

□

3.2.2 Determining the Conditions

Any additive submodule of $\mathbb{Z}^N$ of finite index can be additively generated by $\mathbf{m}_1, \ldots, \mathbf{m}_N$, where

$$\mathbf{m}_i = \sum_{j=i}^{N} m_{i,j} \mathbf{e}_j$$

and $m_{i,j} \in \mathbb{Z}$ for $1 \leq i \leq j \leq N$. If we additionally stipulate that $m_{i,i} > 0$ for all $1 \leq i \leq N$ and $0 \leq m_{i,j} < m_{j,j}$ for $1 \leq i < j \leq N$, then each additive submodule has a unique such generating set. The index of this additive submodule is $\prod_{i=1}^{N} m_{i,i}$, the determinant of the $N \times N$ matrix with the $m_{i,j}$ as entries. We require the index to be a power of p, and this is achieved by ensuring that $m_{i,i}$ is a power of p for each $1 \leq i \leq N$.

For the additive submodule to be an ideal, we must also ensure that

$$[\mathbf{m}_i, \mathbf{e}_j] \in \langle \mathbf{m}_1, \ldots, \mathbf{m}_N \rangle_{\mathbb{Z}} \qquad (3.5)$$

for $1 \leq i, j \leq N$. These requirements give rise to a number of polynomial divisibility conditions amongst the $m_{i,j}$. Our next task is to determine these conditions explicitly.

Lemma 3.8. $m_{i,j} = 0$ *for all* $1 \leq i < j \leq N$, $j > n$.

Proof. We prove by reverse induction on j that $m_{i,j} = 0$ for all $1 \leq i < j$. The base case is $j = N$. Suppose $1 \leq i < N$. By Corollary 3.7,

$$[\mathbf{m}_i, \mathbf{e}_2, \mathbf{e}_n] = m_{i,N} \mathbf{e}_N.$$

This must lie within the $\mathbb{Z}$-span of $\mathbf{m}_1, \ldots, \mathbf{m}_N$, and so $m_{N,N} \mid m_{i,N}$. Since we are assuming $0 \leq m_{i,N} < m_{N,N}$, $m_{i,N} = 0$.

By the inductive hypothesis, assume $m_{i_1,j_1} = 0$ for all i_1, j_1 such that $j < j_1 \leq N$, $1 \leq i_1 < j_1$. In particular, this implies that $\mathbf{m}_j = m_{j,j} \mathbf{e}_j$. Let $(a, b) = f^{-1}(j)$. Since $j > n$, $a < b$. By Corollary 3.7,

$$[\mathbf{m}_i, \mathbf{e}_{a+1}, \mathbf{e}_b] = m_{i,j}\mathbf{e}_j .$$

Since $\mathbf{m}_j = m_{j,j}\mathbf{e}_j$, we once again have $m_{j,j} \mid m_{i,j}$ and $0 \le m_{i,j} < m_{j,j}$, so $m_{i,j} = 0$ for all $1 \le i < j$. This establishes our induction. □

Lemma 3.8 implies that $[\mathbf{m}_i, \mathbf{e}_j] \in \{0, \mathbf{m}_i\}$ for $1 \le i \le N$, $1 \le j \le n$, so we have now satisfied (3.5) for $1 \le j \le n$. It also implies that $\mathbf{m}_i = m_{i,i}\mathbf{e}_i$ for $n+1 \le i \le N$. Recall that each such $\mathbf{e}_i$ is the elementary matrix $E_{j,k}$ where $(j,k) = f^{-1}(i)$. These elementary matrices are more naturally indexed by the pair (j,k) than by the ordering imposed by the bijection f, so we shall relabel the coefficients of the off-diagonal basis elements accordingly. Set $n_{j,k} = m_{f(j,k),f(j,k)}$ for $1 \le j < k \le n$, so that $\mathbf{m}_{f(j,k)} = n_{j,k}E_{j,k}$.

We now determine the conditions the $n_{j,k}$ must satisfy among themselves. These arise from ensuring that

$$[n_{j_1,k_1}E_{j_1,k_1}, E_{j_2,k_2}] \in \langle n_{j,k}E_{j,k} : 1 \le j < k \le n \rangle \tag{3.6}$$

for $1 \le j_1 < k_1 \le n$, $1 \le j_2 < k_2 \le n$.

Lemma 3.9. *Suppose* $1 \le j_1 < k_1 \le n$, $1 \le j_2 < k_2 \le n$. *Then*

$$[E_{j_1,k_1}, E_{j_2,k_2}] = \begin{cases} E_{j_1,k_2} & \text{if } k_1 = j_2 , \\ -E_{j_2,k_1} & \text{if } j_1 = k_2 , \\ 0 & \text{otherwise} . \end{cases}$$

In particular, if $k_1 \ne j_2$ and $j_1 \ne k_2$, $[n_{j_1,k_1}E_{j_1,k_1}, E_{j_2,k_2}] = 0$.

Proof. Routine matrix calculations. □

Lemma 3.10. *Suppose that $1 \le j_1 < k_1 \le n$ and $1 \le j_2 < k_2 \le n$. If either $j_1 = j_2$ and $k_2 < k_1$, or $k_1 = k_2$ and $j_1 < j_2$, then $n_{j_1,k_1} \mid n_{j_2,k_2}$.*

Proof. If $j_1 = j_2$, $k_2 < k_1$, then $[n_{j_2,k_2}E_{j_2,k_2}, E_{k_2,k_1}] = n_{j_2,k_2}E_{j_1,k_1}$, and if $k_1 = k_2$, $j_1 < j_2$, then $[n_{j_2,k_2}E_{j_2,k_2}, E_{j_1,j_2}] = -n_{j_2,k_2}E_{j_1,k_1}$. Either way, to satisfy (3.6) we require $n_{j_1,k_1} \mid n_{j_2,k_2}$. □

Corollary 3.11. *If $1 \le j_1 \le j_2 < k_2 \le k_1 \le n$, then $n_{j_1,k_1} \mid n_{j_2,k_2}$.*

Proof. Using Lemma 3.10 at most twice, $n_{j_1,k_1} \mid n_{j_2,k_1} \mid n_{j_2,k_2}$. □

We therefore require the conditions $n_{j,k} \mid n_{j,k-1}$ for $1 \le j < k-1 < n$ and $n_{j-1,k} \mid n_{j,k}$ for $2 \le j < k < n$. All other conditions that the $n_{j,k}$ satisfy among themselves are implied by these conditions.

Finally, we consider Lie brackets of the form $[\mathbf{m}_i, E_{j,k}]$ for $i \le n$, $1 \le j < k \le n$. By Lemma 3.6, $[\mathbf{m}_i, E_{j,k}] = -\left(\sum_{r=j+1}^{k} m_{i,r}\right)E_{j,k}$, and this gives rise to the condition

$$n_{j,k} \left| \sum_{r=j+1}^{k} m_{i,r} \right. . \tag{3.7}$$

If $k = j + 1$, (3.7) reduces to the monomial condition $n_{j,j+1} \mid m_{i,j+1}$. If $j < k - 1$, the monomial conditions of the form (3.7) and Corollary 3.11 together imply that $n_{j,k} \mid n_{j_1,j_1+1} \mid m_{i,j_1+1}$ for $j \leq j_1 < k$. Hence we shall only need to enforce the conditions (3.7) for $k = j + 1$.

3.2.3 Constructing the Zeta Function

Collecting together all the conditions we derived in the previous section, we have

$$m_{i,i} \text{ is a power of } p \text{ for } 1 \leq i \leq n, \tag{3.8}$$
$$n_{j,k} \text{ is a power of } p \text{ for } 1 \leq j < k \leq n, \tag{3.9}$$
$$0 \leq m_{i,j} < m_{j,j} \text{ for } 1 \leq i < j \leq n, \tag{3.10}$$
$$n_{j,j+1} \mid m_{i,j+1} \text{ for } 1 \leq i \leq j+1 \leq n, \tag{3.11}$$
$$n_{j,k} \mid n_{j,k-1} \text{ for } 1 \leq j < k \leq n, \ k - j \geq 2, \tag{3.12}$$
$$n_{j,k} \mid n_{j+1,k} \text{ for } 1 \leq j < k \leq n, \ k - j \geq 2. \tag{3.13}$$

Substituting $m_{i,i} = p^{A_i}$ and $n_{j,k} = p^{B_{j,k}}$ for $1 \leq i \leq n$, $1 \leq j < k \leq n$, with each $A_i, B_{j,k} \in \mathbb{N}$ eliminates (3.8) and (3.9) and splits (3.11) into two separate sets of conditions (3.15) and (3.16):

$$0 \leq m_{i,j+1} < p^{A_{j}+1} \text{ for } 1 \leq i < j+1 \leq n, \tag{3.14}$$
$$B_{j,j+1} \leq A_{j+1} \text{ for } 1 \leq j \leq n-1, \tag{3.15}$$
$$p^{B_{j,j+1}} \mid m_{i,j+1} \text{ for } 1 \leq i < j+1 \leq n, \tag{3.16}$$
$$B_{j,k} \leq B_{j,k-1} \text{ for } 1 \leq j < k \leq n, \ k - j \geq 2, \tag{3.17}$$
$$B_{j,k} \leq B_{j+1,k} \text{ for } 1 \leq j < k \leq n, \ k - j \geq 2. \tag{3.18}$$

Let W denote the set of all

$$(A_1, \ldots, A_n, m_{1,2}, m_{1,3}, \ldots, m_{n-1,n}, B_{1,2}, B_{1,3}, \ldots, B_{n-1,n}) \in \mathbb{N}^{n^2}$$

satisfying (3.14)–(3.18). We therefore have

$$\zeta_{\mathfrak{tr}_n(\mathbb{Z}),p}^{\triangleleft}(s) = \sum_W \left(\prod_{i=1}^{n} p^{-A_i s} \prod_{1 \leq j < k \leq n} p^{-B_{j,k} s} \right). \tag{3.19}$$

3.2.4 Transforming the Conditions

As they stand, the conditions (3.14)–(3.18) are not sufficient to deduce our results. Some changes of variable are necessary to transform the conditions.

For $1 \leq i < j + 1 \leq n$, set $m_{i,j+1} = m'_{i,j+1} p^{B_{j,j+1}}$. and for $1 \leq j \leq n - 1$, set $A_{j+1} = A'_{j+1} + B_{j,j+1}$. For notational simplicity, we also set $A_1 = A'_1$. These changes eliminate the conditions (3.15) and (3.16), and (3.14) becomes

$$0 \leq m'_{i,j+1} < p^{A'_{j+1}} \text{ for } 1 \leq i < j + 1 \leq n .$$

Since there are no other restrictions on the $m'_{i,j}$, we may sum over the $m'_{i,j}$ for $1 \leq i < j \leq n$ to obtain

$$\zeta^{\triangleleft}_{\mathfrak{tr}_n(\mathbb{Z}),p}(s) = \sum_{W'} \left(\prod_{i=1}^{n} p^{A'_i(i-1-s)} \prod_{1 \leq j < n} p^{-2B_{j,j+1}s} \prod_{\substack{1 \leq j < k \leq n \\ k-j \geq 2}} p^{-B_{j,k}s} \right) ,$$

where W' is the set of all $(A_1, \ldots, A_n, B_{1,2}, B_{1,3}, \ldots, B_{n-1,n}) \in \mathbb{N}^N$ satisfying the conditions (3.17) and (3.18). Furthermore, these conditions are independent of the A'_i, so we sum the A'_i to obtain a factor $\zeta^{\triangleleft}_{\mathbb{Z}^n,p}(s) = \prod_{i=1}^{n} \zeta_p(s - (i-1))$. Thus

$$\zeta^{\triangleleft}_{\mathfrak{tr}_n(\mathbb{Z}),p}(s) = \zeta^{\triangleleft}_{\mathbb{Z}^n,p}(s) \sum_{W''} \left(\prod_{1 \leq j < n} p^{-2B_{j,j+1}s} \prod_{\substack{1 \leq j < k \leq n \\ k-j \geq 2}} p^{-B_{j,k}s} \right) , \qquad (3.20)$$

where W'' is the set of all $(B_{1,2}, B_{1,3}, \ldots, B_{n-1,n}) \in \mathbb{N}^{\frac{1}{2}n(n-1)}$ satisfying (3.17) and (3.18). Set

$$R_n(Y) = \sum_{W''} \left(\prod_{1 \leq j < n} Y^{2B_{j,j+1}} \prod_{\substack{1 \leq j < k \leq n \\ k-j \geq 2}} Y^{B_{j,k}} \right) ,$$

so that $\zeta^{\triangleleft}_{\mathfrak{tr}_n(\mathbb{Z}),p}(s) = \zeta^{\triangleleft}_{\mathbb{Z}^n,p}(s) R_n(p^{-s})$. $R_n(Y)$ is clearly independent of p, and so the first part of Theorem 3.1 now follows from Corollary 3.4.

3.2.5 Deducing the Functional Equation

There is still work to do to prove the functional equation (3.1). The next step is to eliminate the conditions (3.17). For $1 \leq j < k < n$, set $B_{j,k} = B'_{j,k} + B_{j,k+1}$. For the sake of notational simplicity, we also set $B'_{j,n} = B_{j,n}$ for $1 \leq j < n$. Inductively, $B_{j,k} = B'_{j,k} + B'_{j,k+1} + \cdots + B'_{j,n}$. Equation (3.20) becomes

$$\zeta^{\triangleleft}_{\mathfrak{tr}_n(\mathbb{Z}),p}(s) = \zeta^{\triangleleft}_{\mathbb{Z}^n,p}(s) \sum_{W'''} \left(\prod_{1 \leq j < k \leq n} p^{-(k-j+1)B'_{j,k}s} \right) , \qquad (3.21)$$

where W''' is the set of all $(B'_{1,2}, B'_{1,3}, \ldots, B'_{n-1,n}) \in \mathbb{N}^{\frac{1}{2}n(n-1)}$ satisfying

$$B'_{j,k} + B'_{j,k+1} + \cdots + B'_{j,n} \leq B'_{j+1,k} + B'_{j+1,k+1} + \cdots + B'_{j+1,n} \qquad (3.22)$$

for $1 \leq j < k \leq n$, $k - j \geq 2$.

When $k = n$ in (3.22), we have

$$B'_{j,n} \leq B'_{j+1,n} \text{ for } 1 \leq j \leq n - 2 .$$

Set $B'_{j,n} = B''_{j,n} + B'_{j-1,n}$ for $2 \leq j \leq n - 1$ so that $B'_{j,n} = B''_{j,n} + B''_{j-1,n} + \cdots + B''_{1,n}$. For notational simplicity, we also set $B'_{1,n} = B''_{1,n}$ and $B'_{j,k} = B''_{j,k}$ for $1 \leq j < k < n$. Equation (3.21) now becomes

$$\zeta^{\triangleleft}_{\mathfrak{tr}_n(\mathbb{Z}),p}(s) = \zeta^{\triangleleft}_{\mathbb{Z}^n,p}(s) \sum_{W''''} \left(\prod_{1 \leq j < k \leq n} p^{-e_{j,k} B''_{j,k} s} \right) , \qquad (3.23)$$

where

$$e_{j,k} = \begin{cases} k - j + 1 & \text{if } 1 \leq j < k < n, \\ \frac{1}{2}(n - j + 1)(n - j + 2) - 1 & \text{if } k = n, 1 \leq j < n, \end{cases} \qquad (3.24)$$

and W'''' is the set of all solutions $(B''_{1,2}, B''_{1,3}, \ldots, B''_{n-1,n}) \in \mathbb{N}^{\frac{1}{2}n(n-1)}$ satisfying

$$B''_{j,k} + B''_{j,k+1} + \cdots + B''_{j,n-1} \leq B''_{j+1,k} + B''_{j+1,k+1} + \cdots + B''_{j+1,n} \qquad (3.25)$$

for $1 \leq j < k < n$, $k - j \geq 2$.

Each condition in (3.25) has one less term on the inferior side than on the superior. Hence, when these inequalities are replaced with linear equations by adding a slack variable to the inferior side, each such linear equation will have the same number of terms on each side, all with coefficient 1. The all-1 vector $\mathbf{1}$ is always a solution of such systems of linear equations. Hence, by Corollaries 3.4 and 3.5,

$$\zeta^{\triangleleft}_{\mathfrak{tr}_n(\mathbb{Z}),p}(s)\big|_{p \to p^{-1}}$$

$$= \zeta^{\triangleleft}_{\mathbb{Z}^n,p}(s)\big|_{p \to p^{-1}} (-1)^{\frac{1}{2}n(n-1)} \prod_{1 \leq j < k \leq n} p^{-e_{j,k} s} \sum_{W''''} \left(\prod_{1 \leq j < k \leq n} p^{-e_{j,k} B'_{j,k} s} \right)$$

$$= (-1)^n p^{\binom{n}{2} - ns} (-1)^{\frac{1}{2}n(n-1)} \left(\prod_{1 \leq j < k \leq n} p^{-e_{j,k} s} \right) \zeta^{\triangleleft}_{\mathfrak{tr}_n(\mathbb{Z}),p}(s)$$

$$= (-1)^N p^{\binom{n}{2} - ns - \left(\sum_{1 \leq j < k \leq n} e_{j,k} \right) s} \zeta^{\triangleleft}_{\mathfrak{tr}_n(\mathbb{Z}),p}(s) .$$

It remains to evaluate the sum $\sum_{1 \leq j < k \leq n} e_{j,k}$. For $1 \leq j < n$,

$$\sum_{k=j+1}^{n} e_{j,k} = \sum_{k=j+1}^{n-1} (k - j + 1) + \tfrac{1}{2}(n - j + 1)(n - j + 2) - 1$$

$$= \tfrac{1}{2}(n - j)(n - j + 1) + \tfrac{1}{2}(n - j + 1)(n - j + 2) - 2$$

$$= (n - j + 1)^2 - 2 .$$

So

$$\sum_{1 \leq j < k \leq n} e_{j,k} = \sum_{j=1}^{n-1}((n - j + 1)^2 - 2)$$

$$= \left(\sum_{j=1}^{n} j^2 \right) - (2n - 1)$$

$$= \tfrac{1}{6}n(n + 1)(2n + 1) - (2n - 1)$$

$$= \tfrac{1}{6}(2n^3 + 3n^2 - 11n + 6) .$$

Hence

$$\zeta^{\triangleleft}_{\mathfrak{tr}_n(\mathbb{Z}),p}(s)|_{p \to p^{-1}} = (-1)^N p^{\binom{n}{2} - \frac{1}{6}(2n^3 + 3n^2 - 5n + 6)s} \zeta^{\triangleleft}_{\mathfrak{tr}_n(\mathbb{Z}),p}(s) ,$$

and this completes the proof of Theorem 3.1.

3.3 Explicit Examples

Theorem 3.1 gives us some idea of the overall shape of these zeta functions. However, it is worthwhile to calculate a few of them to see what they actually look like. Straightforward calculations give us

Proposition 3.12.

$$\zeta^{\triangleleft}_{\mathfrak{tr}_1(\mathbb{Z}),p}(s) = \zeta_{\mathbb{Z},p}(s) ,$$

$$\zeta^{\triangleleft}_{\mathfrak{tr}_2(\mathbb{Z}),p}(s) = \zeta_{\mathbb{Z}^2,p}(s)\zeta_p(2s) ,$$

$$\zeta^{\triangleleft}_{\mathfrak{tr}_3(\mathbb{Z}),p}(s) = \zeta_{\mathbb{Z}^3,p}(s)\zeta_p(2s)^2\zeta_p(5s) ,$$

$$\zeta^{\triangleleft}_{\mathfrak{tr}_4(\mathbb{Z}),p}(s) = \zeta_{\mathbb{Z}^4,p}(s)\zeta_p(2s)^3\zeta_p(5s)^2\zeta_p(8s)\zeta_p(9s)\zeta_p(10s)^{-1} .$$

The above results can be obtained by hand with little difficulty. However, a computer was used to obtain the following result.

Theorem 3.13.

$$\zeta^{\triangleleft}_{\mathfrak{tr}_5(\mathbb{Z}),p}(s) = \zeta_{\mathbb{Z}^5,p}(s)\zeta_p(2s)^3\zeta_p(5s)^2\zeta_p(8s)\zeta_p(9s) \left(\prod_{k=11}^{14} \zeta_p(ks) \right) P(p^{-s}) ,$$

where

$$P(Y) = 1 + Y^2 + Y^4 + Y^5 + Y^6 + Y^7 + 2Y^8 + 2Y^9 + Y^{10} + 2Y^{11} + 2Y^{12} + Y^{13}$$
$$+ Y^{14} + Y^{15} + Y^{16} + Y^{17} - Y^{20} - Y^{21} - Y^{22} - Y^{23} - Y^{24} - 2Y^{25}$$
$$- 2Y^{26} - Y^{27} - 2Y^{28} - 2Y^{29} - Y^{30} - Y^{31} - Y^{32} - Y^{33} - Y^{35} - Y^{37}.$$

$\zeta^{\triangleleft}_{\mathfrak{tr}_6(\mathbb{Z}),p}(s)$ and $\zeta^{\triangleleft}_{\mathfrak{tr}_7(\mathbb{Z}),p}(s)$ have also been calculated with the help of a computer.

Theorem 3.14.

$$\zeta^{\triangleleft}_{\mathfrak{tr}_6(\mathbb{Z}),p}(s) = \zeta_{\mathbb{Z}^6,p}(s)\zeta_p(2s)^3\zeta_p(5s)^2\zeta_p(8s)\zeta_p(9s)\left(\prod_{k=11}^{20}\zeta_p(ks)\right)W^{\triangleleft}_{\mathfrak{tr}_6(\mathbb{Z})}(p^{-s}),$$

$$\zeta^{\triangleleft}_{\mathfrak{tr}_7(\mathbb{Z}),p}(s) = \zeta_{\mathbb{Z}^7,p}(s)\zeta_p(2s)^3\zeta_p(4s)\zeta_p(5s)^2\zeta_p(9s)\left(\prod_{k=11}^{27}\zeta_p(ks)\right)W^{\triangleleft}_{\mathfrak{tr}_7(\mathbb{Z})}(p^{-s}),$$

where $W^{\triangleleft}_{\mathfrak{tr}_6(\mathbb{Z})}(Y)$ and $W^{\triangleleft}_{\mathfrak{tr}_7(\mathbb{Z})}(Y)$ are given in Appendix A from, p. 188 onwards.

3.4 Variations

3.4.1 Quotients of $\mathfrak{tr}_n(\mathbb{Z})$

Fix some $n \in \mathbb{N}_{>0}$. Let S be a nonempty subset of $\{(j,k) : 1 \leq j < k \leq n\}$ such that if $(j,k) \in S$ and $1 \leq j_1 \leq j < k \leq k_1 \leq n$ then $(j_1,k_1) \in S$. Let $I_S = \langle\{E_{j,k} : (j,k) \in S\}\rangle_{\mathbb{Z}}$, the ideal generated by off-diagonal basis elements of $\mathfrak{tr}_n(\mathbb{Z})$ indexed by S. It is not difficult to see that $I_S \triangleleft \mathfrak{tr}_n(\mathbb{Z})$, so we may consider the quotient of $\mathfrak{tr}_n(\mathbb{Z})$ by I_S.

Quotienting out by I_S does not destroy the uniformity property of the local zeta functions of $\mathfrak{tr}_n(\mathbb{Z})$. Before we prove this, we give the following lemma which is more-or-less an adaptation of Lemma 3.10:

Lemma 3.15. *Suppose that $1 \leq j_1 < k_1 \leq n$, $1 \leq j_2 < k_2 \leq n$ and $(j_1,k_1) \notin S$. If either $j_1 = j_2$ and $k_2 < k_1$, or $k_1 = k_2$ and $j_1 < j_2$, then $n_{j_1,k_1} \mid n_{j_2,k_2}$.*

Proof. The proof is identical to that of Lemma 3.10 once we ensure $E_{j_2,k_2} \notin I_S$ if $j_1 \leq j_2 < k_2 \leq k_1$, $E_{k_2,k_1} \notin I_S$ if $k_2 < k_1$ and $E_{j_1,j_2} \notin I_S$ if $j_1 < j_2$. This follows since, under the above circumstances, $(j_2,k_2),(k_2,k_1),(j_1,j_2) \notin S$. ☐

Theorem 3.16. *There exists a univariate rational function $R_S(Y)$ such that for all primes p,*

$$\zeta^{\triangleleft}_{\mathfrak{tr}_n(\mathbb{Z})/I_S,p}(s) = \zeta^{\triangleleft}_{\mathbb{Z}^n,p}(s)R_S(p^{-s}).$$

Proof. We may apply the proof of Theorem 3.1 with a few modifications. We omit any basis elements $E_{j,k}$ for $(j,k) \in S$, and the bijection f must be adjusted. In particular, it is now a map

$$f : \{\, (j,k) : 1 \leq j < k \leq n, \ (j,k) \notin S \,\} \to \{1, \ldots, N - |S|\}\,.$$

Lemma 3.8 continues to apply, although we must replace Lemma 3.10 with Lemma 3.15. Finally, if $(j, j+1) \in S$, then we do not have $B_{j,j+1} \mid A_{j+1}$, so we set $A_{j+1} = A'_{j+1}$ and $m_{i,j+1} = m'_{i,j+1}$ for $1 \leq i \leq j+1$. We therefore obtain

$$\zeta^{\lhd}_{\mathfrak{tr}_n(\mathbb{Z})/I_S, p}(s) = \zeta^{\lhd}_{\mathbb{Z}^n, p}(s) R_S(p^{-s})\,,$$

where

$$R_S(Y) = \sum_{W''_S} \left(\prod_{\substack{1 \leq j < n \\ (j, j+1) \notin S}} Y^{2B_{j,j+1}} \prod_{\substack{1 \leq j < k \leq n \\ k - j \geq 2 \\ (j,k) \notin S}} Y^{B_{j,k}} \right) \tag{3.26}$$

and W''_S is the set of all $(B_{j,k} : (j,k) \notin S) \in \mathbb{N}^{\frac{1}{2}n(n-1)-|S|}$ satisfying

$$B_{j,k} \leq B_{j,k-1} \text{ for } 1 \leq j < k \leq n, \ k - j \geq 2, (j,k) \notin S\,, \tag{3.27}$$
$$B_{j,k} \leq B_{j+1,k} \text{ for } 1 \leq j < k \leq n, \ k - j \geq 2, (j,k) \notin S\,. \tag{3.28}$$

Hence the result.   □

However it is not always true that a functional equation holds.

Theorem 3.17. *Let $S = \{(1,4), (1,5)\}$. Then $\zeta^{\lhd}_{\mathfrak{tr}_5(\mathbb{Z})/I_S, p}(s)$ satisfies no functional equation of the form (3.1).*

Proof. A computer calculation has shown that

$$\zeta^{\lhd}_{\mathfrak{tr}_5(\mathbb{Z})/I_S, p}(s) = \zeta^{\lhd}_{\mathbb{Z}^5, p}(s) \zeta_p(2s)^3 \zeta_p(5s)^2 \zeta_p(8s) \zeta_p(9s) \zeta_p(11s) \zeta_p(12s) Q(p^{-s})\,,$$

where

$$Q(Y) = 1 + Y^2 + Y^4 + Y^5 + Y^6 + Y^7 + 2Y^8 + Y^9 + Y^{10} + Y^{11} + Y^{12} - Y^{17}$$
$$- Y^{18} - Y^{19} - 2Y^{20} - Y^{21} - Y^{22} - Y^{23} - Y^{24} - Y^{25} - Y^{26} - Y^{28}\,.$$

From this the result follows immediately.   □

Nonetheless, we can prove a theorem giving many cases when a functional equation does hold. For $1 \leq j < n$, let

$$w_j = \min(\{\, k : j < k < n, \ (j, k+1) \in S \,\} \cup \{n\})\,,$$

and for $1 \leq j < n - 1$, put $d_j = w_{j+1} - w_j$. For convenience, set $w_0 = 0$, $d_0 = w_1$.

Definition 3.18. (j, k) *is a* corner *of* S *if* $(j, k) \in S$ *but* $(j, k-1) \notin S$ *and* $(j+1, k) \notin S$, *or equivalently if* $w_j = k - 1$ *and* $w_j < w_{j+1}$.

Definition 3.19. *Suppose* $1 \le j < n - 1$. *A* corner (j, k) *of* S *is* square *if* $j \ge d_j$, $w_{j-d_j} < w_{j-d_j+1}$ *and* $w_{j-d_j+1} = \cdots = w_j$.

Definition 3.20. *A* corner (j, k) *of* S *is* on the mth superdiagonal *if* $k = j + m$.

Example 3.21. Suppose $n = 7$, $S = \{(1, 5), (1, 6), (1, 7), (2, 5), (2, 6), (2, 7), (3, 7)\}$. S has two corners, $(2, 5)$ and $(3, 7)$, both of which are square, since $d_2 = 2$ and $w_2 = w_1 = 4$, and $d_3 = 1$, $w_3 = 6$.

Example 3.22. Suppose $S = \{(j, k) : k - j \ge m\}$, i.e. everything on or beyond the mth superdiagonal. This has $n - m$ corners, $(j, j+m)$ for $1 \le j \le n - m$. Each corner is square since $w_{j+1} = w_j + 1$ for $1 \le j \le n - m$.

Theorem 3.23. *Suppose all the corners of* S *are square or on the first or second superdiagonal. Then*

$$\zeta^{\triangleleft}_{\mathfrak{tr}_n(\mathbb{Z})/I_S, p}(s)|_{p \to p^{-1}} = (-1)^{N - |S|} p^{\binom{n}{2} - C_S s} \zeta^{\triangleleft}_{\mathfrak{tr}_n(\mathbb{Z})/I_S, p}(s)$$

for some $C_S \in \mathbb{N}$.

Proof. From the proof of Theorem 3.16, we have

$$\zeta^{\triangleleft}_{\mathfrak{tr}_n(\mathbb{Z})/I_S, p}(s) = \zeta^{\triangleleft}_{\mathbb{Z}^n, p}(s) \sum_{W''_S} \left(\prod_{\substack{1 \le j < n \\ (j, j+1) \notin S}} p^{-2B_{j, j+1}s} \prod_{\substack{1 \le j < k \le n \\ k - j \ge 2 \\ (j, k) \notin S}} p^{-B_{j, k}s} \right), \quad (3.29)$$

where W''_S is the subset of $(B_{j, k} : (j, k) \notin S) \in \mathbb{N}^{\frac{1}{2}n(n-1) - |S|}$ satisfying the conditions (3.27) and (3.28).

Set $B_{j, k} = B'_{j, k} + B_{j, k+1}$ for all j, k such that $1 \le j < n - 1$, $j < k < w_j$. For completeness, set $B_{j, w_j} = B'_{j, w_j}$ for all $1 \le j < n$. Inductively, this becomes $B_{j, k} = B'_{j, k} + B'_{j, k+1} + \cdots + B'_{j, w_j}$. Doing so eliminates the conditions (3.27), and (3.28) becomes

$$B'_{j, k} + \cdots + B'_{j, w_j} \le B'_{j+1, k} + \cdots B'_{j+1, w_{j+1}} \quad (3.30)$$

for $1 \le j < n - 1$, $j + 1 < k \le w_j$. If $1 \le j < n - 1$ and $w_j = w_{j+1}$, the condition $B_{j, w_j} \le B_{j+1, w_j}$ becomes $B'_{j, w_j} \le B'_{j+1, w_j}$, so we may set $B'_{j+1, w_j} = B''_{j+1, w_j} + B'_{j, w_j}$. For all other $B'_{j, k}$, i.e. those for which $k < w_j$, $j = 1$, or $k = w_j > w_{j-1}$, set $B'_{j, k} = B''_{j, k}$. Inductively, this becomes

$$B'_{j, k} = \sum_{\substack{j' \le j \\ w'_j = w_j}} B''_{j', k}.$$

We split into three cases, depending on j:

1. If $w_j \leq j+1$, there are no conditions of the form (3.30) for this value of j. In particular, this happens if $(j, j+1)$ or $(j, j+2)$ is a corner of S.
2. If $w_j = w_{j+1}$, then the conditions (3.30) become

$$B''_{j,k} + B''_{j,k+1} + \cdots + B''_{j,w_j-1}$$
$$\leq B''_{j+1,k} + B''_{j+1,k+1} + \cdots + B''_{j+1,w_j-1} + B''_{j+1,w_j} \quad (3.31)$$

for $j+1 < k < w_j$. It is clear that the LHS of (3.31) has one less term than the RHS.

3. Now suppose $w_j < w_{j+1}$ and $w_j > j+1$, so (j, w_j+1) is a corner of S not on either of the first two superdiagonals. By assumption, this corner must be square, so $B'_{j,w_j} = B''_{j,w_j} + \cdots + B''_{j-d_j+1,w_j}$. Recall that $d_j = w_{j+1} - w_j$. Equation (3.30) becomes

$$B''_{j,k} + B''_{j,k+1} + \cdots + B''_{j,w_j-1} + B''_{j,w_j} + B''_{j-1,w_j} + \cdots + B''_{j-d_j+1,w_j}$$
$$\leq B''_{j+1,k} + B''_{j+1,k+1} + \cdots + B''_{j+1,w_{j+1}} \quad (3.32)$$

for $j+1 < k < w_j$. The LHS of (3.32) has $w_j - k + d_j$ terms and the RHS has $w_{j+1} - k + 1$, one more than on the LHS.

Hence,

$$\zeta^{\lhd}_{\mathfrak{tr}_n(\mathbb{Z})/I_S,p}(s) = \zeta^{\lhd}_{\mathbb{Z}^n,p}(s) \sum_{W'''_S} \left(\prod_{1 \leq j < k \leq n} p^{-e_{j,k,S} B''_{j,k} s} \right), \quad (3.33)$$

for some positive integers $e_{j,k,S}$, and W'''_S is the set of all $(B_{j,k} : (j,k) \notin S) \in \mathbb{N}^{\frac{1}{2}n(n-1)-|S|}$ satisfying (3.31) and (3.32). The inferior side of each condition in both (3.31) and (3.32) has one less term than the superior, and all terms have coefficient 1. When slack variables are added to the inferior sides, the resulting system of linear equations will have $\mathbf{1}$ as a solution. Put

$$C_S = n + \sum_{\substack{1 \leq j < k \leq n \\ (j,k) \notin S}} e_{j,k,S} \, .$$

The result now follows from Corollaries 3.4 and 3.5. $\qquad\qquad\square$

We have chosen not to give explicit values for the integers $e_{j,k,S}$, as we believe the resulting expression would be fiddly and awkward. For the same reason we do not to give an explicit formula for the constant C_S.

It can be seen from Theorem 3.23 that if $\zeta^{\lhd}_{\mathfrak{tr}_n(\mathbb{Z})/I_S,p}(s)$ is *not* to satisfy a functional equation, then $n \geq 5$. In fact, $\mathfrak{tr}_5(\mathbb{Z})/I_S$ for $S = \{(1,4),(1,5)\}$ or $\{(1,5),(2,5)\}$ are the smallest cases that Theorem 3.23 cannot be applied to. A more general result giving conditions on S such that $\zeta^{\lhd}_{\mathfrak{tr}_n(\mathbb{Z})/I_S,p}(s)$ does not satisfy a functional equation is more difficult to come by. It is not in general true that if $\mathbf{1}$ is not a solution to a system of homogeneous linear equations $\mathbf{E}$, then there is no simple relation between $F(\mathbf{E}; \mathbf{X})$ and $F(\mathbf{E}; 1/\mathbf{X})$.

3.4.2 Counting All Subrings

Zeta functions counting all subrings in a Lie ring L can be defined in an analogous way to those counting ideals. However, it is considerably more difficult to calculate these zeta functions in the case that $L = \mathfrak{tr}_n(\mathbb{Z})$, the main reason being that there is no equivalent of Lemma 3.8. Taylor [57, p. 149] shows that

$$\zeta_{\mathfrak{tr}_2(\mathbb{Z}),p}^{\leq}(s) = \zeta_p(s)\zeta_p(s-1)^2\zeta_p(2s-2)\zeta_p(2s-1)^{-1} .$$

However, even the calculation of $\zeta_{\mathfrak{tr}_3(\mathbb{Z}),p}^{\leq}(s)$ seems infeasible and out of reach.

4

Local Functional Equations

4.1 Introduction

In this chapter, we consider the functional equations that various local zeta functions of groups and Lie rings are known to satisfy. These local functional equations take the form

$$\zeta^*_{L,p}(s)\big|_{p \to p^{-1}} = (-1)^r p^{b-as} \zeta^*_{L,p}(s) \tag{4.1}$$

for $a, b, r \in \mathbb{N}$. Frequently, the values of a, b and r can be given explicitly in terms of various invariants of the group or Lie ring.

4.2 Algebraic Groups

In [21], Lubotzky and the first author consider the zeta function of an algebraic group $\mathfrak{G}$ as defined in the previous chapter. Under certain assumptions on $\mathfrak{G}$, they demonstrate that this zeta function satisfies a functional equation of the form

$$\zeta_{\mathfrak{G},p}(s)\big|_{p \to p^{-1}} = (-1)^n p^{b-as} \zeta_{\mathfrak{G},p}(s)$$

with $a, b, n \in \mathbb{N}$, for all but finitely many primes p. As noted in the Introduction, these zeta functions are counting certain subgroups within a $\mathfrak{T}$-group Γ. A similar functional equation for $\zeta^\wedge_{\Gamma,p}(s)$ then follows.

4.3 Nilpotent Groups and Lie Rings

We describe below a conjecture sketched by the second author concerning local functional equations of these zeta functions of $\mathfrak{T}$-groups and Lie rings. This conjecture offers a potential explanation of the functional equations of zeta functions of nilpotent Lie rings and $\mathfrak{T}$-groups that we have seen to date. It can also explain why the local zeta functions that do not satisfy a functional equation don't.

However, we have been unable to rigorously formulate the conjecture. There are certain technical conditions which must be assumed. Nonetheless, we present this conjecture as it stands and sketch the proof of a significant special case. We spend the rest of the chapter deducing consequences regarding local functional equations from this conjecture.

4.4 The Conjecture

The conjecture we present below is not a direct conjecture about functional equations. Instead, it is a 'reciprocity' conjecture for p-adic integrals, inspired by Theorem 3.3. If the cone data are all monomial, the integral can be expressed as a sum over integral points in a polyhedral cone. Theorem 3.3 can then be applied to deduce a reciprocity result regarding these monomial integrals. The conjecture below is an attempt to generalise this result to cone integrals where the cone data is not monomial.

We recall some definitions from Chap. 2:

Definition 4.1. *Let $\mathbf{x} = (x_1, \ldots, x_n)$ be a vector of variables, and for $i = 0, \ldots, l$, let $f_i(\mathbf{x}), g_i(\mathbf{x}) \in \mathbb{Q}[\mathbf{x}]$ be homogeneous polynomials. The* (closed) *cone integral corresponding to the cone data $\mathcal{D} = \{f_0, g_0, f_1, g_1, \ldots, f_l, g_l\}$ is defined to be*

$$Z_{\mathcal{D}}(s, p) = \int_{W_{\mathcal{D}}} |f_0(\mathbf{x})|_p^s |g_0(\mathbf{x})|_p \, \mathrm{d}\mu \, ,$$

where

$$W_{\mathcal{D}} = \{\, \mathbf{x} \in \mathbb{Z}_p^n : v(f_i(\mathbf{x})) \leq v(g_i(\mathbf{x})) \text{ for } i = 1, \ldots, l \,\} \, .$$

We also define the corresponding open *cone integral to be*

$$Z_{\mathcal{D}}^{\circ}(s, p) = \int_{W_{\mathcal{D}}^{\circ}} |f_0(\mathbf{x})|_p^s |g_0(\mathbf{x})|_p \, \mathrm{d}\mu \, ,$$

where

$$W_{\mathcal{D}}^{\circ} = \{\, \mathbf{x} \in (p\mathbb{Z}_p)^n : v(f_i(\mathbf{x})) < v(g_i(\mathbf{x})) \text{ for } i = 1, \ldots, l \,\} \, .$$

Proposition 2.1 (p. 23) implies that for $* \in \{\lhd, \leq\}$, $\zeta_{L,p}^*(s + d)$ (where $d = \operatorname{rank} L$) can be expressed as a closed cone integral.

We will also be working with resolutions of singularities. Following Sect. 5 of [7], we make the following definitions.

Definition 4.2. *A resolution* (Y, h) *for a homogeneous polynomial* F *over* $\mathbb{Q}$ *consists of a closed integral subscheme* Y *of* $\mathbb{P}^K_{X_{\mathbb{Q}}}$ *(where* $X_{\mathbb{Q}} = \mathrm{Spec}(\mathbb{Q}[\mathbf{x}])$ *and* $\mathbb{P}^K_{X_{\mathbb{Q}}}$ *denotes projective* K*-space over the scheme* $X_{\mathbb{Q}}$*) and the morphism* $h : Y \to X$ *which is the restriction to* Y *of the projection morphism* $\mathbb{P}^K_{X_{\mathbb{Q}}}$*, such that*

1. Y *is smooth over* $\mathrm{Spec}(\mathbb{Q})$*;*
2. *The restriction* $h : Y \setminus h^{-1}(D) \to X \setminus D$ *is an isomorphism, where* $D = \mathrm{Spec}(\mathbb{Q}[\mathbf{x}]/(F)) \subset X_{\mathbb{Q}}$*; and*
3. *The reduced scheme* $(h^{-1}(D))_{\mathrm{red}}$ *associated to* $h^{-1}(D)$ *has only normal crossings (as a subscheme of* Y*).*

Definition 4.3. *Let* E_i*,* $i \in T$ *be the irreducible components of the reduced scheme* $(h^{-1}(D))_{\mathrm{red}}$ *over* $\mathrm{Spec}(\mathbb{Q})$*. For* $i \in T$*, let* N_i *be the multiplicity of* E_i *in the divisor of* $F \circ h$ *on* Y *and let* $\nu_i - 1$ *be the multiplicity of* E_i *in the divisor of* $h^*(\,dx_1 \wedge \cdots \wedge dx_n)$*, The* (N_i, ν_i) *for* $i \in T$ *are called the* numerical data *of the resolution* (Y, h) *for* F*.*

We also recall some necessary facts about reduction of varieties mod p. When $X = X_{\mathbb{Q}} = \mathrm{Spec}(\mathbb{Q}[\mathbf{x}])$ one defines the reduction mod p of a closed integral subscheme Y of $\mathbb{P}^K_{X_{\mathbb{Q}}}$ as follows. Let $\tilde{X} = \mathrm{Spec}(\mathbb{Z}[\mathbf{x}])$ and $\tilde{Y}$ be the scheme-theoretic closure of Y in $\mathbb{P}^K_{\tilde{X}}$. Then the reduction mod p of Y is the scheme $\tilde{Y} \times_{\mathbb{Z}} \mathrm{Spec}(\mathbb{F}_p)$ and we denote it by $\overline{Y}$. Let $\tilde{h} : \tilde{Y} \to \tilde{X}$ be the restriction to $\tilde{Y}$ of the projection morphism $\mathbb{P}^K_{\tilde{X}} \to \tilde{X}_{\mathbb{Q}}$ and $\overline{h} : \overline{Y} \to \overline{X}$ be obtained from $\tilde{h}$ by base extension. Thus

Definition 4.4. *A resolution* (Y, h) *for* F *over* $\mathbb{Q}$ *has* good reduction *mod* p *if*

1. $\overline{Y}$ *is smooth over* $\mathrm{Spec}(\mathbb{F}_p)$*;*
2. $\overline{E_i}$ *is smooth over* $\mathrm{Spec}(\mathbb{F}_p)$ *for each* $i \in T$*, and* $\bigcup_{i \in T} \overline{E_i}$ *has only normal crossings as a subscheme of* $\overline{Y}$*; and*
3. $\overline{E_i}$ *and* $\overline{E_j}$ *have no common irreducible components when* $i \neq j$*.*

Any resolution over $\mathbb{Q}$ has good reduction mod p for almost all primes p ([7, Theorem 2.4]).

We can now roughly state our conjecture:

Conjecture 4.5. Assume the above notation. Let $f_0(\mathbf{x}), g_0(\mathbf{x}), \ldots, f_l(\mathbf{x}), g_l(\mathbf{x})$ be homogeneous polynomials and put $F = \prod_{i=0}^{l} f_i(\mathbf{x}) g_i(\mathbf{x})$. Suppose that $\mu(W_{\mathcal{D}}^{\circ}) > 0$, the resolution (Y, h) of F has good reduction mod p, and some as-yet-undetermined conditions hold. Then, for all but finitely many primes p,

$$Z_{\mathcal{D}}(s, p)|_{p \to p^{-1}} = p^n Z_{\mathcal{D}}^{\circ}(s, p) \, . \tag{4.2}$$

It is clear that we need the resolution to have good reduction mod p. In [24], du Sautoy and Taylor calculate $\zeta_{\overline{\mathfrak{sl}_2(\mathbb{Z})}, p}^{\leq}(s)$ for all p. In this case, the resolution of the polynomial F has bad reduction at $p = 2$, and indeed $\zeta^{\leq}_{\mathfrak{sl}_2(\mathbb{Z}),2}(s)$ satisfies no local functional equation.

Remark 4.6. To show that other conditions are necessary, consider the integral

$$Z_{\mathcal{D}}(s,p) = \int_{x|y} |xy(x+y)|^s_p \, \mathrm{d}\mu \ .$$

It is not difficult to calculate that

$$Z_{\mathcal{D}}(s,p) = (1 - p^{-1})\zeta_p(s+1)\zeta_p(3s+2)(1 - 2p^{-1} + p^{-1-s}) \ ,$$
$$Z^{\circ}_{\mathcal{D}}(s,p) = (1 - p^{-1})^2 p^{-3-4s}\zeta_p(s+1)\zeta_p(3s+2) \ .$$

In this case the polynomial $F = x^2 y^2(x+y)$ has good reduction at all primes p, and it is clear that $\mu(W^{\circ}_{\mathcal{D}}) > 0$, but $Z_{\mathcal{D}}(s,p)$ and $Z^{\circ}_{\mathcal{D}}(s,p)$ do not satisfy (4.2). The polynomial $xy(x+y)$ requires a single blow-up to resolve the singularity at $(0,0)$. Normally, we would split into two cases, $v(x) \leq v(y)$ and $v(x) > v(y)$, but it is clear that the condition $x \mid y$ under the integral renders the second case inconsistent. Set $y = xy'$, so that $xy(x+y) = x^3 y'(1+y')$.

Any point $\{(x,y,x',y') \in \mathbb{Z}^2_p \times \mathbb{P}^1(\mathbb{Q}_p) : xy' = x'y\}$ on the variety $(1+y')$ maps to the point $(x,-x)$ under the resolution h. In $W^{\circ}_{\mathcal{D}}$, however, there are no such points, since $v(x) < v(y)$. The correspondence fails this time because $|x+y|_p = |x|_p$ for all $(x,y) \in W^{\circ}_{\mathcal{D}}$, but the same is not true for all $(x,y) \in W_{\mathcal{D}}$.

4.5 Special Cases Known to Hold

If all the f_i and g_i are monomial, $Z_{\mathcal{D}}(s,p)$ and $Z^{\circ}_{\mathcal{D}}(s,p)$ reduce to cone sums of integer points in polyhedral cones in $\mathbb{R}^n_{\geq 0}$ with a coefficient $(1-p^{-1})^n$. This special case of the conjecture follows easily from Stanley's Theorem (Theorem 3.3), with no technical conditions.

Theorem 4.7. *Conjecture 4.5 holds if all the f_i and g_i are monomial.*

Proof. If f_i and g_i are monomial for $0 \leq i \leq l$, whether a point $\mathbf{x} \in W_{\mathcal{D}}$ (or $W^{\circ}_{\mathcal{D}}$) depends only on the p-adic valuations of the x_i. So, let $a_i = v(x_i)$. The polynomial divisibility conditions defining $W_{\mathcal{D}}$ and $W^{\circ}_{\mathcal{D}}$ become linear inequalities in the a_i. By adding slack variables we obtain a system of linear equations $\mathbf{E}$. We take each X_i to be a suitable monomial in p and p^{-s}, and then

$$Z_{\mathcal{D}}(s,p) = (1 - p^{-1})^n F(\mathbf{E};\mathbf{X}) \quad \text{and} \quad Z^{\circ}_{\mathcal{D}}(s,p) = (1 - p^{-1})^n \bar{F}(\mathbf{E};\mathbf{X}) \ .$$

If $\mu(W^{\circ}_{\mathcal{D}}) > 0$ then $\bar{\mathbf{S}} \neq \varnothing$. The corank of the system of equations $\mathbf{E}$ is n, since there are $n + l$ variables and l linearly independent linear equations. By our choice of $\mathbf{X}$,

$$F(\mathbf{E}; \mathbf{X})|_{p \to p^{-1}} = F(\mathbf{E}; 1/\mathbf{X}) = (-1)^d \bar{F}(\mathbf{E}; \mathbf{X}) .$$

Hence

$$Z_{\mathcal{D}}(s, p)|_{p \to p^{-1}} = p^d Z_{\mathcal{D}}^{\circ}(s, p) ,$$

as required. $\square$

This conjecture also generalises in part a theorem of Denef and Meuser [8]. Denef and Meuser work over any finite extension K of $\mathbb{Q}_p$ with R_K the ring of integers of K. They assume that $l = 0$ (i.e. the integral is over the whole of R_K^n) and $g_0(\mathbf{x}) = 1$. Indeed, we follow the strategy used in their proof below when we (attempt to) prove a special case of this conjecture.

The most important special case of Conjecture 4.5 is the following. It is in some ways parallel to Corollary 3.5 of the previous chapter. We believe this special case requires no extra conditions such as those which would exclude the example described in Remark 4.6.

Corollary 4.8. *Suppose the cone data satisfies*

$$\deg f_i(\mathbf{x}) + 1 = \deg g_i(\mathbf{x}) \text{ for } i = 1, \dots, l . \tag{4.3}$$

Assume Conjecture 4.5 holds. Then

$$Z_{\mathcal{D}}(s, p)|_{p \to p^{-1}} = p^{-\deg g_0 - s \deg f_0} Z_{\mathcal{D}}(s, p) . \tag{4.4}$$

Proof. Since f_i and g_i are homogeneous, $f_i(p\mathbf{x}) = p^{\deg f_i} f_i(\mathbf{x})$ and $g_i(p\mathbf{x}) = p^{\deg g_i} g_i(\mathbf{x})$, and this implies $W_{\mathcal{D}}^{\circ} = pW_{\mathcal{D}}$. It is then routine to see that $Z_{\mathcal{D}}^{\circ}(s, p) = p^{-n - \deg g_0 - s \deg f_0} Z_{\mathcal{D}}(s, p)$. Hence, by Conjecture 4.5,

$$Z_{\mathcal{D}}(s, p)|_{p \to p^{-1}} = p^n Z_{\mathcal{D}}^{\circ}(s, p) = p^{-\deg g_0 - s \deg f_0} Z_{\mathcal{D}}(s, p)$$

for all but finitely many primes p. $\square$

4.6 A Special Case of the Conjecture

In this section we present an almost-complete proof of a significant special case of Conjecture 4.5. We hope that sketching a proof of a substantial case can lay the groundwork for future attempts at the conjecture.

We focus on the case $\deg f_i(\mathbf{x}) = \deg g_i(\mathbf{x})$ for $1 \le i \le l$, because it can be 'projectivised' easily. If $\mathbf{x}$ is a solution to a set of polynomial conditions of the form $v(f_i(\mathbf{x})) \le v(g_i(\mathbf{x}))$ with $\deg f_i(\mathbf{x}) = \deg g_i(\mathbf{x})$ for $i = 1, \dots, l$, then so is $\lambda \mathbf{x}$ for any $\lambda \in \mathbb{Z}_p$. Note that this case includes the integral presented in Remark 4.6. We therefore need to make an assumption in the proof to exclude this exceptional case.

We follow the proof of the functional equation of the Igusa local zeta function given in [8], using certain results of du Sautoy and Grunewald [17] in place of corresponding results of Denef.

Let

$$Z_{\mathcal{D}}(s,p) = \int_{W_{\mathcal{D}}} |f_0(\mathbf{x})|_p^s |g_0(\mathbf{x})|_p \, \mathrm{d}\mu \ ,$$

where

$$W_{\mathcal{D}} = \{ \mathbf{x} \in \mathbb{Z}_p^n : v(f_i(\mathbf{x})) \le v(g_i(\mathbf{x})) \text{ for } i = 1, \dots, l \}$$

and $f_i(\mathbf{x})$, $g_i(\mathbf{x})$ for $0 \le i \le l$ are homogeneous polynomials, with $\deg f_i(\mathbf{x}) = \deg g_i(\mathbf{x})$ for $1 \le i \le l$. Let $d = \deg f_0(\mathbf{x})$ and $d' = \deg g_0(\mathbf{x})$. For clarity, we shall drop the subscript $\mathcal{D}$ from $W_{\mathcal{D}}$. Also, let

$$\hat{Z}_{\mathcal{D}}(s,p) = \int_{\hat{W}} |f_0(\mathbf{x})|_p^s |g_0(\mathbf{x})|_p \, \mathrm{d}\mu \ ,$$

where

$$\hat{W} = \{ \mathbf{x} \in \mathbb{Z}_p^n : v(f_i(\mathbf{x})) + \deg(f_i) \le v(g_i(\mathbf{x})) + \deg(g_i) \text{ for } i = 1, \dots, l \} \ .$$

It is easy to see that $W^\circ = p\hat{W}$, and hence

$$Z_{\mathcal{D}}^\circ(s,p) = p^{-ds-d'-n} \hat{Z}_{\mathcal{D}}(s,p) \ ,$$

so we are left with proving that

$$Z_{\mathcal{D}}(s,p)|_{p \to p^{-1}} = p^{-ds-d'} \hat{Z}_{\mathcal{D}}(s,p) \ .$$

We split the proof as it stands up into a number of steps in the hope that it makes it easier to follow.

4.6.1 Projectivisation

As in [8], we start by projectivising the integral. Let

$$\tilde{W} = \{ \mathbf{x} \in W : x_i \in \mathbb{Z}_p^* \text{ for some } i \} \ ,$$

i.e. points in W with at least one unit coordinate, and let

$$\tilde{Z}_{\mathcal{D}}(s,p) = \int_{\tilde{W}} |f_0(\mathbf{x})|_p^s |g_0(\mathbf{x})|_p \, \mathrm{d}\mu \ .$$

It is easy to see that

$$Z_{\mathcal{D}}(s,p) = \sum_{k=0}^{\infty} \int_{\substack{W \\ \min_i v(x_i) = k}} |f_0(\mathbf{x})|_p^s |g_0(\mathbf{x})|_p \, \mathrm{d}\mu \ ,$$

and a change of variables $\mathbf{x} = p^k \mathbf{u}$ gives

$$Z_{\mathcal{D}}(s,p) = \sum_{k=0}^{\infty} p^{-k(ds+d'+n)} \int_{\substack{W \\ \min_i v(u_i)=0}} |f_0(\mathbf{u})|_p^s |g_0(\mathbf{u})|_p \, d\mu$$

$$= \zeta_p(ds + d' + n) \tilde{Z}_{\mathcal{D}}(s,p) \ .$$

Note that $\mathbf{x} \in W$ if and only if $\mathbf{u} \in W$ since the powers of p appearing on each side of each condition $f_i(\mathbf{x}) \mid g_i(\mathbf{x})$ cancel out.

Next, we write $\tilde{W}$ as the disjoint union of $\tilde{W}_1, \ldots, \tilde{W}_n$, where

$$\tilde{W}_r = \{ \mathbf{u} \in W : u_i \in p\mathbb{Z}_p \text{ for } 1 \leq i < r \text{ and } u_r \in \mathbb{Z}_p^* \} \ .$$

Then

$$\tilde{Z}_{\mathcal{D}}(s,p) = \sum_{r=1}^{n} \int_{\tilde{W}_r} |f_0(\mathbf{u})|_p^s |g_0(\mathbf{u})|_p \, d\mu \ .$$

There is an obvious map $\gamma : \tilde{W} \to \mathbb{P}^{n-1}(\mathbb{Q}_p)$, and we note that $|f(\mathbf{u})|_p$, $|g(\mathbf{u})|_p$ and whether $u \in W$ depends only on $\gamma(\mathbf{u})$. Let ω be the Haar measure on $\mathbb{P}^{n-1}(\mathbb{Q}_p)$ which induces the Haar measure on the unit ball $\mathbb{Z}_p^{n-1}$ of each affine chart satisfying $\omega(a + (p\mathbb{Z}_p)^{n-1}) = p^{-(n-1)}$. We see that

$$\int_{\tilde{W}_r} |f(\mathbf{u})|_p^s |g(\mathbf{u})|_p \, d\mu = (1 - p^{-1}) \int_{V_r} |f(\mathbf{u})|_p^s |g(\mathbf{u})|_p \, d\omega \ , \tag{4.5}$$

where $V_r = \gamma(\tilde{W}_r)$.

4.6.2 Resolution

At this point in [8], Denef and Meuser use a previous result proved by Denef which in essence tells them what the integral looks like on the other side of a resolution of singularities. Our analogue is provided by results of du Sautoy and Grunewald in [17], which we modify slightly.

Let $\mathbf{x} = (x_1, \ldots, x_m)$, $X_{\mathbb{Q}} = \text{Spec}(\mathbb{Q}[\mathbf{x}])$, and let (Y, h) be a resolution for $F = \prod_{i=0}^{l} f_i(\mathbf{x}) g_i(\mathbf{x})$. Let T be an indexing set for the irreducible components E_i, and $t = |T|$. For $i \in T$, let E_i be the irreducible components of the reduced scheme $(h^{-1}(D))_{\text{red}}$ over $\text{Spec}(\mathbb{Q})$. Set $\mathcal{D} = \{f_0, g_0, \ldots, f_l, g_l\}$, and let integral

$$Z_{\mathcal{D}}(s,p) = \int_W |f_0(\mathbf{x})|_p^s |g_0(\mathbf{x})|_p \, d\mu \ .$$

Suppose that the resolution (Y, h) has good reduction mod p. Then, du Sautoy and Grunewald show that

$$Z_{\mathcal{D}}(s,p) = (1 - p^{-1})^m \sum_{I \subseteq T} c_{p,I} \mathcal{J}_I(s,p) \ ,$$

where

$$c_{p,I} = |\{\, a \in \overline{Y}(\mathbb{F}_p) : a \in \overline{E}_i \text{ if and only if } i \in I \,\}| \,,$$

and $\mathcal{J}_I(s,p)$ is a rational function in p^{-1} and p^{-s} given by

$$\mathcal{J}_I(s,p) = \frac{1}{(p-1)^{m-|I|}} \sum_{(k_1,\ldots,k_{|I|}) \in \Lambda_I} p^{-\sum_{j=1}^{|I|} k_j (A_{j,I}s + B_{j,I})} \,,$$

where $I = \{i_1, \ldots, i_{|I|}\}$,

$$\Lambda_I = \left\{ (k_1, \ldots, k_{|I|}) \in \mathbb{N}_{>0}^{|I|} : \sum_{j=1}^{|I|} N_{i_j}(f_i)k_j \leq \sum_{j=1}^{|I|} N_{i_j}(g_i)k_j \text{ for } i = 1, \ldots, l \right\}$$

$$(4.6)$$

and $N_{i_j}(f_i)$, $N_{i_j}(g_i)$, $A_{j,I}$ and $B_{j,I}$ are some constants depending on the numerical data of the resolution (Y, h).

du Sautoy and Grunewald's formula can easily be modified to evaluate integrals over a union of cosets mod $(p\mathbb{Z}_p)^m$. If U is such a union of cosets, then we define

$$Z_{\mathcal{D},U}(s,p) = \int_{W \cap U} |f_0(\mathbf{x})|_p^s |g_0(\mathbf{x})|_p \, d\mu \,,$$

and a simple modification of du Sautoy and Grunewald's proof gives us

$$Z_{\mathcal{D},U}(s,p) = (1 - p^{-1})^n \sum_{I \subseteq T} c_{p,I,U} \mathcal{J}_I(s,p) \,,$$

where $\overline{U}$ denotes the reduction of U mod p and

$$c_{p,I,U} = |\{\, a \in \overline{Y}(\mathbb{F}_p) : a \in \overline{E}_i \text{ iff } i \in I, \text{ and } \overline{h}(a) \in \overline{U} \,\}| \,.$$

Summing gives us

$$\tilde{Z}_{\mathcal{D}}(s,p) = \sum_{r=1}^{n} (1 - p^{-1})^{n-1} \sum_{I \subseteq T} c_{p,I,V_r} \mathcal{J}_I(s,p)$$

$$= (1 - p^{-1})^{n-1} \sum_{I \subseteq T} \left(\sum_{r=1}^{n} c_{p,I,V_r} \right) \mathcal{J}_I(s,p)$$

$$= (1 - p^{-1})^{n-1} \sum_{I \subseteq T} c_{p,I} \mathcal{J}_I(s,p) \,,$$

so that

$$Z_{\mathcal{D}}(s,p) = \frac{(1 - p^{-1})^n}{1 - p^{-(ds+d'+n)}} \sum_{I \subseteq T} c_{p,I} \mathcal{J}_I(s,p) \,.$$

Similarly,

$$\hat{Z}_{\mathcal{D}}(s,p) = \frac{(1-p^{-1})^n}{1-p^{-(ds+d'+n)}} \tilde{\hat{Z}}_{\mathcal{D}}(s,p) \, ,$$

where

$$\tilde{\hat{Z}}_{\mathcal{D}}(s,p) = \sum_{I \subseteq T} c_{p,I} \hat{\mathcal{J}}_I(s,p)$$

and

$$\hat{\mathcal{J}}_I(s,p) = \frac{1}{(p-1)^{n-1-|I|}} \sum_{(k_1,\dots,k_{|I|}) \in \hat{\Lambda}_I} p^{-\sum_{j=1}^{|I|} k_j (A_{j,I}s + B_{j,I})} \, ,$$

where

$$\hat{\Lambda}_I = \left\{ (k_1,\dots,k_{|I|}) \in \mathbb{N}_{>0}^{|I|} : \sum_{j=1}^{|I|} N_{i_j}(f_i)k_j < \sum_{j=1}^{|I|} N_{i_j}(g_i)k_j \text{ for } i=1,\dots,l \right\}.$$

Note that the definition of $\hat{\Lambda}_I$ is identical to that of Λ_I (4.6), except that the linear inequalities are strict.

4.6.3 Manipulating the Cone Sums

At this point, we have to take into account the polyhedral cones, and the rational functions $\mathcal{J}_I(s,p)$ counting integer points within them.

Now

$$\left. \frac{(1-p^{-1})^n}{1-p^{-(ds+d'+n)}} \right|_{p \to p^{-1}} = (-1)^{n-1} p^{-ds-d'} \frac{(1-p^{-1})^n}{1-p^{-(ds+d'+n)}} \, ,$$

so it is sufficient to prove that

$$\left. \sum_{I \subseteq T} c_{p,I} \hat{\mathcal{J}}_I(s,p) \right|_{p \to p^{-1}} = (-1)^{n-1} \sum_{I \subseteq T} c_{p,I} \mathcal{J}_I(s,p) \, . \tag{4.7}$$

The $\hat{\mathcal{J}}_I(s,p)$s that are nonzero are $(p-1)^{|I|-n+1}$ times a cone sum of the 'open cone' form required for Stanley's theorem to hold. For all sets I with $\hat{\mathcal{J}}_I(s,p) \neq 0$, Stanley's theorem tells us that

$$\hat{\mathcal{J}}_I(s,p)|_{p \to p^{-1}} = (-1)^{|I|}(-p)^{n-1-|I|} \mathcal{J}_I^{\perp}(s,p) \, , \tag{4.8}$$

where

$$\mathcal{J}_I^\perp(s,p) = \frac{1}{(p-1)^{n-1-|I|}} \sum_{(k_1,\dots,k_{|I|})\in\Lambda_I^\perp} p^{-\sum_{j=1}^{|I|} k_j(A_{j,I}s+B_{j,I})}$$

and

$$\Lambda_I^\perp = \left\{ (k_1,\dots,k_{|I|}) \in \mathbb{N}^{|I|} : \sum_{j=1}^{|I|} N_{i_j}(f_i)k_j \le \sum_{j=1}^{|I|} N_{i_j}(g_i)k_j \text{ for } i = 1,\dots,l \right\}.$$

$\Lambda_I^\perp$ is essentially the closed cone corresponding to $\hat{\Lambda}_I$.

For $I \subseteq T$, define

$$b_{p,I} = |\{ a \in \overline{Y}(\mathbb{F}_p) : a \in \overline{E}_i \text{ if } i \in I \}|.$$

The relationships between the $b_{p,I}$s and the $c_{p,I}$s are

$$b_{p,I} = \sum_{I\subseteq J\subseteq T} c_{p,J} \tag{4.9}$$

and

$$c_{p,I} = \sum_{I\subseteq J\subseteq T} (-1)^{|J|-|I|} b_{p,J}. \tag{4.10}$$

The $b_{p,I}$s are counting points on smooth projective varieties; we shall need to use the properties of the Weil zeta function of such varieties.

If $b_{p,I}$ is a polynomial in p for all p, then it is clear what we mean by $b_{p,I}|_{p\to p^{-1}}$. However, we don't want to restrict ourselves to such limited cases. If we define $b_{p^e,I}$ to be the number of $\mathbb{F}_{p^e}$-rational points on $V := \bigcap_{i\in I} E_i$, then the analogue of the Riemann Hypothesis for the Weil zeta function of V implies that the function $N_V(e)$ defined by

$$N_V : \mathbb{N}_{>0} \to \mathbb{N}$$
$$e \mapsto |\{ \mathbb{F}_{p^e}\text{-rational points of } V \}|$$

has a unique extension to $\mathbb{Z}$. We shall then take $b_{p,I}|_{p\to p^{-1}}$ to be $N_V(-1)$. It then follows from the functional equation of the Weil zeta function of V that

$$b_{p,I}|_{p\to p^{-1}} = p^{|I|-n+1} b_{p,I}. \tag{4.11}$$

We also need the following combinatorial lemma, an immediate consequence of Lemma 1 in [8]. For I fixed,

Lemma 4.9.

$$\sum_{I\subseteq J\subseteq T} b_{p,J}(-p)^{|J|-|I|} = \sum_{I\subseteq K\subseteq T} c_{p,K}(1-p)^{|K|-|I|}. \tag{4.12}$$

We say that a subset I of T is *good* if $\hat{\mathcal{J}}_I(s,p) \neq 0$, *quasi-good* if $\hat{\mathcal{J}}_I(s,p) = 0$ but $\mathcal{J}_I(s,p) \neq 0$, and *bad* otherwise. Finally, we note that

$$\mathcal{J}_I^{\perp}(s,p) = \sum_{H \subseteq I} (p-1)^{|I|-|H|} \mathcal{J}_H(s,p) . \tag{4.13}$$

We now have enough to guide us through the following piece of algebraic manipulation. We have

$$\sum_{I \subseteq T} c_{p,I} \hat{\mathcal{J}}_I(s,p) = \sum_{J \subseteq T} b_{p,J} \sum_{I \subseteq J} (-1)^{|J|-|I|} \hat{\mathcal{J}}_I(s,p) ,$$

so, using (4.10), (4.11), (4.8), (4.12) and (4.13) in that order,

$$\sum_{I \subseteq T} c_{p,I} \hat{\mathcal{J}}_I(s,p) \Bigg|_{p \to p-1}$$

$$= \left[\sum_{J \subseteq T} b_{p,J} \left(\sum_{I \subseteq J} (-1)^{|J|-|I|} \hat{\mathcal{J}}_I(s,p) \right) \right]\Bigg|_{p \to p-1}$$

$$= \sum_{J \subseteq T} p^{|J|-n+1} b_{p,J} \left(\sum_{\substack{I \subseteq J \\ I \text{ good}}} (-1)^{|J|-|I|} (-p)^{n-1-|I|} (-1)^{|I|} \mathcal{J}_I^{\perp}(s,p) \right)$$

$$= (-1)^{n-1} \sum_{J \subseteq T} b_{p,J} \left(\sum_{\substack{I \subseteq J \\ I \text{ good}}} (-p)^{|J|-|I|} \mathcal{J}_I^{\perp}(s,p) \right)$$

$$= (-1)^{n-1} \sum_{\substack{I \subseteq T \\ I \text{ good}}} \mathcal{J}_I^{\perp}(s,p) \left(\sum_{I \subseteq J \subseteq T} b_{p,J} (-p)^{|J|-|I|} \right)$$

$$= (-1)^{n-1} \sum_{\substack{I \subseteq T \\ I \text{ good}}} \mathcal{J}_I^{\perp}(s,p) \left(\sum_{I \subseteq K \subseteq T} c_{p,K} (1-p)^{|K|-|I|} \right)$$

$$= (-1)^{n-1} \sum_{\substack{H \subseteq I \subseteq K \subseteq T \\ I \text{ good}}} (p-1)^{|I|-|H|} \mathcal{J}_H(s,p) c_{p,K} (1-p)^{|K|-|I|}$$

$$= (-1)^{n-1} \sum_{\substack{H \subseteq I \subseteq K \subseteq T \\ I \text{ good}}} (p-1)^{|K|-|H|} \mathcal{J}_H(s,p) c_{p,K} (-1)^{|K|-|I|} . \tag{4.14}$$

4.6.4 Cones and Schemes

From (4.7) and (4.14), it now suffices to show that

$$c_{p,H} = \sum_{\substack{H \subseteq I \subseteq K \subseteq T \\ I \text{ good}}} (p-1)^{|K|-|H|} c_{p,K}(-1)^{|K|-|I|} \tag{4.15}$$

for all good or quasi-good H. If H is bad, $\mathcal{J}_H(s,p) = 0$, so the coefficient of $\mathcal{J}_H(s,p)$ in (4.14) is irrelevant. This statement is independent of the cone sums; all we have left to work with are the schemes over $\mathbb{F}_p$ and the cones themselves.

We may assume from now on that $|H| < n$. $c_{p,I} = 0$ for all I with $|I| \geq n$, and if $|H| \geq n$, (4.15) trivially holds.

Remark 4.10. We assume at this point that the whole set T is good. This assumption is certainly necessary, since it rules out the integral presented in Remark 4.6. However, we do not know what conditions to impose on the integral to ensure that T is always good.

We let $\overline{D_T}$ be the polyhedral cone mentioned in Chap. 3 of [17]. It is defined as follows:

$$\overline{D_T} = \left\{ (z_1, \ldots, z_t) \in \mathbb{R}_{\geq 0}^t : \sum_{j=1}^{t} N_j(f_i) z_j \leq \sum_{j=1}^{t} N_j(g_i) z_j \text{ for } i = 1, \ldots, l \right\},$$

where $t = |T|$. $\overline{D_T}$ is a cone in $\mathbb{R}^t$, with each dimension corresponding to one of the varieties of the resolution. For any subset of $\mathbb{R}^m$, we define its dimension to be the dimension of its $\mathbb{R}$-linear span.

A *wall* of $\overline{D_T}$ is a face of codimension 1. Walls of $\overline{D_T}$ are of two forms:

1. Walls of the form $x_k = 0$ for some j (*coordinate walls*)
2. Walls of the form

$$\sum_{j=1}^{t} N_j(f_i) z_j = \sum_{j=1}^{t} N_j(g_i) z_j$$

for some i (*non-coordinate walls*)

For $I \subseteq T$, we define

$$\text{face}(I) = \{ (z_1, \ldots, z_t) \in \overline{D_T} : z_i > 0 \iff i \in I \}.$$

If I is bad, $\text{face}(I) = \varnothing$. Otherwise, $\text{face}(I)$ is a face of $\overline{D_T}$ of dimension $|I|$. Since T is good, the dimension of $\overline{D_T}$ is $|T|$. A set I is quasi-good if and only if $\text{face}(I)$ is contained in at least one non-coordinate wall.

Our first lemma in this section shows that good sets are 'upwards-closed':

Lemma 4.11. *If $I \subseteq J \subseteq T$ and I is good, then J is good.*

Proof. It suffices to assume $|J| = |I| + 1$. Let $\mathbf{u}$ be any integer point in the open face of $\overline{D_T}$ corresponding to I.

Each of the inequalities defining $\overline{D_T}$ will be satisfied strictly. Pick $j \in T \setminus I$, and for $k \neq j$, set $u'_k = u_k$. Let u'_j be a rational number which can be chosen to be small enough so that $\mathbf{u}'$ still strictly satisfies the inequalities. $\mathbf{u}'$ may not be an integer point, but if not, a suitable integer multiple of it will be integral. Clearing denominators if necessary will give us an integer point in $\hat{A}_{I \cup \{j\}}$, so $I \cup \{j\}$ is good. $\qquad \square$

Proposition 4.12. *Let $I' \supseteq H$ be good. Then*

$$\sum_{\substack{I' \subseteq I \subseteq K \subseteq T \\ I \text{ good}}} (p-1)^{|K|-|H|} c_{p,K} (-1)^{|K|-|I|} = (p-1)^{|I'|-|H|} c_{p,I'} \; .$$

Proof. Since I' is good and $I' \subseteq I$, I is good. Thus we have

$$\sum_{I' \subseteq I \subseteq K \subseteq T} (p-1)^{|K|-|H|} c_{p,K} (-1)^{|K|-|I|}$$

$$= \sum_{I' \subseteq K \subseteq T} (p-1)^{|K|-|H|} c_{p,K} \sum_{I' \subseteq I \subseteq T} (-1)^{|K|-|I|}$$

$$= \sum_{I' \subseteq K \subseteq T} (p-1)^{|K|-|H|} c_{p,K} (1-1)^{|K|-|I'|}$$

$$= (p-1)^{|I'|-|H|} c_{p,I'} \; .$$

$\qquad \square$

Corollary 4.13. *Equation (4.15) holds if H is good.*

Proof. Proposition 4.12 with $I' = H$. $\qquad \square$

4.6.5 Quasi-Good Sets

The case with H good could be easily dealt with. We didn't need to know anything about the $c_{p,I}$s; everything follows as an identity in the $c_{p,I}$s. We are now left with proving (4.15) for H quasi-good. For this we require the terminology of convex polytopes (see, for example, [30]).

A *convex polytope* $\mathcal{P}$ is a bounded subset of $\mathbb{R}^m$ defined as the intersection of finitely many closed half-spaces. The *dimension* of a convex polytope is the dimension of its linear span as a vector space over $\mathbb{R}$. We suppose $\dim \mathcal{P} = m$. A *wall* of $\mathcal{P}$ is an $(m-1)$-dimensional hyperplane which does not bisect $\mathcal{P}$ but whose intersection with $\mathcal{P}$ is $(m-1)$-dimensional. A *face* F of $\mathcal{P}$ is a nonempty intersection of $\mathcal{P}$ with a number of walls. We do not consider the empty set to be a face of $\mathcal{P}$, but we do consider the whole polytope $\mathcal{P}$ to be a face of itself. A face F is *proper* if $\dim F < m$.

Let $\mathcal{F}(\mathcal{P})$ denote the set of nonempty faces of $\mathcal{P}$, and for F a face of $\mathcal{P}$, let $\mathcal{F}(\mathcal{P}; F)$ denote the set of faces of $\mathcal{P}$ containing F. The most important result concerning this counting function is the following variation on the Euler characteristic, namely

$$\sum_{F' \in \mathcal{F}(\mathcal{P};F)} (-1)^{\dim F'} = 0 \qquad (4.16)$$

for all proper faces F of $\mathcal{P}$ [30, p. 137].

A *polyhedral cone* C is a subset of $\mathbb{R}^m_{\geq 0}$ defined as the intersection of finitely many closed half-spaces, with the property that $\lambda \mathbf{x} \in C$ for all $\mathbf{x} \in C$, $\lambda \in \mathbb{R}_{>0}$. In particular, all the boundary hyperplanes must pass through the origin $\mathbf{0}$. We define walls and faces of a polyhedral cone in an analogous way as for convex polytopes.

Proposition 4.14. *Let F be a proper face of the polyhedral cone C. Then*

$$\sum_{F' \in \mathcal{F}(C;F)} (-1)^{\dim F'} = 0 \, .$$

Proof. Let $\mathcal{H}$ denote the half plane $\{\, \mathbf{x} \in \mathbb{R}^m \mid \sum_i x_i = 1 \,\}$ and $\mathcal{H}^-$ the closed half-space $\{\, \mathbf{x} \in \mathbb{R}^m \mid \sum_i x_i \leq 1 \,\}$. $\mathcal{P} := C \cap \mathcal{H}^-$ is a convex polytope since it is clearly convex and is contained within the m-dimensional unit hypercube. There is a bijective dimension-preserving correspondence between the faces of C containing F and faces of $\mathcal{P}$ containing $F \cap \mathcal{H}^-$, from which the result follows.   □

We shall prove the following combinatorial theorem, which gives us some idea about how the quasi-good faces behave. Let $W_1, \ldots, W_l$ denote the walls of the cone C.

Theorem 4.15. *Suppose there exists $\mathbf{x}_0 \in \text{face}(T)$ such that $\mathbf{x} \in \bigcap_{i=1}^l W_i$. Then for all quasi-good sets $H \subset T$,*

$$\sum_{\substack{H \subseteq I \subseteq T \\ I \ good}} (-1)^{|T|-|I|} = 0 \, .$$

We shall prove this result after proving a number of combinatorial lemmas. Fix a subset $\Sigma \subseteq \{1, \ldots, l\}$ and a face $F \subsetneq \bigcap_{i \in \Sigma} W_i$.

Lemma 4.16. *For $S \subseteq \Sigma$, set $W_S^{\cap} = \cap_{i \in S} W_i$. Then*

$$\sum_{F' \in \mathcal{F}(W_S^{\cap};F)} (-1)^{\dim F'} = 0 \, .$$

Proof. Proposition 4.14 with $C = W_S^{\cap}$.   □

Lemma 4.17. *For $S \subseteq \Sigma$, set $\dot{W}_S^\cap = \bigcap_{i \in S} W_i \setminus \bigcup_{j \in \Sigma \setminus S} W_j$. Then*

$$\sum_{F' \in \mathcal{F}(\dot{W}_S^\cap; F)} (-1)^{\dim F'} = 0 \, .$$

Proof. By reverse induction on $|S|$. The case $S = \Sigma$ is clear. Otherwise, we note that

$$W_S^\cap = \dot{\bigcup_{S'}} \dot{W}_{S'}^\cap$$

for some combination of sets S' with $S \subsetneq S' \subseteq \Sigma$. By induction, the alternating sum over each $\dot{W}_{S'}^\cap$ is zero, and the sets $\dot{W}_{S'}^\cap$ are disjoint, so the result follows. □

Lemma 4.18. *For $S \subseteq \Sigma$, set $W_S^\cup = \bigcup_{i \in S} W_i$. Then*

$$\sum_{F' \in \mathcal{F}(C;F) \setminus \mathcal{F}(W_S^\cup; F)} (-1)^{\dim F'} = 0 \, .$$

Proof. $W_S^\cup$ is the disjoint union of terms of the form $\dot{W}_{S'}^\cap$ for various $S' \subseteq \Sigma$. The alternating sum over each $\dot{W}_{S'}^\cap$ is zero, hence

$$\sum_{\mathcal{F}(W_S^\cup; F)} (-1)^{\dim F'} = 0 \, .$$

The result follows from this and Proposition 4.14. □

Proof. (of Theorem 4.15) Set $F = \text{face}(H)$. It is clear that there is then a bijective correspondence between faces of C containing F and good or quasi-good subsets of T containing H, with $\dim(\text{face}(I)) = |I|$.

Let $\Sigma = \{ i \mid 1 \leq i \leq l, W_i \not\supseteq F \}$, the indexing set of the hyperplanes strictly containing F. Since F contains only quasi-good points, $\Sigma \neq \varnothing$. It is clear that $F \subseteq \bigcap_{i \in \Sigma} W_i$, but there cannot be equality since $\mathbf{x}_0 \in W_i$ for all i but $\mathbf{x}_0 \notin F$. There is a bijective correspondence between faces contained in $W_\Sigma^\cup$ and quasi-good subsets of T containing H, and hence between faces containing F but not contained in $W_\Sigma^\cup$ and good subsets of T containing H. The result then follows from Lemma 4.18. □

4.6.6 Quasi-Good Sets: The Monomial Case

If we additionally assume that the cone data is monomial, we can show that (4.15) holds.

Proposition 4.19. *Suppose that the cone data* $\mathcal{D} = \{f_0, g_0, \ldots, f_l, g_l\}$ *are monomial, and that* $H \subsetneq T$. *Then (4.15) holds.*

Proof. Since the cone data are monomial, $c_{p,K} = (p-1)^{n-1-|K|}$ if $K \subsetneq T$, with $c_{p,T} = 0$. We now substitute this into the RHS of (4.15):

$$\sum_{\substack{H \subseteq I \subseteq K \subseteq T \\ I \text{ good}}} (p-1)^{|K|-|H|} c_{p,K} (-1)^{|K|-|I|}$$

$$= (p-1)^{n-1-|H|} \sum_{\substack{H \subseteq I \subseteq K \subsetneq T \\ I \text{ good}}} (-1)^{|K|-|I|}$$

$$= c_{p,H} \left(\sum_{\substack{H \subseteq I \subseteq K \subseteq T \\ I \text{ good}}} (-1)^{|K|-|I|} - \sum_{\substack{H \subseteq I \subseteq T \\ I \text{ good}}} (-1)^{|T|-|I|} \right)$$

$$= c_{p,H} \left(\sum_{\substack{H \subseteq I \subseteq T \\ I \text{ good}}} (1-1)^{|T|-|I|} - \sum_{\substack{H \subseteq I \subseteq T \\ I \text{ good}}} (-1)^{|T|-|I|} \right)$$

$$= c_{p,H} (1 - 0)$$

$$= c_{p,H} ,$$

using Lemma 4.15 with $\mathbf{x}_0 = (1, 1, \ldots, 1)$ for the penultimate step. □

Note that Proposition 4.19 does not prove anything new. In fact, Theorem 4.7 is more general and has a much simpler proof. Nonetheless it demonstrates one case where we can follow the proof through to completion. To go any further without assuming the cone data is monomial seems to require intimate knowledge of the coefficients $c_{p,I}$.

4.7 Applications of Conjecture 4.5

We now look at applying Conjecture 4.5 to functional equations of local zeta functions. We adopt a cavalier attitude to the incompleteness of Conjecture 4.5; the aim here is to demonstrate that what it predicts agrees with calculations made to date. Before we do this, we take the time to consider the conjectures that have so far been formulated concerning these functional equations.

For brevity, we shall refer to a Lie ring additively isomorphic to $\mathbb{Z}^d$ (or $\mathbb{Z}_p^d$) as a $\mathbb{Z}$-Lie ring (or $\mathbb{Z}_p$-Lie ring) of rank d, where we assume $d \in \mathbb{N}$. All calculated examples at nilpotency class 2 satisfy the following conjectures, which were also stated in [15] and [57]:

Conjecture 4.20. Let L be a class-2-nilpotent $\mathbb{Z}$-Lie ring of rank d. Then, for all but finitely many primes p,

$$\zeta_{L,p}^{\triangleleft}(s)\big|_{p\to p^{-1}} = (-1)^d p^{\binom{d}{2}-(d+n)s}\zeta_{L,p}^{\triangleleft}(s) , \qquad (4.17)$$

where $n = \operatorname{rank}(L/Z(L))$.

Conjecture 4.21. Let L be a class-2-nilpotent $\mathbb{Z}$-Lie ring of rank d. Then, for all but finitely many primes p,

$$\zeta_{L,p}^{\leq}(s)\big|_{p\to p^{-1}} = (-1)^d p^{\binom{d}{2}-ds}\zeta_{L,p}^{\leq}(s) . \qquad (4.18)$$

Both conjectures have been confirmed by Voll [59].

For Lie rings of higher nilpotency class, Taylor makes the following

Conjecture 4.22 ([57]). Let L be a nilpotent $\mathbb{Z}$-Lie ring of rank d. Then, for all but finitely many primes p,

$$\zeta_{L,p}^{\leq}(s)\big|_{p\to p^{-1}} = (-1)^d p^{\binom{d}{2}-ds}\zeta_{L,p}^{\leq}(s) . \qquad (4.19)$$

This conjecture generalises Conjecture 4.21 and has also been confirmed by Voll [59]. In fact, Voll proves that Conjecture 4.22 holds for *any* not-necessarily-associative ring L.

A similar conjecture for the zeta functions counting ideals was formulated by the second author in his thesis. Let $\sigma_i(L)$ denote the ith term of the upper-central series of L, which we recall is defined by $\sigma_0(L) = \{0\}$, $\sigma_1(L) = Z(L)$ and $\sigma_i(L)/\sigma_{i-1}(L) = Z(L/\sigma_{i-1}(L))$. All examples in Chap. 2 at nilpotency classes 3 and 4 satisfy the following conjecture.

Conjecture 4.23. Let L be a class-c-nilpotent $\mathbb{Z}$-Lie ring of rank d. One of the following two alternatives holds:

• For all but finitely many primes p,

$$\zeta_{L,p}^{\triangleleft}(s)\big|_{p\to p^{-1}} = (-1)^d p^{\binom{d}{2}-Ns}\zeta_{L,p}^{\triangleleft}(s) ,$$

where $N = \sum_{i=0}^{c} \operatorname{rank}(L/\sigma_i(L))$.
• For all but finitely many primes p, $\zeta_{L,p}^{\triangleleft}(s)$ satisfies no such functional equation.

In [57], Taylor also considered the ideal zeta functions of the Lie rings

$$M_n = \langle z, x_1, x_2, \ldots, x_n : [z, x_i] = x_{i+1} \text{ for } i = 1, \ldots, n-1 \rangle$$

for $n \in \mathbb{N}$, and made the following conjecture.

Conjecture 4.24. Let $L = M_n$ of rank $d = n + 1$. Then, for all but finitely many primes p,

$$\zeta_{L,p}^{\triangleleft}(s)\big|_{p\to p^{-1}} = (-1)^d p^{\binom{d}{2}-\left(\binom{d+1}{2}-1\right)s}\zeta_{L,p}^{\triangleleft}(s) ,$$

where $d = n + 1$.

The exponent of p^{-s} agrees with that predicted by Conjecture 4.23; Taylor conjectures that such a functional equation always holds.

Finally, we briefly move away from nilpotent Lie rings. Suppose L arises as the $\mathbb{Z}$-span of a Chevalley basis corresponding to a simple Lie algebra over $\mathbb{C}$. du Sautoy asks the following question on p. 219 of [12]:

Question 4.25. Does $\zeta_{L\otimes\mathbb{Z}_p}^{\leq}(s)$ satisfy a functional equation of the form

$$\zeta_{L\otimes\mathbb{Z}_p}^{\leq}(s)\big|_{p\to p^{-1}} = (-1)^n p^{-as+b}\zeta_{L\otimes\mathbb{Z}_p}^{\leq}(s)$$

for all but perhaps finitely many primes p?

Voll's work on the local zeta functions counting all subrings of a $\mathbb{Z}$-Lie ring L of rank d answers this question in the affirmative.

In the following sections we will see that Conjecture 4.5 implies all of Conjectures 4.20–4.24 as well as a positive answer to Question 4.25. In particular, it agrees with Voll's results.

Let L be a $\mathbb{Z}$-Lie ring of rank d, $* \in \{\leq, \triangleleft\}$. From Proposition 2.2, there exists a set of cone data $\mathcal{D}^*$ such that

$$\zeta_{L,p}^*(s) = (1 - p^{-1})^{-d}Z_{\mathcal{D}^*}(s - d, p) . \tag{4.20}$$

Conjecture 4.5 relates $Z_{\mathcal{D}^*}(s,p)|_{p\to p^{-1}}$ and $Z_{\mathcal{D}^*}^{\circ}(s,p)$. It is clear from (4.20) that $(1-p^{-1})^{-d}Z_{\mathcal{D}^*}(s-d,p)$ is counting additive submodules of L with certain properties, i.e. ideals or subrings of L. Our applications are founded on the observation that $(1-p^{-1})^{-d}Z_{\mathcal{D}^*}^{\circ}(s-d,p)$ is also counting additive submodules, and on determining what these additive submodules are.

Definition 4.26. *Let L be a $\mathbb{Z}$-Lie ring, p a rational prime. An ideal $I \trianglelefteq L$ is a p-ideal of L if:*

1. *$I \subseteq pL$, and*
2. *For all $x \in L$, $y \in I$, $[x,y] \in pI$.*

If I is a p-ideal of L, we write $I \triangleleft_p L$.

Definition 4.27. *Let L be a $\mathbb{Z}$-Lie ring, p a rational prime. A subring $H \leq L$ is a p-subring of L if:*

1. *$H \subseteq pL$, and*
2. *For all $x, y \in L$, $[x,y] \in pH$.*

If H is a p-subring of L, we write $H <_p L$.

p-ideals and p-subrings can also be defined for $\mathbb{Z}_p$-Lie rings in a similar way.

Lemma 4.28. *Let L be a $\mathbb{Z}$-Lie ring and $* \in \{\leq, \lhd\}$. By Proposition 2.2, there exist sets of cone data $\mathcal{D}^\lhd$, $\mathcal{D}^\leq$ such that*

$$\zeta_{L,p}^\lhd(s) = (1 - p^{-1})^{-d} Z_{\mathcal{D}^\lhd}(s - d, p) ,$$

$$\zeta_{L,p}^\leq(s) = (1 - p^{-1})^{-d} Z_{\mathcal{D}^\leq}(s - d, p) .$$

Define

$$\zeta_{L,p}^{\lhd_p}(s) = \sum_{n=0}^\infty a_{p^n}^{\lhd_p}(L) p^{-ns} ,$$

$$\zeta_{L,p}^{<_p}(s) = \sum_{n=0}^\infty a_{p^n}^{<_p}(L) p^{-ns} ,$$

where $a_{p^n}^{\lhd_p}(L)$ is the number of p-ideals of index p^n in L and $a_{p^n}^{<_p}(L)$ is the number of p-subrings of index p^n in L. Then

$$\zeta_{L,p}^{\lhd_p}(s) = (1 - p^{-1})^{-d} Z_{\mathcal{D}^\lhd}^\circ(s - d, p) ,$$

$$\zeta_{L,p}^{<_p}(s) = (1 - p^{-1})^{-d} Z_{\mathcal{D}^\leq}^\circ(s - d, p) .$$

Proof. Let $\mathcal{B} = (\mathbf{e}_1, \ldots, \mathbf{e}_d)$ be a basis for L, and $\mathbf{m}_1, \ldots, \mathbf{m}_d$ a set of additive generators for an additive submodule H of L. The polynomial divisibility conditions $v(f_k(\mathbf{x})) \leq v(g_k(\mathbf{x}))$ ensure that Lie brackets $[\mathbf{e}_i, \mathbf{m}_j]$ are in the $\mathbb{Z}_p$-span of $\{\mathbf{m}_1, \ldots, \mathbf{m}_d\}$ for $1 \leq i, j \leq d$. Indeed, when we express

$$[\mathbf{m}_i, \mathbf{e}_j] = \sum_{r=1}^d \lambda_{i,j,r} \mathbf{m}_r ,$$

each nonzero coefficient $\lambda_{i,j,r}$ is of the form $g_k(\mathbf{x})/f_k(\mathbf{x})$ for some k. Enforcing $v(f_i(\mathbf{x})) < v(g_i(\mathbf{x}))$ will then ensure that each $[\mathbf{m}_i, \mathbf{e}_j]$ is in the $p\mathbb{Z}_p$-span of $\{\mathbf{m}_1, \ldots, \mathbf{m}_d\}$ for $1 \leq i, j \leq d$, i.e. is in pH.

A similar argument works when we count all subrings instead of just ideals. We must consider the Lie brackets $[\mathbf{m}_i, \mathbf{m}_j]$ for $1 \leq i < j \leq d$ instead of $[\mathbf{m}_i, \mathbf{e}_j]$ for $1 \leq i, j \leq d$.

Finally, since $Z_{\mathcal{D}^*}^\circ(s, p)$ integrates over a subset of $(p\mathbb{Z}_p)^{\binom{d+1}{2}}$, it is clear that it is only counting additive submodules contained in pL. $\qquad\square$

Corollary 4.29. *Let L be a $\mathbb{Z}$-Lie ring. Assume Conjecture 4.5. Then, for all primes p outside a finite set dependent on L,*

$$\zeta_{L,p}^\lhd(s)\big|_{p \to p^{-1}} = (-1)^d p^{\binom{d}{2}} \zeta_{L,p}^{\lhd_p}(s) , \tag{4.21}$$

$$\zeta_{L,p}^\leq(s)\big|_{p \to p^{-1}} = (-1)^d p^{\binom{d}{2}} \zeta_{L,p}^{<_p}(s) . \tag{4.22}$$

Proof. $Z_{\mathcal{D}^\lhd}(s,p)$ and $Z_{\mathcal{D}^\leq}(s,p)$ are both integrals over $\binom{d+1}{2}$ variables. So, since we are assuming Conjecture 4.5,

$$Z_{\mathcal{D}^\lhd}(s,p)\big|_{p \to p^{-1}} = p^{\binom{d+1}{2}} Z_{\mathcal{D}^\lhd}^\circ(s,p) \,,$$

$$Z_{\mathcal{D}^\leq}(s,p)\big|_{p \to p^{-1}} = p^{\binom{d+1}{2}} Z_{\mathcal{D}^\leq}^\circ(s,p) \,,$$

for all but finitely many primes p. Also, it is straightforward that

$$(1 - p^{-1})^{-d}\big|_{p \to p^{-1}} = (-1)^d p^{-d}(1 - p^{-1})^{-d} \,.$$

The result now follows from Lemma 4.28.   □

Corollary 4.29 suggests that the functional equations we have seen arise from correspondences between ideals and p-ideals, or subrings and p-subrings. This therefore motivates us to study these correspondences.

4.8 Counting Subrings and p-Subrings

The correspondence between subrings and p-subrings is encapsulated by the following lemma.

Lemma 4.30. *Let L be a $\mathbb{Z}$-Lie ring, and H be an additive submodule of L. Then H is a subring of L if and only if pH is a p-subring of L.*

Proof. Clearly $pH \subseteq pL$. For all $x, y \in H$, $[x, y] \in H$ if and only if $[px, py] \in p(pH)$.   □

Theorem 4.31. *Let L be a $\mathbb{Z}$-Lie ring. Assume Conjecture 4.5. Then, for all but finitely many primes p,*

$$\zeta_{L,p}^\leq(s)\big|_{p \to p^{-1}} = (-1)^d p^{\binom{d}{2}-ds} \zeta_{L,p}^\leq(s) \,. \tag{4.23}$$

Proof. Lemma 4.30 implies that the multiplication-by-p map is a bijective correspondence between the subrings of p-power index in L and the p-subrings of p-power index in L, for all primes p. Hence

$$\zeta_{L,p}^{<_p}(s) = p^{-ds} \zeta_{L,p}^\leq(s) \tag{4.24}$$

for all primes p. Combining (4.24) with (4.22), we obtain that Conjecture 4.5 implies the functional equation

$$\zeta_{L,p}^\leq(s)\big|_{p \to p^{-1}} = (-1)^d p^{\binom{d}{2}-ds} \zeta_{L,p}^\leq(s) \,.$$

This gives us the result.   □

A second proof of Theorem 4.31 follows from Corollary 4.8:

Proof. By Proposition 2.2, there exists a set of cone data $\mathcal{D} = \{f_0, g_0, \ldots, f_l, g_l\}$ such that

$$\zeta^{\leq}_{\overline{L},p}(s + d) = (1 - p^{-1})^{-d} Z_\mathcal{D}(s, p) .$$

By Proposition 2.1, the cone data satisfy $\deg f_i(\mathbf{x}) + 1 = \deg g_i(\mathbf{x})$ for $1 \leq i \leq l$. Assuming Conjecture 4.5, Corollary 4.8 implies that $Z_\mathcal{D}(s, p)$ satisfies the functional equation

$$Z_\mathcal{D}(s, p)\big|_{p \to p^{-1}} = p^{-\deg g_0 - s \deg f_0} Z_\mathcal{D}(s, p) .$$

Proposition 2.2 additionally implies that $\deg f_0 = d$ and $\deg g_0 = \binom{d}{2}$. Hence

$$\zeta^{\leq}_{\overline{L},p}(s + d)\bigg|_{p \to p^{-1}} = (-p)^{-d} p^{-\binom{d}{2} - ds} \zeta^{\leq}_{\overline{L},p}(s + d) ,$$

and this easily rearranges into the functional equation (4.23).   □

In both proofs we can relax the assumption that L is a Lie ring. L can instead be any not-necessarily associative ring provided it is additively isomorphic to $\mathbb{Z}^d$ for some $d \in \mathbb{N}$. Conjectures 4.21 and 4.22 are thus mere special cases of Conjecture 4.5, as is a positive answer to Question 4.25. However, if L is not a nilpotent Lie ring we do not obtain a corresponding result counting subgroups in a $\mathfrak{T}$-group.

The correspondence between subrings and p-subrings is rather trivial. It is encapsulated in one tiny lemma (Lemma 4.30). Furthermore, there is a more direct proof of Theorem 4.31 which avoids introducing the concept of a p-subring. However, it does illustrate the approach we shall apply to local zeta functions counting ideals, where such a direct proof does not exist.

4.9 Counting Ideals and p-Ideals

Let L be a Lie ring additively isomorphic to $\mathbb{Z}$. By Proposition 2.2, there exists a set of cone data $\mathcal{D} = \{f_0, g_0, \ldots, f_l, g_l\}$ such that

$$Z_\mathcal{D}(s, p) = (1 - p^{-1})^d \zeta^{\triangleleft}_{L,p}(s + d) .$$

Proposition 2.1 establishes that the cone data satisfy $\deg f_i(\mathbf{x}) = \deg g_i(\mathbf{x})$ for $1 \leq i \leq l$. In this case we cannot apply Corollary 4.8.

Remark 4.6 notes that some unspecified extra conditions for Conjecture 4.5 are necessary in the case where $\deg f_i(\mathbf{x}) = \deg g_i(\mathbf{x})$ for all $1 \leq i \leq l$. We are not sure what these conditions are but we do believe that they will be satisfied by the p-adic integrals representing local ideal zeta functions of nilpotent Lie rings. Although we cannot formulate Conjecture 4.5 rigorously, we can still study the correspondence between ideals and p-ideals that it has led us to.

It can be seen that the soluble Lie rings $\mathfrak{tr}_n(\mathbb{Z})$ for $n \geq 2$ have no p-ideals of finite index for all primes p. In this case we cannot deduce the functional equations we deduced in Chap. 3 from Conjecture 4.5. Therefore we must also make the assumption that p-ideals exist. We shall see later that if L is nilpotent, L has p-ideals of finite index for all p. For this reason we assume from now on that L is nilpotent.

Furthermore, we shall only consider Lie rings. It is likely that similar results could be obtained for more general torsion-free rings with nilpotent multiplication. For simplicity, this is a route we have chosen not to follow.

We will frequently find ourselves working over $\mathbb{Z}_p$ instead of over $\mathbb{Z}$. For a $\mathbb{Z}$-Lie ring L, we shall denote $L \otimes \mathbb{Z}_p$ by L_p. We shall also identify an element $x \in L$ with its image $x \otimes 1_{\mathbb{Z}_p}$ in L_p. In particular, if we have a basis $\mathcal{B} = (\mathbf{e}_1, \ldots, \mathbf{e}_d)$ for a $\mathbb{Z}$-Lie ring L, we shall also write $\mathcal{B}$ for the corresponding basis $(\mathbf{e}_1 \otimes 1_{\mathbb{Z}_p}, \ldots, \mathbf{e}_d \otimes 1_{\mathbb{Z}_p})$ of L_p.

4.9.1 Heights, Cocentral Bases and the π-Map

Nilpotent groups and rings have a notion of the *weight* of an element, i.e. how far down the lower-central series a given element lies. We require an analogue of this notion for the upper-central series.

Definition 4.32. *Let L be a nilpotent Lie ring of class c. For $0 \leq i \leq c$, let $\sigma_i(L)$ denote the ith term of the upper-central series of L. The* height *of an element $x \in L$ is defined by $\mathrm{ht}(x) := \min\{\, i : x \in \sigma_i(L) \,\}$.*

For clarity we shall write the left-normed Lie bracket $[[\ldots[[z_1, z_2], z_3], \ldots], z_m]$ as $[z_1, z_2, z_3, \ldots, z_m]$.

Proposition 4.33. *Let L be a nilpotent Lie ring, $x \in L$. $\mathrm{ht}(x) > h$ if and only if there exist $z_1, \ldots, z_h \in L$ such that $[x, z_1, \ldots, z_h] \neq 0$.*

Proof. By induction on $h \geq 0$. $\mathrm{ht}(x) > 0$ if and only if $x \neq 0$, so the base case is clear.

$\mathrm{ht}(x) > h$ if and only if $x + \sigma_{h-1}(L) \notin Z(L/\sigma_{h-1}(L))$, i.e. if and only if there exists z_1 such that $[x + \sigma_{h-1}(L), z_1 + \sigma_{h-1}(L)] \neq \sigma_{h-1}(L)$, i.e. if and only if $[x, z_1] \notin \sigma_{h-1}(L)$. This is equivalent to $\mathrm{ht}([x, z_1]) > h - 1$, and by our inductive hypothesis, equivalent to there existing $z_2, \ldots, z_h$ such that $[[x, z_1], z_2, \ldots, z_h] \neq 0$. This establishes the induction. $\square$

Proposition 4.34. *Let L be a nilpotent Lie ring, $x, y \in L$, $x, y \neq 0$. Then $\mathrm{ht}([x, y]) < \min(\mathrm{ht}(x), \mathrm{ht}(y))$.*

Proof. Straightforward. $\square$

Proposition 4.35. *Let L be a nilpotent Lie ring with basis $\mathcal{B} = (\mathbf{e}_1, \ldots, \mathbf{e}_d)$. Suppose $\mathrm{ht}(x) = h > 1$. There exists j with $1 \leq j \leq d$ such that $\mathrm{ht}([x, \mathbf{e}_j]) = h - 1$.*

Proof. For a contradiction, suppose $\mathrm{ht}([x, \mathbf{e}_j]) \leq h - 2$ for all $\mathbf{e}_j$. Since x has height h, there must exist $z_1, \ldots, z_{h-1} \in L$ such that

$$[x, z_1, \ldots, z_{h-1}] \neq 0 . \tag{4.25}$$

By our supposition, $\mathrm{ht}([x, z_1]) \leq h - 2$. By (4.25), $[x, z_1, \ldots, z_k] \neq 0$ for all $1 \leq k \leq h-1$. Since $\mathrm{ht}([x, z_1, \ldots, z_k]) > \mathrm{ht}([\mathbf{e}_i, z_1, \ldots, z_{k+1}])$ for $1 \leq k \leq h-2$, this implies that $\mathrm{ht}([x, z_1, \ldots, z_{h-1}]) = 0$. Clearly this contradicts (4.25). $\square$

Proposition 4.36. *Let L be a torsion-free nilpotent Lie ring. Then, for all $1 \leq i \leq c$, $\sigma_i(L)/\sigma_{i-1}(L)$ is torsion-free.*

Proof. For a contradiction, suppose $x \notin \sigma_{i-1}(L)$ but $mx \in \sigma_{i-1}(L)$ for some $m \in \mathbb{N}_{>0}$. Thus $\mathrm{ht}(x) \geq i$ and $\mathrm{ht}(mx) < i$. Proposition 4.33 then implies that there exist $z_1, \ldots, z_{i-1}$ such that

$$\alpha := [x, z_1, \ldots, z_{i-1}] \neq 0 . \tag{4.26}$$

However, since $\mathrm{ht}(mx) < i$,

$$m\alpha = [mx, z_1, \ldots, z_{i-1}] = 0 . \tag{4.27}$$

Since L is torsion-free, (4.26) and (4.27) contradict one another. $\square$

Definition 4.37. *Let L be a torsion-free nilpotent Lie ring of rank d. For $1 \leq i \leq c$, let $\rho_i = \mathrm{rank}(\sigma_i(L))$. A basis $\mathcal{B} = (\mathbf{e}_1, \ldots, \mathbf{e}_d)$ for L is said to be cocentral if, for all $1 \leq i \leq c$, $\sigma_i(L) = \langle \mathbf{e}_{d-\rho_i+1}, \ldots, \mathbf{e}_d \rangle$.*

Remark 4.38. Proposition 4.36 guarantees that for a nilpotent Lie ring of rank d, a cocentral basis $\mathcal{B}$ for L exists.

Lemma 4.39. *Let L be a class-c-nilpotent $\mathbb{Z}$-Lie ring, with cocentral basis $\mathcal{B} = (\mathbf{e}_1, \ldots, \mathbf{e}_d)$. Let $1 \leq h < c$, $1 \leq k_1, k_2, \ldots, k_h \leq d$ and suppose $\mathrm{ht}(\mathbf{e}_i) > h$. Let $\mathbf{m}_i = m_{i,i}\mathbf{e}_i + \cdots + m_{i,d}\mathbf{e}_d$, where $m_{i,i}, \ldots, m_{i,d}$ are independent indeterminates. Let $R_{i,h}$ be the free $\mathbb{Z}$-module with basis $\{m_{i,i}, \ldots, m_{i,d-\rho_h}\}$. Then the coefficients of basis elements in iterated Lie brackets of the form*

$$[\mathbf{m}_i, \mathbf{e}_{k_1}, \mathbf{e}_{k_2}, \ldots, \mathbf{e}_{k_h}] \tag{4.28}$$

generate a $\mathbb{Z}$-module of finite index in $R_{i,h}$.

Proof. By Proposition 4.33, the coefficients of basis elements in (4.28) are linear polynomials over $\mathbb{Z}$ in $m_{i,i}, \ldots, m_{i,d-\rho_h}$. It suffices to show that over $\mathbb{Q}$ they generate a $(d - \rho_h - i + 1)$-dimensional vector space. Suppose for a contradiction that they do not. We may if necessary perform an invertible change of variables $m_{i,j} = \sum_{r=1}^{d-\rho_h} c_{r,j} m'_j$ for $c_{r,j} \in \mathbb{Q}$, so that all the Lie brackets of the form (4.28) are independent of, say, $m'_{i,l}$. Now

$$\mathbf{m}_i = \sum_{j=1}^{n} m_{i,j} \mathbf{e}_j$$

$$= \sum_{r=1}^{n} \left(\sum_{j=1}^{d-\rho_h} c_{r,j} m'_{i,j} \right) \mathbf{e}_r$$

$$= \sum_{j=1}^{n} m'_{i,j} \left(\sum_{r=1}^{d-\rho_h} c_{r,j} \mathbf{e}_r \right) . \tag{4.29}$$

If $[\mathbf{m}_i, \mathbf{e}_{k_1}, \ldots, \mathbf{e}_{k_h}]$ has no term in $m'_{i,l}$ for all $\mathbf{e}_{k_1}, \ldots, \mathbf{e}_{k_h} \in L$, then (4.29) implies that $\sum_{r=1}^{d-\rho_h} c_{r,l} \mathbf{e}_r \in \sigma_{h-1}(L)$. Since the change of variables is invertible, we cannot have $c_{r,l} = 0$ for all $1 \le r \le d - \rho_h$. This contradicts $\mathcal{B}$ being cocentral. $\qquad\square$

Definition 4.40. *Let p be a prime and L a nilpotent $\mathbb{Z}_p$-Lie ring with cocentral basis $\mathcal{B} = (\mathbf{e}_1, \ldots, \mathbf{e}_d)$. We define the linear map $\pi_{\mathcal{B}} : L \to L$ by $\pi_{\mathcal{B}}(\mathbf{e}_i) = p^{\mathrm{ht}(\mathbf{e}_i)} \mathbf{e}_i$.*

Lemma 4.41. *Let L be a nilpotent $\mathbb{Z}$-Lie ring with cocentral basis $\mathcal{B} = (\mathbf{e}_1, \ldots, \mathbf{e}_d)$. For all but finitely many primes p, $P \lhd_p L_p$ implies $P \subseteq \pi_{\mathcal{B}}(L_p)$.*

Proof. Let P be a p-ideal with additive generators $\mathbf{m}_1, \ldots, \mathbf{m}_d$ where $\mathbf{m}_i = m_{i,i} \mathbf{e}_i + \cdots + m_{i,d} \mathbf{e}_d$ for $m_{i,j} \in \mathbb{Z}_p$. We must prove that $p^{\mathrm{ht}(\mathbf{e}_j)} \mid m_{i,j}$ for all $1 \le i \le j \le d$.

Suppose $\mathrm{ht}(\mathbf{e}_i) > h$. By Lemma 4.39, the coefficients of basis elements in (4.28) generate a submodule of finite index in $R_{i,h}$, the free $\mathbb{Z}$-module with basis $\{m_{i,i}, \ldots, m_{i,d-\rho_h}\}$. For all primes p not dividing this finite index, the coefficients $m_{i,i}, \ldots, m_{i,d-\rho_h}$ generate the free $\mathbb{Z}_p$-module $R_{i,h} \otimes \mathbb{Z}_p$.

Since P is a p-ideal,

$$[\mathbf{m}_i, \mathbf{e}_{k_1}, \mathbf{e}_{k_2}, \ldots, \mathbf{e}_{k_h}] \in p^h P \subseteq p^{h+1} L .$$

This implies that p^{h+1} divides all the linear combinations of $m_{i,i}, \ldots, m_{i,d-\rho_h}$ that arise as coefficients of basis elements in the above Lie brackets. But since these span $R_{i,h} \otimes \mathbb{Z}_p$, we must have that $p^{h+1} \mid m_{i,j}$ for $1 \le i \le j \le d - \rho_h$.

Hence $p^{\mathrm{ht}(\mathbf{e}_i)} \mid m_{i,j}$, and thus $P \subseteq \pi_{\mathcal{B}}(L_p)$. $\qquad\square$

Corollary 4.42. *Let L be a nilpotent $\mathbb{Z}$-Lie ring with cocentral basis $\mathcal{B} = (\mathbf{e}_1, \ldots, \mathbf{e}_d)$. For all but finitely many primes p, the unique p-ideal of minimal index in L_p is $\pi_{\mathcal{B}}(L_p)$.*

Proof. It is easy to see that $\pi_{\mathcal{B}}(L_p)$ is a p-ideal. By Lemma 4.41 it contains all other p-ideals of finite index, hence its index must be minimal. $\qquad\square$

Remark 4.43. If $\pi_{\mathcal{B}}(L_p)$ is the unique p-ideal of L_p of minimal index, it is clear that $\pi_{\mathcal{B}}^{-1}(P)$ is an additive submodule of L_p for all p-ideals $P \lhd_p L_p$. However, there is no guarantee at all that $\pi_{\mathcal{B}}^{-1}(P)$ is an ideal, nor even a subring, of L_p.

We can now deduce Conjecture 4.23 from Conjecture 4.5:

Theorem 4.44. *Let L be a $\mathbb{Z}$-Lie ring of nilpotency class c with cocentral basis $\mathcal{B} = (\mathbf{e}_1, \ldots, \mathbf{e}_d)$. Assume Conjecture 4.5. Suppose $\zeta_{L,p}^{\triangleleft}(s)$ satisfies a functional equation of the form*

$$\zeta_{L,p}^{\triangleleft}(s)\big|_{p \to p^{-1}} = (-1)^r p^{b-as} \zeta_{L,p}^{\triangleleft}(s) \tag{4.30}$$

for almost all primes p, with $a, b, r \in \mathbb{Z}$. Then $a = N$, $b = \binom{d}{2}$ and $r \equiv d$ (mod 2), i.e.

$$\zeta_{L,p}^{\triangleleft}(s)\big|_{p \to p^{-1}} = (-1)^d p^{\binom{d}{2} - Ns} \zeta_{L,p}^{\triangleleft}(s), \tag{4.31}$$

where

$$N = \sum_{i=0}^{c-1} \mathrm{rank}(L/\sigma_i(L)) = \sum_{i=1}^{d} \mathrm{ht}(\mathbf{e}_i). \tag{4.32}$$

Proof. By Proposition 2.2, there exists a set of cone data $\mathcal{D}$ such that $\zeta_{L,p}^{\triangleleft}(s) = (1 - p^{-1})^{-d} Z_{\mathcal{D}}(s - d, p)$. Assuming Conjecture 4.5, Corollary 4.29 implies that

$$\zeta_{L,p}^{\triangleleft}(s)\big|_{p \to p^{-1}} = (-1)^d p^{\binom{d}{2}} \zeta_{L,p}^{\triangleleft_p}(s)$$

for p outside a finite set of exceptional primes.

Excluding at most finitely many primes p, Corollary 4.42 establishes that the minimal index of a p-ideal in L_p is

$$p^{\sum_{i=1}^{d} \mathrm{ht}(\mathbf{e}_i)} = p^N,$$

and that there is precisely one p-ideal of this index. Hence if $\zeta_{L_p}^{\triangleleft}(s)$ satisfies a functional equation of the form (4.30), then $\zeta_{L_p}^{\triangleleft_p}(s) = p^{-Ns} \zeta_{L_p}^{\triangleleft}(s)$. Hence $\zeta_{L_p}^{\triangleleft}(s) = \zeta_{L,p}^{\triangleleft}(s)$ satisfies (4.31). $\qquad\square$

4.9.2 Property (†)

We are unable to give a general condition which decides whether the p-local ideal zeta functions of a nilpotent Lie ring should or should not satisfy the functional equation (4.31). The idea behind our next definition is to define an interesting subset of Lie rings within which we can prove a necessary and sufficient condition for this functional equation to be satisfied, assuming Conjecture 4.5.

Definition 4.45. *Let L be a nilpotent $\mathbb{Z}$-Lie ring or $\mathbb{Z}_p$-Lie ring. A cocentral basis $\mathcal{B} = (\mathbf{e}_1, \ldots, \mathbf{e}_d)$ for L has Property (†) if there exists a function $\lambda : \mathcal{B} \to \mathbb{N}_{>0}$ such that for all $1 \leq i, j \leq d$, $[\mathbf{e}_i, \mathbf{e}_j]$ is in the span of basis elements of height $\mathrm{ht}(\mathbf{e}_i) - \lambda(\mathbf{e}_j)$. L has Property (†) if there exists a cocentral basis $\mathcal{B}$ satisfying Property (†).*

The definition of Property (†) is rather abstract, in that it gives no indication what the constants $\lambda(\mathbf{e}_j)$ should be. However, it is in some sense the most general possible definition that allows us to prove the following lemma:

Lemma 4.46. *Let L be a nilpotent $\mathbb{Z}_p$-Lie ring and let $\mathcal{B} = (\mathbf{e}_1, \ldots, \mathbf{e}_d)$ be a cocentral basis for L with (†). Let H be the $\mathbb{Z}_p$-linear span of $\mathbf{m}_1, \ldots, \mathbf{m}_d$ and let $f_{i,j,k}(\mathbf{x})$ and $g_{i,j,k}(\mathbf{x})$ be coprime polynomials in the indeterminates $\mathbf{x} = (m_{1,1}, m_{1,2}, \ldots, m_{d,d})$ such that*

$$[\mathbf{m}_i, \mathbf{e}_j] = \sum_{k=1}^{d} \frac{g_{i,j,k}(\mathbf{x})}{f_{i,j,k}(\mathbf{x})} \mathbf{m}_k . \tag{4.33}$$

1. The conditions that must be satisfied if $H \trianglelefteq L_p$ are

$$f_{i,j,k}(\mathbf{x}) \mid g_{i,j,k}(\mathbf{x}) \tag{4.34}$$

for all $1 \le i, j, k \le d$.
2. The conditions that must be satisfied if $\pi_{\mathcal{B}}(H) \triangleleft_p L_p$ are

$$f_{i,j,k}(\mathbf{x}) \mid p^{\lambda(\mathbf{e}_j)-1} g_{i,j,k}(\mathbf{x}) \tag{4.35}$$

for all $1 \le i, j, k \le d$.

Proof. 1. Clear from (4.33).
2. Put $\hat{m}_{i,j} = p^{\mathrm{ht}(\mathbf{e}_j)} m_{i,j}$ and $\hat{\mathbf{m}}_i = \pi_{\mathcal{B}}(\mathbf{m}_i) = \hat{m}_{i,i}\mathbf{e}_i + \cdots + \hat{m}_{i,d}\mathbf{e}_d$. It is clear that $\hat{m}_{i,j} \in p\mathbb{Z}_p$, so the additive submodule generated by $\hat{\mathbf{m}}_1, \ldots, \hat{\mathbf{m}}_d$ is clearly contained within pL.
Equating the basis elements of height h in (4.33) gives us

$$\sum_{\mathrm{ht}(\mathbf{e}_r)=\lambda(\mathbf{e}_j)+h} m_{i,r}[\mathbf{e}_r, \mathbf{e}_j] = \sum_{\mathrm{ht}(\mathbf{e}_k)\ge h} \frac{g_{i,j,k}(\mathbf{x})}{f_{i,j,k}(\mathbf{x})} \left(\sum_{\mathrm{ht}(\mathbf{e}_t)=h} m_{k,t}\mathbf{e}_t \right) .$$

Multiply both sides by $p^{\lambda(\mathbf{e}_j)+h}$, to give us

$$\sum_{\mathrm{ht}(\mathbf{e}_r)=\lambda(\mathbf{e}_j)+h} \hat{m}_{i,r}[\mathbf{e}_r, \mathbf{e}_j] = \sum_{\mathrm{ht}(\mathbf{e}_k)\ge h} \frac{p^{\lambda(\mathbf{e}_j)} g_{i,j,k}(\mathbf{x})}{f_{i,j,k}(\mathbf{x})} \left(\sum_{\mathrm{ht}(\mathbf{e}_t)=h} \hat{m}_{k,t}\mathbf{e}_t \right) .$$

Summing both sides over h yields

$$[\hat{\mathbf{m}}_i, \mathbf{e}_j] = \sum_{k=1}^{d} \frac{p^{\lambda(\mathbf{e}_j)} g_{i,j,k}(\mathbf{x})}{f_{i,j,k}(\mathbf{x})} \hat{\mathbf{m}}_k .$$

For $\pi_{\mathcal{B}}(H)$ to be a p-ideal, we require the coefficient $p^{\lambda(\mathbf{e}_j)} g_{i,j,k}(\mathbf{x})/f_{i,j,k}(\mathbf{x})$ to be an element of pL. This is true if and only if (4.35) holds.   □

Corollary 4.47. *Let L be a nilpotent $\mathbb{Z}$-Lie ring and let $\mathcal{B} = (\mathbf{e}_1, \ldots, \mathbf{e}_d)$ be a cocentral basis for L with (†). For all primes p, and for all ideals $I \trianglelefteq L_p$ of finite index in L_p, $\pi_{\mathcal{B}}(I) \triangleleft_p L_p$ and $|L_p : \pi_{\mathcal{B}}(I)| = p^N |L_p : I|$.*

Proof. It is clear that (4.34) implies (4.35), hence $I \trianglelefteq L_p$ implies $\pi_{\mathcal{B}}(I) \triangleleft_p L_p$. Since $|L_p : I| = |m_{1,1} \ldots m_{d,d}|^{-1}$ and $|L_p : \pi_{\mathcal{B}}(I)| = |\hat{m}_{1,1} \ldots \hat{m}_{d,d}|^{-1}$ it is also clear that $|L_p : \pi_{\mathcal{B}}(I)| = p^N |L_p : I|$. □

Corollary 4.47 now gives us an obvious criterion for the local ideal zeta functions of L to satisfy the functional equation (4.31):

Corollary 4.48. *Let L be a nilpotent $\mathbb{Z}$-Lie ring with a cocentral basis $\mathcal{B} = (\mathbf{e}_1, \ldots, \mathbf{e}_d)$ having (†). Suppose additionally that for all but finitely many primes p, $\pi_{\mathcal{B}}^{-1}(P) \trianglelefteq L_p$ for all p-ideals P of finite index in L_p. Assume Conjecture 4.5. Then, for all but finitely many primes p, $\zeta_{L,p}^{\triangleleft}(s)$ satisfies (4.31).*

Proof. Lemma 4.47 and our assumption that $\pi_{\mathcal{B}}^{-1}(P) \trianglelefteq L_p$ for all p-ideals P of p-power index in L together imply that there is a bijective correspondence between ideals of L and p-ideals of L. Under this correspondence, an ideal of index p^r corresponds to a p-ideal of index p^{r+N}. Hence the result. □

It is useful to classify Lie rings that satisfy Property (†). For more general rings there may not be a similar classification.

Definition 4.49. *Let L be a $\mathbb{Z}$-Lie ring of nilpotency class c. We define the depth of an element $x \in L$ to be $\mathrm{dep}(x) = c + 1 - \mathrm{ht}(x)$.*

Definition 4.50. *Let L be a nilpotent ring additively isomorphic to $\mathbb{Z}^d$. A cocentral basis $\mathcal{B} = (\mathbf{e}_1, \ldots, \mathbf{e}_d)$ for L is stepped if $\mathcal{B}$ has (†) and we may take $\lambda(\mathbf{e}_j) = \mathrm{dep}(\mathbf{e}_j)$ for $1 \leq j \leq d$. L is stepped if there exists a stepped basis for L.*

Lemma 4.51. *Let L be a Lie ring, $a, b, c \in L$. If c commutes with a and b, c commutes with $[a, b]$.*

Proof. Follows immediately from the Jacobi identity. □

Lemma 4.52. *Let L be a $\mathbb{Z}$-Lie ring and let $\mathcal{B} = (\mathbf{e}_1, \ldots, \mathbf{e}_d)$ be a cocentral basis for L having (†). Suppose $\mathbf{e}_i$, $\mathbf{e}_j$ are non-commuting basis elements. Then*

$$\mathrm{ht}(\mathbf{e}_i) + \lambda(\mathbf{e}_i) = \mathrm{ht}(\mathbf{e}_j) + \lambda(\mathbf{e}_j) . \tag{4.36}$$

Proof. From the definition of Property (†), $[\mathbf{e}_i, \mathbf{e}_j]$ is a linear combination of basis elements of height $\mathrm{ht}(\mathbf{e}_i) - \lambda(\mathbf{e}_j)$. Since the Lie bracket is antisymmetric, $\mathrm{ht}(\mathbf{e}_i) - \lambda(\mathbf{e}_j) = \mathrm{ht}(\mathbf{e}_j) - \lambda(\mathbf{e}_i)$. Rearranging gives the result. □

Theorem 4.53. *Let L be a nonabelian nilpotent Lie ring. Then L has (†) if and only if L is a direct product (perhaps with central amalgamation) of stepped Lie rings.*

Proof. Define the relation $\sim$ on $\mathcal{B}$ by $\mathbf{e}_i \sim \mathbf{e}_j$ if $[\mathbf{e}_i, \mathbf{e}_j] \neq 0$. Clearly $\sim$ is symmetric. Let $\approx$ denote the transitive closure of $\sim$. $\approx$ is an equivalence relation on the set of non-central basis elements. By Lemma 4.52, the function $\mathbf{e}_i \mapsto \mathrm{ht}(\mathbf{e}_i) + \lambda(\mathbf{e}_i)$ is constant on the equivalence classes $\mathcal{C}_1, \ldots, \mathcal{C}_r$ of $\approx$.

Let L_i be the subring generated by $\mathcal{C}_i$. Consider two distinct subrings L_i and L_j, $i \neq j$. For all $a, b \in L_i$, $c \in L_j$, Lemma 4.51 implies that c commutes with $[a, b]$. Hence the subrings L_i and L_j commute, and so their intersection must lie in the centre. Furthermore, $Z(L_i), Z(L_j) \subseteq Z(L)$. It is then clear that $L_1 + Z(L), L_2, L_3, \ldots, L_r$ generate L.

We now claim that each subring L_i is stepped. For $1 \leq i \leq r$, let t_i be the constant value of $\mathrm{ht}(\mathbf{e}_j) + \lambda(\mathbf{e}_j)$ for $\mathbf{e}_j \in \mathcal{C}_i$. Proposition 4.35 implies there exists $\mathbf{e}_j \in L_i$ such that $\lambda(\mathbf{e}_j) = 1$ and thus $\mathrm{ht}(\mathbf{e}_j) = t_i - 1$. Since $Z(L_i) \subseteq Z(L)$, L_i contains elements of height 1, so the nilpotency class of L_i is $t_i - 1$. By the definition of t_i, $\lambda(\mathbf{e}_j) = t_i - \mathrm{ht}(\mathbf{e}_j) = \mathrm{dep}_{L_i}(\mathbf{e}_j)$. Thus L_i is stepped. Clearly $L_1 + Z(L)$ is also stepped if L_1 is.

Conversely it is easy to see that any direct product of stepped Lie rings $L_1, \ldots, L_r$, perhaps with central amalgamation, has (†). For a central basis element $\mathbf{e}_j$, the value of $\lambda(\mathbf{e}_j)$ is arbitrary. For a noncentral basis element $\mathbf{e}_j$, $\mathbf{e}_j \in L_i$ for some unique $1 \leq i \leq r$, and we take $\lambda(\mathbf{e}_j) = \mathrm{dep}_{L_i}(\mathbf{e}_j)$.   □

Remark 4.54. It is not always true that $L_1, \ldots, L_r$ generate L. In particular, this happens if L has an abelian direct factor.

Remark 4.55. The concept of a stepped Lie ring is similar to the concept of a graded Lie ring. Indeed, by taking L_i to be the linear span of the basis elements of depth i, $L_1, \ldots, L_c$ is a grading of a stepped Lie ring. Direct products of such rings are clearly graded as well, and it is easy to see that a direct product with central amalgamation of graded Lie rings is also graded. However, graded Lie rings do not necessarily have (†). Fil_4 (p. 53) is graded – take $L_1 = \langle z \rangle$, $L_{i+1} = \langle x_i \rangle$ for $i = 1, 2, 3, 4$, then $[L_i, L_j] \subseteq L_{i+j}$ – but does not have (†).

Corollary 4.48 is obvious, but it is awkward to use. We now develop a more useful equivalent notion.

Definition 4.56. *Let L be a $\mathbb{Z}$-Lie ring with cocentral basis $\mathcal{B} = (\mathbf{e}_1, \ldots, \mathbf{e}_d)$ having* (†)*. $\mathcal{B}$ has* (∗) *if, for all $1 \leq j, k \leq d$ with $\lambda(\mathbf{e}_j) > 1$,*

$$[\mathbf{e}_k, \mathbf{e}_j] \in \left\langle \left\{ [\mathbf{e}_k, \mathbf{e}_{j_1}, \ldots, \mathbf{e}_{j_l}] : \begin{array}{l} 1 \leq l \leq c, \\ \lambda(\mathbf{e}_{j_1}) = \cdots = \lambda(\mathbf{e}_{j_l}) = 1 \end{array} \right\} \right\rangle_{\mathbb{Q}}. \tag{4.37}$$

This is a somewhat technical definition and is more difficult to understand than Corollary 4.48. However, determining whether a basis has (∗) is a finite calculation, not something offered by Corollary 4.48.

Note also that we use the $\mathbb{Q}$-span in (4.37), rather than the $\mathbb{Z}$-span. (4.37) holds if and only if

$$[\mathbf{e}_k, \mathbf{e}_j] \in \left\langle \left\{ [\mathbf{e}_k, \mathbf{e}_{j_1}, \ldots, \mathbf{e}_{j_l}] : \begin{array}{c} 1 \le l \le c, \\ \lambda(\mathbf{e}_{j_1}) = \cdots = \lambda(\mathbf{e}_{j_l}) = 1 \end{array} \right\} \right\rangle_{\mathbb{Z}_p}$$

for all but finitely many primes p. The same is not true with the $\mathbb{Z}$-span.

Example 4.57. The Lie ring $\mathfrak{g}_{6,17}$ has presentation

$$\langle \mathbf{e}_1, \ldots, \mathbf{e}_6 : [\mathbf{e}_1, \mathbf{e}_2] = \mathbf{e}_3, [\mathbf{e}_1, \mathbf{e}_3] = \mathbf{e}_5, [\mathbf{e}_1, \mathbf{e}_5] = \mathbf{e}_6, [\mathbf{e}_2, \mathbf{e}_4] = \mathbf{e}_6 \rangle .$$

The basis $\mathcal{B} = (\mathbf{e}_1, \ldots, \mathbf{e}_6)$ is cocentral and has (†), with $\lambda(\mathbf{e}_1) = \lambda(\mathbf{e}_2) = 1$, $\lambda(\mathbf{e}_3) = 2$, $\lambda(\mathbf{e}_4) = \lambda(\mathbf{e}_5) = 3$. Now

$$[\mathbf{e}_1, \mathbf{e}_3] = \mathbf{e}_5 , \qquad [\mathbf{e}_1, \mathbf{e}_4] = 0 , \qquad [\mathbf{e}_1, \mathbf{e}_5] = \mathbf{e}_6 ,$$

and

$$[\mathbf{e}_1, \mathbf{e}_2, \mathbf{e}_1] = -\mathbf{e}_5 , \qquad [\mathbf{e}_1, \mathbf{e}_2, \mathbf{e}_1, \mathbf{e}_1] = \mathbf{e}_6 ,$$

so (4.37) holds for $k = 1$. For $k = 2$, we need only check that

$$[\mathbf{e}_2, \mathbf{e}_3] = 0 , \qquad [\mathbf{e}_2, \mathbf{e}_4] = \mathbf{e}_6 , \qquad [\mathbf{e}_2, \mathbf{e}_5] = 0 ,$$

and

$$[\mathbf{e}_2, \mathbf{e}_1, \mathbf{e}_1, \mathbf{e}_1] = -\mathbf{e}_6 .$$

There is nothing to check for $k \ge 3$ since $\mathbf{e}_3, \mathbf{e}_4, \mathbf{e}_5, \mathbf{e}_6$ all commute. Hence $\mathcal{B}$ has $(*)$. Turning to p. 61, we find that $\zeta_{\mathfrak{g}_{6,17},p}^{\triangleleft}(s)$ does indeed satisfy the expected functional equation.

Example 4.58. The Lie ring $\mathfrak{g}_{6,6}$ has presentation

$$\langle \mathbf{e}_1, \ldots, \mathbf{e}_6 : [\mathbf{e}_1, \mathbf{e}_2] = \mathbf{e}_4, [\mathbf{e}_1, \mathbf{e}_3] = \mathbf{e}_5, [\mathbf{e}_1, \mathbf{e}_4] = \mathbf{e}_6, [\mathbf{e}_2, \mathbf{e}_3] = \mathbf{e}_6 \rangle .$$

The basis $\mathcal{B} = (\mathbf{e}_1, \ldots, \mathbf{e}_6)$ is cocentral and has (†), with $\lambda(\mathbf{e}_1) = \lambda(\mathbf{e}_2) = 1$ and $\lambda(\mathbf{e}_3) = \lambda(\mathbf{e}_4) = 2$. Now $[\mathbf{e}_1, \mathbf{e}_3] = \mathbf{e}_5$ but $[\mathbf{e}_1, \mathbf{e}_2, \mathbf{e}_1] = -\mathbf{e}_6$ and $[\mathbf{e}_1, \mathbf{e}_2, \mathbf{e}_2] = 0$. Hence (4.37) does not hold with $j = 3$, $k = 1$, so $\mathfrak{g}_{6,6}$ does not have $(*)$. $\zeta_{\mathfrak{g}_{6,6},p}^{\triangleleft}(s)$ does not satisfy a functional equation, as we see on p. 56.

Definition 4.59. *Let L be a $\mathbb{Z}$-Lie ring with cocentral basis $\mathcal{B}$ having (†). Let $f_{i,j,k}(\mathbf{x}) \le g_{i,j,k}(\mathbf{x})$ for $1 \le k \le d$ be the polynomial divisibility conditions in (4.34). A condition $f_{i,j,k}(\mathbf{x}) \le g_{i,j,k}(\mathbf{x})$ is primary if $\lambda(\mathbf{e}_j) = 1$, and is secondary otherwise.*

From (4.34) and (4.35), the primary conditions are necessary conditions for H to be an ideal and for $\pi_{\mathcal{B}}(H)$ to be a p-ideal. The secondary conditions for $\pi_{\mathcal{B}}(H)$ to be a p-ideal are weaker than those for H to be an ideal. This suggests that $\pi_{\mathcal{B}}(P) \triangleleft L_p$ for all p-ideals P of p-power index if and only if the secondary conditions are redundant. Our next aim is to show that this is in fact the case.

Lemma 4.60. *Let L be a nilpotent $\mathbb{Z}$-Lie ring or $\mathbb{Z}_p$-Lie ring, $\mathcal{B} = (\mathbf{e}_1, \ldots, \mathbf{e}_d)$ a basis for L with (†), and $\mathbf{m}_1, \ldots, \mathbf{m}_d$ be additive generators for an ideal $I \trianglelefteq L$. Suppose that*

$$[\mathbf{m}_i, \mathbf{e}_j] \in \langle \mathbf{m}_1, \ldots, \mathbf{m}_d \rangle$$

for all $\mathbf{e}_j$ with $\lambda(\mathbf{e}_j) = 1$. Then, for all $j_1, \ldots, j_r$ with $\lambda(\mathbf{e}_{j_1}) = \cdots = \lambda(\mathbf{e}_{j_r}) = 1$,

$$[\mathbf{m}_i, \mathbf{e}_{j_1}, \mathbf{e}_{j_2}, \ldots, \mathbf{e}_{j_r}] \in \langle \mathbf{m}_1, \ldots, \mathbf{m}_d \rangle .$$

Proof. If $[\mathbf{m}_i, \mathbf{e}_j] \in \langle \mathbf{m}_1, \ldots, \mathbf{m}_d \rangle$ for all $\mathbf{e}_j$ with $\lambda(\mathbf{e}_j) = 1$, then $[\mathbf{u}, \mathbf{e}_j] \in \langle \mathbf{m}_1, \ldots, \mathbf{m}_d \rangle$ for all $\mathbf{u} \in \langle \mathbf{m}_1, \ldots, \mathbf{m}_d \rangle$ and all $\mathbf{e}_j$ such that $\lambda(\mathbf{e}_j) = 1$. Setting $\mathbf{u} = \mathbf{m}_i, [\mathbf{m}_i, \mathbf{e}_{j_1}], [\mathbf{m}_i, \mathbf{e}_{j_1}, \mathbf{e}_{j_2}], \ldots, [\mathbf{m}_i, \mathbf{e}_{j_1}, \mathbf{e}_{j_2}, \ldots, \mathbf{e}_{j_{r-1}}]$ in turn gives the result. $\square$

Theorem 4.61. *Let L be a $\mathbb{Z}$-Lie ring with cocentral basis $\mathcal{B}$ having (†) and (∗). For all but finitely many primes p, $H \trianglelefteq L_p$ if and only if $\pi_{\mathcal{B}}(H) \triangleleft_p L_p$.*

Proof. For all but finitely many primes p, (4.37) implies that

$$[\mathbf{e}_k, \mathbf{e}_j] \in \left\langle \left\{ [\mathbf{e}_k, \mathbf{e}_{j_1}, \ldots, \mathbf{e}_{j_l}] : \begin{array}{c} 1 \leq l \leq c, \\ \lambda(\mathbf{e}_{j_1}) = \cdots = \lambda(\mathbf{e}_{j_l}) = 1 \end{array} \right\} \right\rangle_{\mathbb{Z}_p} \quad (4.38)$$

for all $\mathbf{e}_j, \mathbf{e}_k$ with $\lambda(\mathbf{e}_j) > 1$. If we set $\mathbf{m}_i = m_{i,i}\mathbf{e}_i + \cdots + m_{i,d}\mathbf{e}_d$ for $1 \leq i \leq d$, it can then be seen that (4.38) is equivalent to

$$[\mathbf{m}_i, \mathbf{e}_j] \in \left\langle \left\{ [\mathbf{m}_i, \mathbf{e}_{j_1}, \ldots, \mathbf{e}_{j_l}] : \begin{array}{c} 1 \leq l \leq c, \\ \lambda(\mathbf{e}_{j_1}) = \cdots = \lambda(\mathbf{e}_{j_l}) = 1 \end{array} \right\} \right\rangle_{\mathbb{Z}_p} \quad (4.39)$$

for all $1 \leq i, j \leq d$ with $\lambda(\mathbf{e}_j) > 1$.

For fixed i, j with $\lambda(\mathbf{e}_j) > 1$, the secondary conditions $f_{i,j,k}(\mathbf{x}) \mid g_{i,j,k}(\mathbf{x})$ for $1 \leq k \leq d$ are 'redundant' (in the sense that they are implied by the primary conditions) for all but finitely many primes p if

$$[\mathbf{m}_i, \mathbf{e}_j] \in \left\langle \left\{ [\mathbf{m}_i, \mathbf{e}_{j_1}, \ldots, \mathbf{e}_{j_r}] : \begin{array}{c} 1 \leq l \leq c, \\ \lambda(\mathbf{e}_{j_1}) = \cdots = \lambda(\mathbf{e}_{j_l}) = 1 \end{array} \right\} \right\rangle_{\mathbb{Z}_p} . \quad (4.40)$$

By Lemma 4.60, $[\mathbf{m}_i, \mathbf{e}_{j_1}, \ldots, \mathbf{e}_{j_r}] \in \langle \mathbf{m}_1, \ldots, \mathbf{m}_d \rangle$ for all $j_1, \ldots, j_r$ with $\lambda(\mathbf{e}_{j_1}) = \cdots = \lambda(\mathbf{e}_{j_r}) = 1$ if and only if $[\mathbf{m}_i, \mathbf{e}_j] \in \langle \mathbf{m}_1, \ldots, \mathbf{m}_d \rangle$ for all j such that $\lambda(\mathbf{e}_j) = 1$. Clearly if (4.40) holds for all i, j with $1 \leq i, j \leq d$ and $\lambda(\mathbf{e}_j) > 1$, then the conditions $f_{i,j,k}(\mathbf{x}) \mid g_{i,j,k}(\mathbf{x})$ for $1 \leq k \leq d$ are implied by those where $\lambda(\mathbf{e}_j) = 1$. Hence $H \trianglelefteq L_p$ if and only if $\pi_{\mathcal{B}}(H) \triangleleft_p L_p$. $\square$

Theorem 4.62. *Let L be a nilpotent $\mathbb{Z}$-Lie ring with basis $\mathcal{B} = (\mathbf{e}_1, \ldots, \mathbf{e}_d)$ which has (†) but does not have (∗). There exists an additive submodule H such that $H \ntrianglelefteq L_p$ but $\pi_{\mathcal{B}}(H) \triangleleft_p L_p$.*

Proof. Since $\mathcal{B}$ does not have $(*)$, there exist basis elements $\mathbf{e}_j, \mathbf{e}_k$ with $\lambda(\mathbf{e}_j) > 1$ such that

$$[\mathbf{e}_k, \mathbf{e}_j] \notin \left\langle \left\{ [\mathbf{e}_k, \mathbf{e}_{j_1}, \ldots, \mathbf{e}_{j_l}] : \begin{array}{c} 1 \leq l \leq c, \\ \lambda(\mathbf{e}_{j_1}) = \cdots = \lambda(\mathbf{e}_{j_l}) = 1 \end{array} \right\} \right\rangle_{\mathbb{Q}}, \qquad (4.41)$$

and hence, for all primes p,

$$[\mathbf{e}_k, \mathbf{e}_j] \notin \left\langle \left\{ [\mathbf{e}_k, \mathbf{e}_{j_1}, \ldots, \mathbf{e}_{j_l}] : \begin{array}{c} 1 \leq l \leq c, \\ \lambda(\mathbf{e}_{j_1}) = \cdots = \lambda(\mathbf{e}_{j_l}) = 1 \end{array} \right\} \right\rangle_{\mathbb{Z}_p}. \qquad (4.42)$$

We construct a suitable additive submodule H additively generated by $\mathbf{m}_1, \ldots, \mathbf{m}_d$. It suffices to assume $m_{i,j} = 0$ for $1 \leq i < j \leq d$, so that $\mathbf{m}_i = m_{i,i} \mathbf{e}_i$ and thus

$$[\mathbf{m}_i, \mathbf{e}_j] = m_{i,i} \sum_{\mathrm{ht}(\mathbf{e}_r) = \mathrm{ht}(\mathbf{e}_i) - \lambda(\mathbf{e}_j)} c_{i,j,r} \mathbf{e}_r$$

for some nonzero integers $c_{i,j,r}$. We may assume p divides none of the nonzero $c_{i,j,r}$. Since we have chosen a 'diagonal' set of generators for H, the requirement that $[\mathbf{m}_i, \mathbf{e}_j] \in \langle \mathbf{m}_1, \ldots, \mathbf{m}_d \rangle$ becomes $m_{r,r} \mid m_{i,i}$ for each r with $c_{i,j,r} \neq 0$.

Choose $\mathbf{e}_k, \mathbf{e}_j$ to satisfy (4.41). Choose $\mathbf{e}_t$ such that $c_{k,j,t} \neq 0$, so that $m_{t,t} \mid m_{k,k}$ if $\langle \mathbf{m}_1, \ldots, \mathbf{m}_d \rangle$ is to be an ideal. Set $m_{t,t} = p$, and if any conjunction of primary conditions implies that $m_{t,t} \mid m_{i,i}$, set $m_{i,i} = p$ too. Set all other $m_{i,i} = 1$. By our construction, all primary conditions are satisfied.

By our choice of $\mathbf{e}_k$ and $\mathbf{e}_j$, $m_{t,t} \mid m_{k,k}$ cannot be a conjunction of primary conditions, so we must have $m_{k,k} = 1$. This clearly implies that $H = \langle \mathbf{m}_1, \ldots, \mathbf{m}_d \rangle$ is not an ideal of L_p, since one of the non-primary conditions does not hold.

We now prove that $\pi_{\mathcal{B}}(H)$ is a p-ideal. Recall that

$$\pi_{\mathcal{B}}(H) = \langle p^{\mathrm{ht}(\mathbf{e}_1)} m_{1,1}, \ldots, p^{\mathrm{ht}(\mathbf{e}_d)} m_{d,d} \rangle_{\mathbb{Z}_p}.$$

For $1 \leq i, l \leq d$, we must show that $[\mathbf{m}_i, \mathbf{e}_l] \in \langle p\mathbf{m}_1, \ldots, p\mathbf{m}_d \rangle$. Now

$$[\mathbf{m}_i, \mathbf{e}_l] = \sum_{\mathrm{ht}(\mathbf{e}_r) = \mathrm{ht}(\mathbf{e}_i) - \lambda(\mathbf{e}_l)} c_{i,l,r} p^{\mathrm{ht}(\mathbf{e}_r) + \lambda(\mathbf{e}_l)} m_{i,i} \mathbf{e}_r,$$

so for each r with $c_{i,l,r} \neq 0$,

$$m_{r,r} \mid p^{\lambda(\mathbf{e}_l) - 1} m_{i,i}. \qquad (4.43)$$

If $\lambda(\mathbf{e}_l) = 1$, (4.43) reduces to $m_{r,r} \mid m_{i,i}$, one of the primary conditions that we know H satisfies. If $\lambda(\mathbf{e}_l) > 1$, (4.43) trivially holds since $m_{r,r} \mid p$ for all $1 \leq r \leq d$. Hence $\pi_{\mathcal{B}}(H) \lhd_p L_p$. $\qquad \square$

Corollary 4.63. *Let L be a nilpotent $\mathbb{Z}$-Lie ring with basis $\mathcal{B} = (e_1, \dots, e_d)$ which has (†). Assume Conjecture 4.5. Then, for all but finitely many primes p, $\zeta_{L,p}^{\triangleleft}(s)$ satisfies the functional equation (4.31) if and only if $\mathcal{B}$ has (∗).*

Proof. Corollary 4.48 and Theorems 4.61 and 4.62.   □

Corollary 4.64. *Let L be a ring with (†), i.e. a direct product (possibly with central amalgamation) of stepped rings $L_1, \dots, L_r$. Then L has (∗) if and only if for each $1 \le i \le r$, L_i has (∗).*

Proof. If $e_j \in L_i$, then $[\mathbf{m}_k, e_j]$ only has nonzero coefficients of other basis elements in L_i. If all the direct factors have (∗), it is clear that L will. If (with no loss of generality) L_1 does not have (∗), we will still be able to construct an non-ideal $H \triangleleft L_p$ such that $\pi_{\mathcal{B}}(H) \triangleleft_p L$.   □

Theorems 4.61 and 4.62 give us a way of determining whether ideals and p-ideals correspond. However, we can further cut down the work we need to do. Our next lemma essentially allows us to reorder the Lie brackets within (4.37). Its proof is routine manipulation of the Jacobi identity.

Lemma 4.65. *Let L be a Lie ring and $\mathfrak{b}$ be any iterated Lie bracket of elements in L, $y \in L$. Let $S \subseteq L$ be any set of elements such that $\mathfrak{b}$ is in the subring generated by S. Then*

$$[\mathfrak{b}, y] \in \langle \{ [y, z_1, z_2, \dots, z_r] \mid 1 \le r \le c, z_i \in S \} \rangle_{\mathbb{Z}} .$$

Remark 4.66. We may assume y and all the z_i are distinct by considering them as indeterminates and follow the method outlined in the proof to obtain an identity expressing $[\mathfrak{b}, y]$ as a linear combination of left-normed brackets.

Proof. Let the number of Lie brackets within $\mathfrak{b}$ be b. Define the altitude of y in $\mathfrak{b}$, $\text{alt}_{\mathfrak{b}}(y)$, to be the number of Lie brackets it is contained within. We proceed by reverse induction on $\text{alt}_{\mathfrak{b}}(y)$. If $\text{alt}_{\mathfrak{b}}(y) = b$, then $\mathfrak{b}$ is left-normed and the result is trivially true. Initially, we have $\text{alt}_{\mathfrak{b}}(y) = 0$.

If $\text{alt}_{\mathfrak{b}}(y) < b$, then there exists a Lie bracket within $\mathfrak{b}$ not containing y. Let $[[A, B], C]$ be the innermost Lie bracket containing y and this other Lie bracket, with y somewhere within C. The Jacobi identity implies that

$$[[A, B], C] = [[C, B], A] + [[A, C], B] . \tag{4.44}$$

Using (4.44) we may express $\mathfrak{b}$ as $\mathfrak{b} = \mathfrak{b}_1 + \mathfrak{b}_2$. For $i = 1, 2$ either $\mathfrak{b}_i = 0$ or $\text{alt}_{\mathfrak{b}_i}(y) = \text{alt}_{\mathfrak{b}}(y) + 1$ and there are b brackets within $\mathfrak{b}_i$. By our inductive hypothesis, $\mathfrak{b}_1$ and $\mathfrak{b}_2$ are expressible as some $\mathbb{Z}$-linear combination of left-normed Lie brackets, so $\mathfrak{b}$ is too.   □

Definition 4.67. *Let L be a $\mathbb{Z}$-Lie ring or $\mathbb{Z}_p$-Lie ring with basis $\mathcal{B}$ having (†). Denote by $\Gamma_{\mathcal{B}}(L)$ the subring generated by all $e_j \in \mathcal{B}$ such that $\lambda(e_j) = 1$.*

Corollary 4.68. *Let L be a $\mathbb{Z}_p$-Lie ring with basis $\mathcal{B}$ having (†), $\mathbf{e}_j \in \mathcal{B}$. If $\mathbf{e}_j \in \Gamma_{\mathcal{B}}(L)$ then (4.37) holds for all $\mathbf{e}_k \in \mathcal{B}$.*

Proof. Lemma 4.65 with $S = \{\, z \in \mathcal{B} \mid \lambda(z) = 1 \,\}$. □

Corollary 4.69. *Let L be a $\mathbb{Z}$-Lie ring with basis $\mathcal{B}$ having (†). Suppose $|L : \Gamma_{\mathcal{B}}(L)| < \infty$. Assume Conjecture 4.5. Then $\zeta_{L,p}^{\triangleleft}(s)$ satisfies (4.31) for all but finitely many primes p.*

Proof. For all primes p not dividing $|L : \Gamma_{\mathcal{B}}(L)|$, $\Gamma_{\mathcal{B}}(L_p) = L_p$. Hence $\mathcal{B} \subset \Gamma_{\mathcal{B}}(L_p)$ and by Corollary 4.68, $\mathcal{B}$ has (∗). Thus Corollary 4.63 implies the result. □

This next corollary provides a useful quick check for the correspondence to fail, and hence predict that local ideal zeta function does not satisfy a functional equation.

Corollary 4.70. *Let L be a $\mathbb{Z}$-Lie ring with basis $\mathcal{B} = (\mathbf{e}_1, \ldots, \mathbf{e}_d)$ having (†). Suppose $\Gamma_{\mathcal{B}}(L)$ is not of finite index in L. Suppose that there exist $\mathbf{e}_k, \mathbf{e}_j \in \mathcal{B}$ such that $\mathbf{e}_k \in \Gamma_{\mathcal{B}}(L)$, but for all primes p, $\mathbf{e}_j \notin \Gamma_{\mathcal{B}}(L_p)$ and $[\mathbf{e}_k, \mathbf{e}_j] \notin \Gamma_{\mathcal{B}}(L_p)$. Then $\mathcal{B}$ does not have (∗). Hence, assuming Conjecture 4.5, $\zeta_{L,p}^{\triangleleft}(s)$ satisfies no functional equation for all but finitely many primes p.*

Proof. If $\mathbf{e}_k \in \Gamma_{\mathcal{B}}(L_p)$, $\mathbf{e}_j \notin \Gamma_{\mathcal{B}}(L_p)$ and $[\mathbf{e}_k, \mathbf{e}_j] \notin \Gamma_{\mathcal{B}}(L_p)$, then (4.41) clearly holds. □

Having laid the groundwork above, we can now prove results about when local ideal zeta functions of Lie rings should satisfy functional equations.

Proposition 4.71. *Let L be a class-2 nilpotent Lie ring. Assume Conjecture 4.5. Then, for all but finitely many primes p, $\zeta_{L,p}^{\triangleleft}(s)$ satisfies (4.31). In other words, Conjecture 4.5 implies Conjecture 4.20.*

Proof. Since L has nilpotency class 2 it is clear that L is stepped. L trivially has (∗) since there are no nontrivial secondary conditions. The result now follows from Theorem 4.61. □

Lemma 4.72. *For $c, d \geq 2$, let $F_{c,d}$ denote the free class-c-nilpotent Lie ring on d generators. Then $F_{c,d}$ is stepped.*

Proof. Let $\mathbf{e}_1, \ldots, \mathbf{e}_d$ be free generators of $F_{c,d}$, and for $1 \leq i \leq c$, set $r_i = \operatorname{rank}(\gamma_i(F_{c,d})/\gamma_{i+1}(F_{c,d}))$. Set $s_i = \operatorname{rank}(\gamma_i(F_{c,d})/\gamma_{i+1}(F_{c,d})) = r_1 + \cdots + r_i$. For convenience set $s_0 = 0$.

$F_{c,d}$ has a basis $\mathcal{B} = (\mathbf{e}_1, \ldots, \mathbf{e}_{s_c})$ with the property that for $s_{i-1}+1 \leq l \leq s_i$, $\mathbf{e}_l$ is a left-normed Lie bracket of length i in the free generators $\mathbf{e}_1, \ldots, \mathbf{e}_d$. Define L_i to be the $\mathbb{Z}$-span of $\mathbf{e}_{s_{i-1}+1}, \ldots, \mathbf{e}_{s_i}$. Then

$$F_{c,d} = L_1 \oplus L_2 \oplus \cdots \oplus L_c$$

as $\mathbb{Z}$-module direct sums. Furthermore, Lemma 4.65 implies that any Lie bracket with $i + j$ elements, left-normed or otherwise, can be rewritten as a $\mathbb{Z}$-linear combination of left-normed Lie brackets of length $i + j$. Hence $[L_i, L_j] \subseteq L_{i+j}$, so $F_{c,d}$ is graded with respect to $L_1, \ldots, L_c$.

If $s_{i-1} + 1 \leq l \leq s_i$, then $\mathbf{e}_l \in L_i$. $[\mathbf{e}_l, \underbrace{\mathbf{e}_1, \ldots, \mathbf{e}_1}_{c-i}]$ is a left-normed Lie bracket of length c. By the freeness of $F_{c,d}$, this Lie bracket is nonzero. Hence, by Proposition 4.33, $\mathrm{ht}(\mathbf{e}_l) > c - i$.

For all $z_1, \ldots, z_{c-i+1} \in F_{c,d}$, $[\mathbf{e}_l, z_1, \ldots, z_{c-i+1}]$ is a left-normed Lie bracket of length $c + 1$, so is zero. Thus $\mathrm{ht}(\mathbf{e}_l) = c - i + 1$, so $\mathrm{dep}(\mathbf{e}_l) = i$. Hence $F_{c,d}$ is stepped. $\qquad\square$

Theorem 4.73. *Suppose $c, d \geq 2$. Assume Conjecture 4.5. Then, for all but finitely many primes p,*

$$\left. \zeta_{F_{c,d},p}^{\triangleleft}(s) \right|_{p \to p^{-1}} = (-1)^{N_1} p^{\binom{N_1}{2} - N_2 s} \zeta_{F_{c,d},p}^{\triangleleft}(s) , \tag{4.45}$$

where

$$N_1 = \sum_{i=1}^{c} \frac{1}{i} \sum_{j \mid i} \mu(j) d^{i/j} = \mathrm{rank}(F_{c,d}) ,$$

$$N_2 = \sum_{i=1}^{c} \sum_{j \mid i} \mu(j) d^{i/j} .$$

and μ is the Möbius function.

Proof. Lemma 4.72 implies that $F_{c,d}$ has a stepped basis $\mathcal{B}$. The formula

$$r_i = \frac{1}{i} \sum_{j \mid i} \mu(j) d^{i/j}$$

is due to Witt, see for example Theorem 5.11 of [45]. It is also clear that $\Gamma_{\mathcal{B}}(F_{c,d}) = F_{c,d}$, so by Corollary 4.69, (4.45) holds. $\qquad\square$

Proposition 4.74. *Let L be a 2-generated stepped $\mathbb{Z}$-Lie ring. Then $\zeta_{L,p}^{\triangleleft}(s)$ satisfies (4.31) for all but finitely many primes p.*

Proof. We shall prove that $\mathrm{rank}(L/\sigma_{c-1}(L)) = 2$, the result will then follow from Corollary 4.69. Since $L/\sigma_{c-1}(L)$ is abelian,

$$\mathrm{rank}(L/\sigma_{c-1}(L)) \leq \mathrm{rank}(L/[L, L]) = 2 .$$

Choose some $x \notin \sigma_{c-1}(L)$. Since $x \notin \sigma_{c-1}(L)$, there must exist $y \in L$ such that $y \notin \sigma_{c-2}(L)$ and $[x, y] \notin \sigma_{c-2}(L)$. This implies $y \notin \sigma_{c-1}(L)$ and since $[x, y] \notin \sigma_{c-2}(L)$, x and y must be linearly independent modulo $\sigma_{c-1}(L)$. Hence $\mathrm{rank}(L/\sigma_{c-1}(L)) = 2$. $\qquad\square$

The following proposition demonstrates infinitely many Lie rings which (conjecturally) have no functional equation:

Proposition 4.75. *For $r \in \mathbb{N}_{>0}$, let $\mathbf{h} = (h_1, \ldots, h_r)$ be a finite sequence of natural numbers satisfying $h_1 \geq h_2 \geq \cdots \geq h_r \geq 2$. Let $L_\mathbf{h}$ be the Lie ring on generators $\{z\} \cup \{\mathbf{e}_{i,j} \mid 1 \leq i \leq r, 1 \leq j \leq h_i\}$ with the only nontrivial Lie brackets (up to antisymmetry) being $[z, \mathbf{e}_{i,j}] = \mathbf{e}_{i,j+1}$ for $1 \leq i \leq r, 1 \leq j < h_i$. Set $c = h_1$ and $d = 1 + h_1 + \cdots + h_r$. Assume Conjecture 4.5. If $h_1 = \cdots = h_r$,*

$$\zeta^{\triangleleft}_{L_\mathbf{h},p}(s)|_{p \to p^{-1}} = (-1)^d p^{\binom{d}{2} - Ns} \zeta^{\triangleleft}_{L_\mathbf{h},p}(s) , \tag{4.46}$$

for all but finitely many primes p, where $N = c + \frac{1}{2}c(c+1)r$. If $h_r < h_1$, $\zeta^{\triangleleft}_{L_\mathbf{h},p}(s)$ satisfies no such functional equation.

Proof. Firstly, it is clear that $\mathrm{ht}(\mathbf{e}_{i,j}) = h_i + 1 - j$, so $L_\mathbf{h}$ is clearly stepped with respect to any basis $\mathcal{B}$ listing the generators in descending order of height.

If $h_1 = \cdots = h_r$, then $z, \mathbf{e}_{1,1}, \ldots, \mathbf{e}_{r,1}$ are the basis elements of depth 1, and it is clear that they generate $L_\mathbf{h}$. Hence $\Gamma_\mathcal{B}(L_\mathbf{h}) = L_\mathbf{h}$, and by Corollary 4.69, (4.46) holds.

If $h_r < h_1$, then $\mathrm{dep}(\mathbf{e}_{r,1}) > 1$. Hence $\mathbf{e}_{r,1}, \mathbf{e}_{r,2} \notin \Gamma_\mathcal{B}(L_\mathbf{h})$. Now $z \in \Gamma_\mathcal{B}(L)$ and $[z, \mathbf{e}_{r,1}] = \mathbf{e}_{r,2}$, hence, by Corollary 4.70, $\zeta^{\triangleleft}_{L_\mathbf{h},p}(s)$ satisfies no functional equation. $\square$

We can deduce Taylor's conjecture about the maximal class Lie rings M_n (Conjecture 4.24) as special cases of both of these last two propositions:

Corollary 4.76. *Assume Conjecture 4.5. Then, for all but finitely many primes p,*

$$\zeta^{\triangleleft}_{M_n,p}(s)|_{p \to p^{-1}} = (-1)^d p^{\binom{d}{2} - (\binom{d+1}{2} - 1)s} \zeta^{\triangleleft}_{M_n,p}(s) ,$$

where $d = n + 1$.

Proof. M_n is 2-generated and also $M_n = L_\mathbf{h}$ for the singleton sequence $\mathbf{h} = (n)$. $\square$

Corollary 4.70 is a quick way of verifying the lack of functional equation. Although it is useful, it is not universal. Consider the Lie ring L with presentation

$$\left\langle \mathbf{e}_1, \ldots, \mathbf{e}_9 : \begin{array}{l} [\mathbf{e}_1, \mathbf{e}_2] = \mathbf{e}_3, [\mathbf{e}_1, \mathbf{e}_3] = \mathbf{e}_6, [\mathbf{e}_2, \mathbf{e}_3] = \mathbf{e}_7, [\mathbf{e}_1, \mathbf{e}_4] = \mathbf{e}_7, \\ [\mathbf{e}_2, \mathbf{e}_5] = \mathbf{e}_6, [\mathbf{e}_4, \mathbf{e}_5] = \mathbf{e}_9, [\mathbf{e}_2, \mathbf{e}_6] = \mathbf{e}_8, [\mathbf{e}_1, \mathbf{e}_7] = \mathbf{e}_8 \end{array} \right\rangle .$$

L has nilpotency class 4 and has stepped basis $\mathcal{B} = (\mathbf{e}_1, \ldots, \mathbf{e}_9)$, with $\Gamma_\mathcal{B}(L) = \langle \mathbf{e}_1, \mathbf{e}_2, \mathbf{e}_3, \mathbf{e}_6, \mathbf{e}_7, \mathbf{e}_8 \rangle$. Corollary 4.70 cannot be applied since $[\mathbf{e}_k, \mathbf{e}_j] \notin \Gamma_\mathcal{B}(L)$ implies $\{\mathbf{e}_k, \mathbf{e}_j\} = \{\mathbf{e}_4, \mathbf{e}_5\}$, and $\mathbf{e}_4, \mathbf{e}_5 \notin \Gamma_\mathcal{B}(L)$. However, it follows from Corollary 4.48 that $\zeta^{\triangleleft}_{L,p}(s)$ is predicted to satisfy no functional equation. For each prime p, $H = \langle e_1, e_2, \ldots, e_8, pe_9 \rangle_{\mathbb{Z}_p}$ is clearly not an ideal (nor even a subring!) of L_p, but $\pi_B(H) \vartriangleleft_p L_p$.

We conclude this section by referring back to Chap. 2. The majority of Lie rings considered have (†). The exceptions are Fil_4, $\mathrm{Fil}_4 \times \mathbb{Z}$, $\mathfrak{g}_{6,15}$, $\mathfrak{g}_{137B}$, $\mathfrak{g}_{137D}$, $\mathfrak{g}_{1357B}$, $\mathfrak{g}_{1357C}$ and $\mathfrak{g}_{1457B}$. All the others have (†) and all of them satisfy a functional equation if and only if they have (∗). The local ideal zeta functions of those with (∗) satisfy the functional equation (4.31), and Corollary 4.70 applies to each Lie ring without (∗).

Many of the examples in Chap. 2 arose from taking the $\mathbb{Z}$-span of a nilpotent Lie algebra of dimension 6 or 7 over $\mathbb{C}$, classified in [44] and [26]. Whilst we have been able to complete many of the calculations, there are still a number of Lie rings arising from these classifications for which we were not able to calculate the ideal zeta function. It therefore seems worthwhile listing such Lie rings, and for those that have (†), whether Corollary 4.63 implies a functional equation satisfied by the ideal zeta function.

The only such Lie rings of rank 6 whose ideal zeta functions are not given in Chap. 2 are $\mathfrak{g}_{6,n}$ for $n \in \{2, 11, 18, 19, 20, 21, 22\}$. Of these, $\mathfrak{g}_{6,2} \cong M_5$, $\mathfrak{g}_{6,18}$ and $\mathfrak{g}_{6,21}$ are also stepped and Corollary 4.63 predicts the ideal zeta functions of all three to satisfy (4.31). The remaining four Lie rings do not have (†).

Amongst Lie rings of rank 7 there are many calculations of ideal zeta functions which have yet to be done. In the table below, we list these Lie rings and whether or not they have (†). For each set of upper-central series dimensions, we list the suffixes of Lie algebras as used in [26]. There are six infinite families indexed by a single parameter. These are denoted by an asterisk. In all six cases, whether the Lie ring has (†) is independent of the parameter. Note that there are gaps in the suffixes: the Lie algebras (147C), (1357K), (12457M), (13457H) and (123457G) do not exist.

Dimensions	Has (†)	Doesn't have (†)
147	D,E*	–
247	C-K	L-R
257	D,I,J	E-H,L
357	A-C	–
1357	D,M*,O-R,S*	E,F,I,J,L,N*
2357	–	A-D
2457	A-C,L,M	D-K
12357	A	B,C
12457	A,C,H,L	B,D-G,I-K,N*
13457	A,C	B,D-G,I
23457	C	A,B,D-G
123457	A	B-F,H,I*

Of those that have (†), all except 257D, 357A-C, 1357D and 2457A-C satisfy (4.37). For each of these eight exceptions, their predicted lack of functional equation can be deduced from Corollary 4.70.

4.9.3 Lie Rings Without (†)

We have focused on Lie rings with (†) since we can apply Proposition 4.47. If we wish to determine whether the local ideal zeta function satisfies a functional equation, we only need to look for a p-ideal that doesn't correspond to an ideal, or show that no such p-ideal exists. If we instead consider Lie rings without (†) we can no longer do this. There may well be ideals that don't correspond to p-ideals under the π-map, as well as p-ideals that don't correspond to ideals, and some cancellation may take place.

Fil$_4$, defined by the presentation

$$\langle z, x_1, x_2, x_3, x_4 : [z, x_1] = x_2, [z, x_2] = x_3, [z, x_3] = x_4, [x_1, x_2] = x_4 \rangle ,$$

is in some sense the 'simplest' Lie ring without (†). There certainly aren't any of smaller rank. Let p be any prime and put $\mathcal{B} = (z, x_1, x_2, x_3, x_4)$, $L_p = $ Fil$_4 \otimes \mathbb{Z}_p$ and $N = 5 + 4 + 3 + 2 = 14$. Then set

$$H = \langle p^2 z + p x_1 + x_2, p^2 x_1, p x_2 + x_3, p x_3 + x_4, p x_4 \rangle ,$$
$$P = \pi_{\mathcal{B}}(H) = \langle p^6 z + p^5 x_1 + p^3 x_2, p^6 x_1, p^4 x_2 + p^2 x_3, p^3 x_3 + p x_4, p^2 x_4 \rangle .$$

It is a routine task to verify that $H \lhd L_p$ but $P \not\lhd_p L_p$. However, it turns out that this ideal not associated to a p-ideal, and all others of index no more than 10, are 'cancelled out' by p-ideals P' of index $\leq 10 + N$ such that $\pi_{\mathcal{B}}^{-1}(P') \not\lhd L_p$. There are, however, more p-ideals of index p^{11+N} than ideals of index p^{11}. This phenomenon can be observed in the numerator polynomial of $\zeta_{\text{Fil}_4, p}^{\lhd}(s)$ (p. 54). For each term $c X^a Y^b$ with $b \leq 10$, there exists a term $c X^{23-a} Y^{42-b}$, but this doesn't happen for the term $-X^6 Y^{11}$.

It is perhaps worth noting that $\mathfrak{g}_{6,15}$, $\mathfrak{g}_{137B}$ and $\mathfrak{g}_{137D}$ are the only Lie rings without (†) whose ideal zeta functions we have calculated and which satisfy a functional equation. In all three cases the Lie ring is isospectral to a Lie ring with (†) – $\mathfrak{g}_{6,17}$, $M_3 \times_\mathbb{Z} M_3$ and $\mathfrak{g}_{137C}$ respectively – so it raises the following:

Question 4.77. Let L be a $\mathbb{Z}$-Lie ring without (†). Suppose $\zeta_{L,p}^{\lhd}(s)$ satisfies (4.31) for all but finitely many primes p. Does there always exist a $\mathbb{Z}$-Lie ring L_1 with (†) such that $\zeta_{L,p}^{\lhd}(s) = \zeta_{L_1,p}^{\lhd}(s)$ for all but finitely many primes p?

We suspect the answer is 'no', but on the scant evidence we have it would be foolish to elevate this to a conjecture.

5

Natural Boundaries I: Theory

5.1 A Natural Boundary for $\zeta_{\mathrm{GSp}_6}(s)$

We begin this chapter with an explicit demonstration that the global zeta function of the algebraic group GSp_6 has a natural boundary.

Theorem 5.1. *Let* $Z(s) = \prod_p (1 + (p + p^2 + p^3 + p^4)p^{-s} + p^{5-2s})$, *where the product is taken over all primes* p. *Then*

1. $Z(s)$ *converges on* $\{\, s \in \mathbb{C} : \Re(s) > 5 \,\}$;
2. $Z(s)$ *can be meromorphically continued to* $\{\, s \in \mathbb{C} : \Re(s) > 4 \,\}$;
3. $\{\, s \in \mathbb{C} : \Re(s) = 4 \,\}$ *is a natural boundary for* $Z(s)$.

Proof. 1. An infinite product $\prod_{n \in I}(1 + a_n)$ converges absolutely if the corresponding sum $\sum_{n \in I} |a_n|$ converges. Now $\sum_{p \text{ prime}} |p^{-s}|$ converges on $\{\, s \in \mathbb{C} : \Re(s) > 1 \,\}$. Hence we see that in our infinite product $Z(s)$ it is the term p^{4-s} which is the limit of convergence. Hence $Z(s)$ converges on $\{\, s \in \mathbb{C} : \Re(s) > 5 \,\}$.

2. To meromorphically continue the function to $\{\, s \in \mathbb{C} : \Re(s) > 4 \,\}$, we have to produce a function meromorphic on $\{\, s \in \mathbb{C} : \Re(s) > 4 \,\}$ which coincides with $Z(s)$ on $\{\, s \in \mathbb{C} : \Re(s) > 5 \,\}$.
Let

$$
F(s) = \prod_p (1 + (p + p^2 + p^3 + p^4)p^{-s} + p^{5-2s}) \frac{(1 - p^{-s+4})}{(1 - p^{-2s+8})} \ .
$$

We claim that this converges on $\{\, s \in \mathbb{C} : \Re(s) > 4 \,\}$, in which case it coincides with $Z(s).\frac{\zeta(2s-8)}{\zeta(s-4)}$ on $\{\, s \in \mathbb{C} : \Re(s) > 5 \,\}$. Since $\frac{\zeta(s-4)}{\zeta(2s-8)}$ is meromorphic on the whole complex plane, we can meromorphically continue $Z(s)$ from $\{\, s \in \mathbb{C} : \Re(s) > 5 \,\}$ to $\{\, s \in \mathbb{C} : \Re(s) > 4 \,\}$ by setting $Z(s) = F(s).\frac{\zeta(s-4)}{\zeta(2s-8)}$ for s with $\Re(s) > 4$.

The terms in the infinite product $F(s)$ can be rewritten

$$(1 + (p + p^2 + p^3 + p^4)p^{-s} + p^{5-2s})\frac{(1 - p^{-s+4})}{(1 - p^{-2s+8})}$$

$$= 1 + \frac{(p + p^2 + p^3)p^{-s} + p^{5-2s}}{1 + p^{4-s}}.$$

To ascertain the radius of convergence of the infinite product of such expressions we again look at the absolute convergence of sums. The limit of convergence is given now by the term $\sum_p \left|\frac{p^{3-s}}{1+p^{4-s}}\right|$ which converges for s with $\Re(s) > 4$. We have therefore continued $Z(s)$ successfully to $\{ s \in \mathbb{C} : \Re(s) > 4 \}$ by using the Riemann zeta function to pass the pole at $\Re(s) = 5$.

3. However $\Re(s) = 4$ is as far as we can go. We show now how we can realise every point on this boundary as the limit point of zeros from $\{ s \in \mathbb{C} : \Re(s) > 4 \}$. To do this we consider solutions of the equation $1 + (X + X^2 + X^3 + X^4)Y + X^5Y^2 = 0$. We will be interested in the value of s for solutions of the form $(X, Y) = (p, p^{-s})$.
We make the substitution $U = X^4Y$ and $V = X^{-1}$. Hence we want to consider the equation

$$F(V, U) = U^2V^3 + U(1 + V + V^2 + V^3) + 1 = 0.$$

This has a trivial solution at $(V, U) = (0, -1)$. The partial derivatives at this point are then given by

$$F_V(V, U)|_{(0,-1)} = 3U^2V^2 + U(1 + 2V + 3V^2)|_{(0,-1)} = -1 \,,$$

$$F_U(V, U)|_{(0,-1)} = 2UV^3 + 1 + V + V^2 + V^3|_{(0,-1)} = 1 \,.$$

We can therefore use the Implicit Function Theorem to expand U as a function of V in the neighbourhood around the solution $(0, -1)$ to get

$$U = -1 - \frac{F_V}{F_U}\bigg|_{(0,-1)} V + \Omega(V)$$

$$= -1 + V + \Omega(V) \,,$$

where $\Omega(V)$ is a power series in V starting with V^2 or some higher term. So for p large enough at the point $V = p^{-1}$ we get the solution $U = -1 + p^{-1} + \Omega(p^{-1})$. Setting $U = p^{4-s}$ we therefore get a solution of $(1 + (p + p^2 + p^3 + p^4)p^{-s} + p^{5-2s}) = 0$ for values of s satisfying

$$p^{4-s} = -1 + p^{-1} + \Omega(p^{-1}) \,.$$

Hence for all $n \in \mathbb{Z}$ we have a solution of the form

$$s = 4 - \frac{\log(1 - p^{-1} + \Omega(p^{-1}))}{\log p} + \frac{(2n - 1)\pi i}{\log p} \,.$$

Now

$$\delta_p = -\frac{\log(1 - p^{-1} + \Omega(p^{-1}))}{\log p} \to 0$$

as $p \to \infty$. If we fix some point $A = 4 + ai$ on the boundary $\Re(s) = 4$ then we can arrange some sequence of integers n_p for each prime p so that

$$\frac{(2n_p - 1)\pi}{\log p} \to a$$

as $p \to \infty$. Hence each point A on the boundary is a limit point of zeros. Finally we must check that the zeros are on the right-hand side of this boundary which follows since $\delta_p > 0$ for large enough p. We therefore cannot continue $Z(s)$ beyond its natural boundary at $\Re(s) = 4$. This proves Theorem 5.1. $\qquad\square$

Corollary 5.2. *The global zeta function $Z_{\mathrm{GSp}_6}(s) = \prod_p Z_{\mathrm{GSp}_6,p}(s)$ has a natural boundary at $\Re(s) = 4$.*

Proof. It was established in [36] that

$$Z_{\mathrm{GSp}_6,p}(s) = \frac{1 + (p + p^2 + p^3 + p^4)p^{-s} + p^{5-2s}}{(1 - p^{-s})(1 - p^{3-s})(1 - p^{5-s})(1 - p^{6-s})}.$$

The result is now immediate from Theorem 5.1. $\qquad\square$

Remark 5.3. Notice that we had a helpful minor miracle during the course of the proof of the natural boundary in the fact that $\delta_p > 0$ which forced the zeros to lie on the right-hand side of the boundary. If they had been on the left-hand side they would not have been helpful as there may have been a way to continue the function to avoid picking up these zeros. However since there is a unique way to analytically continue a function, once the zeros have appeared, we are stuck with them. As we shall see in the next section, we shall make this a hypothesis of our general result on natural boundaries. It means that we are unable to completely answer the conjecture mentioned in the introduction of a generalisation of Estermann's result to two variables. For example the method above would fail for the polynomial $1 + (X + X^2 - X^3 + X^4)Y + X^5Y^2$.

What is perhaps extraordinary is that all the examples of zeta functions of Lie rings that have been calculated involve polynomials that also satisfy this minor miracle.

5.2 Natural Boundaries for Euler Products

We see in this section how far we can take the methodology employed in the previous section. Let

$$W(X,Y) = 1 + \sum_{k=1}^{l}(a_{0,k} + a_{1,k}X + \cdots + a_{n_k,k}X^{n_k})Y^k ,$$

where $a_{i,k} \in \mathbb{Z}$ and we assume that $a_{n_k,k} \neq 0$. Put $\deg_X W = d_X$, say.
We consider the analytic behaviour of the function

$$Z(s) = \prod_{p \text{ prime}} W(p, p^{-s}) .$$

We shall prove in this section Conjecture 1.11 under certain conditions that
we shall explain during the course of the discussion.

Lemma 5.4. Let $\alpha = \max\{\frac{1+n_k}{k} : k = 1,\ldots,l\}$. Then $Z(s)$ converges on
$\{s \in \mathbb{C} : \Re(s) > \alpha\}$.

Proof. The sum $\sum_p |a_{i,k}p^{i-ks}|$ converges on $\{s \in \mathbb{C} : \Re(s) > \alpha\}$. Hence the
infinite product $Z(s)$ also converges on $\{s \in \mathbb{C} : \Re(s) > \alpha\}$.   □

Where and whether we can meromorphically continue $Z(s)$ is going to
depend on the zeros of $W(X,Y)$. These zeros are determined by the Puiseux
power series and the corresponding Newton diagrams. The Puiseux power
series as we shall see are just a more sophisticated version of the implicit
function theorem that we used in the previous section.

We first make a substitution into our polynomial so that we are considering
the behaviour of a polynomial as one of the variables tends to zero rather than
infinity. Let $\beta = \max\{\frac{n_k}{k} : k \in I\}$. Let $j \in I$ be as small as possible with the
property that $\frac{n_j}{j} = \beta$ and put

$$U = X^\beta Y ,$$
$$V = X^{-1/j} .$$

Then setting $J = \{(k,i) : i = 0,\ldots,n_k \text{ and } k = 1,\ldots,l\}$,

$$W(X,Y) = F(V,U) = 1 + \sum_{(k,i) \in J} a_{i,k}V^{(n_j k - ij)}U^k$$

$$= 1 + \sum_{(k,i) \in K} b_{i,k}V^i U^k .$$

Note that $n_j k - ij \geq 0$.

The theory of the Puiseux power series guarantees us the existence of
power series Ω_i for $i = 1,\ldots,l$ such that

$$\Omega_i(V^{1/q}) = c_i \left(V^{1/q}\right)^{e_i} + \Omega_{i,1}(V^{1/q}) ,$$

where $\Omega_{i,1}(V^{1/q})$ is a power series in $V^{1/q}$ starting with $\left(V^{1/q}\right)^{e_i+1}$ or some
higher term and $e_i \in \mathbb{Z}$. Since the coefficients of $F(V,U)$ as a polynomial in U

are polynomials in V, the power series $\Omega_i(V^{1/q})$ converge for all V and define the l zeros of the polynomial $F(V, U) = W(X, Y)$.

We can read off the initial term of these Puiseux power series from the Newton polygon of $W(X, Y)$. Let $\mathcal{L}_W$ denote the set of lattice points (n, m) which correspond to the occurrence of a non-zero term $X^n Y^m$ in $W(X, Y)$. Let $\mathfrak{l}_1, \ldots, \mathfrak{l}_r$ be the lines starting from the point $(n_0, 0)$ and ascending to the line with end point (n_l, l) marking out the right-hand convex hull which we call $\mathcal{N}$.

For each line $\mathfrak{l}_i$, let $m_i = \min\{m : (n, m) \in \mathfrak{l}_i \cap \mathcal{L}_W\}$. Let u_i and v_i be coprime with the property that

$$\frac{v_i}{u_i} = \frac{(m_{i+1} - m_i)}{(n_{m_{i+1}} - n_{m_i})} ,$$

the gradient of the line $\mathfrak{l}_i$. Consider the quasi-homogeneous polynomial made up of all the homogeneous components of $W(X, Y)$ sitting on the line $\mathfrak{l}_i$:

$$\sum_{(n,m)\in \mathfrak{l}_i \cap \mathcal{L}_W} a_{n,m} X^n Y^m = X^{n_{m_i}} Y^{m_i} \widetilde{W}_i(X^{u_i} Y^{v_i}) ,$$

where $\widetilde{W}_i(Z)$ is a polynomial of degree $(m_{i+1} - m_i)/v_i$ in one variable with non-zero constant coefficient $a_{n_{m_i}, m_i}$.

In [18] we defined the ghost polynomial of $W(X, Y)$ as follows:

$$\widetilde{W}(X, Y) = X^{d_X - n_l} \prod_{i=1}^{r} \widetilde{W}_i(X^{u_i} Y^{v_i})$$

$$= X^{d_X - n_l} \prod_{i=1}^{l} (X^{e_i/q} Y - c_i) .$$

It has the property as explained in [18] of picking out the leading term of the Puiseux power series in which we are interested. We have made an additional change of variable to that in [18]. Note that β is the inverse of the value of the minimal gradient in the Newton polygon. Hence

$$\widetilde{W}(X, Y) = V^{(n_l - d_X)j} \prod_{i=1}^{r} \widetilde{W}_i(V^{s_i} U^{t_i})$$

$$= V^{(n_l - d_X)j} \widetilde{W}_1(U^{t_1}) \prod_{i=2}^{r} \widetilde{W}_i(V^{s_i} U^{t_i}) .$$

It is the zeros defined by the Puiseux power series corresponding to the first piece of the Newton polygon that will be important to us. That is except for one case which we can remove from the outset. If $\widetilde{W}_1(Z)$ is a product of cyclotomic polynomials and $\widetilde{W}_1(X^{u_i} Y^{v_i})$ is actually a factor of $W(X, Y)$, then $\prod_p \widetilde{W}_1(p^{u_i} p^{-s v_i})$ is meromorphic on the whole of $\mathbb{C}$. Hence we can just factor it off and suppose we are in a situation in which either $\widetilde{W}_1(Z)$ is a not a product of cyclotomic polynomials or $\widetilde{W}_1(X^{u_i}Y^{v_i})$ is not a factor of $W(X,Y)$. Note that if we keep on doing this and are left with a constant then of course $Z(s)$ is meromorphic and of the shape predicated by the conjecture. And conversely if $W(X,Y)$ is built out of products of cyclotomic polynomials in one variable then we will be able to remove each section corresponding to sides of the Newton polygon until we are left with a constant.

So we want to show that in the case that we are left with terms of the Newton polygon, $Z(s)$ will to have a natural boundary at $\Re(s) = \beta$, the inverse of the first gradient.

We shall divide into a number of cases. First though, we need to show that we can meromorphically continue to $\Re(s) > \beta$.

Lemma 5.5. $Z(s)$ can be meromorphically continued to $\Re(s) > \beta$.

Proof. There is a unique way to write $W(X,Y)$ formally as a product:

$$W(X,Y) = \prod_{(n,m)\in\mathbb{N}^2} (1 - X^n Y^m)^{c_{n,m}} . \tag{5.1}$$

To see how to find such an expression for a general bivariate polynomial, one clears each term with $\mathbb{N}^2$ ordered lexicographically from the right. To clear a term $(-1)^{\varepsilon_{n,m}} e_{n,m} X^n Y^m$ where $\varepsilon_{n,m} = 0$ or 1 and $e_{n,m} > 0$ one introduces a factor $(1 - X^n Y^m)^{e_{n,m}}$ if $\varepsilon_{n,m} = 1$ or two terms $(1 - X^{2n}Y^{2m})^{e_{n,m}}(1 - X^n Y^m)^{-e_{n,m}}$ if $\varepsilon_{n,m} = 0$. This only introduces terms higher up the lexicographical ordering which will be cleared later. For each fixed m there will only be a finite number of terms $(-1)^{\varepsilon_{n,m}} e_{n,m} X^n Y^m$ that we will ever have to clear so the procedure does approximate $W(X,Y)$ modulo polynomials starting in higher and higher degrees of Y. The uniqueness is clear since each $c_{n,m}$ will be recursively defined from terms lower in the lexicographical ordering.

The next claim is that if we have

$$W(X,Y) = \prod_{m \leq M} (1 - X^n Y^m)^{c_{n,m}} + \sum_{m > M} e_{n,m} X^n Y^m , \tag{5.2}$$

if $e_{n,m} \neq 0 \neq c_{n,m}$ then $n/m \leq \beta$. This follows because the pairs (n,m) appearing are all generated additively by $\{(n,m) : a_{n,m} \neq 0\}$ and if $n_1/m_1 \leq \beta$ and if $n_2/m_2 \leq \beta$ then $(n_1 + n_2)/(m_1 + m_2) \leq \beta$. For each fixed M, we set $\beta_M = \max\{(n+1)/m : m > M$ and $e_{n,m} \neq 0\}$. Then for $\Re(s) > \beta_M$ the following infinite product converges absolutely:

$$W_M(s) = \prod_p \left(1 + \frac{\sum_{m>M} e_{n,m} p^{n-ms}}{\prod_{m \leq M}(1 - p^{n-ms})^{c_{n,m}}} \right) .$$

Hence on $\Re(s) > \beta_M$ we continue $Z(s)$ by defining

$$Z(s) = \prod_{\substack{(n,m)\in\mathbb{N}^2 \\ m\leq M}} \zeta(ms-n)^{-c_{n,m}} W_M(s).$$

To prove that this is a meromorphic continuation of $Z(s)$ we have to check that on the region of convergence of $Z(s)$, the two expressions for $Z(s)$ agree. For this we just need to use the fact that if $\prod a_n$, $\prod b_n$ and $\prod a_n b_n$ all converge absolutely then $(\prod a_n \times \prod b_n) = \prod a_n b_n$. By letting $M \to \infty$, this continues $Z(s)$ up to $\Re(s) > \beta$ since β_M has a limit which is β or smaller as $M \to \infty$.

If we want to see the zeros of one of the local factors $W(p, p^{-s})$ we shall use the following identity in $\Re(s) > \beta_M$:

$$Z(s) = \prod_{\substack{(n,m)\in\mathbb{N}^2 \\ m\leq M}} \zeta(ms-n)^{-c_{n,m}} W_M(s)$$

$$= W(p, p^{-s}) \prod_{\substack{(n,m)\in\mathbb{N}^2 \\ m\leq M}} \zeta_p(ms-n)^{c_{n,m}} \prod_{\substack{(n,m)\in\mathbb{N}^2 \\ m\leq M}} \zeta(ms-n)^{-c_{n,m}} W_{M,p}(s),$$

where

$$W_{M,p}(s) = \prod_{q\neq p}\left(1 + \frac{\sum_{m>M} e_{n,m} q^{n-ms}}{\prod_{m\leq M}(1-q^{n-ms})^{c_{n,m}}}\right).$$

Note that since a convergent product of non-zero factors is not zero, we shall get that $W_M(s)$ is non-zero except for zeros of $W(p, p^{-s})$.   □

We now consider several case distinctions for our polynomial $W(X, Y)$.

Case 1: $\widetilde{W}_1(U^{t_1})$ is not cyclotomic. This is equivalent to there existing infinitely many (n, m) with $c_{n,m} \neq 0$ and $n/m = \beta$. In this case there exists a corresponding Puiseux power series with $|c_i| < 1$. To prove this we use the same argument as in the proof of Estermann's result. Let c_i ($i = 1, \ldots, d_1 = \deg \widetilde{W}_1$) be all the roots of $\widetilde{W}_1$. Now we know that $c_1 \cdots c_{d_1} = 1$, the constant term of $\widetilde{W}_1$. Suppose that $|c_i| \geq 1$ for all i. Then this would imply $|c_i| = 1$. But if all the c_i lie on the unit circle then $\widetilde{W}_1$ is cyclotomic contrary to our assumption. Hence there must be some i with $|c_i| < 1$.

We shall prove that this case has a lot of zeros of $W(p, p^{-s})$ on the right of $\Re(s) = \beta$ which can't get cancelled. We shall call a polynomial that falls under this case of *Type I*.

Case 2: $\widetilde{W}_1(U^{t_1})$ is cyclotomic (hence there are finitely many (n, m) with $c_{n,m} \neq 0$ and $n/m = \beta$), and in addition there are only finitely many pairs

(n, m) with $c_{n,m} > 0$ and $(n + 1)/m > \beta$ but there exists a corresponding Puiseux power series

$$\Omega_i(V^{1/q}) = c_i + \Omega_{i,1}(V^{1/q})$$
$$= c_i + c_{i,1}\left(V^{1/q}\right)^{\gamma_i} + \Omega_{i,2}(V^{1/q})$$

with

$$|\Omega_i(V^{1/q})| < 1 \tag{5.3}$$

for small enough V. This case also has a lot of zeros of $W(p, p^{-s})$ on the right of $\Re(s) = \beta$ which can't get cancelled. This will be called a polynomial of *Type II*.

Case 3: $\widetilde{W}_1(U^{t_1})$ is cyclotomic (hence there are finitely many (n, m) with $c_{n,m} > 0$ and $n/m = \beta$), there are infinitely many pairs (n, m) with $c_{n,m} > 0$ and $(n + 1)/m > \beta$, and there exists a corresponding Puiseux power series

$$\Omega_i(V^{1/q}) = c_i + \Omega_{i,1}(V^{1/q})$$
$$= c_i + c_{i,1}\left(V^{1/q}\right)^{\gamma_i} + \Omega_{i,2}(V^{1/q})$$

satisfying (5.3) for V small enough. Polynomials in this case we call of *Type III*.

Type III polynomials require an assumption that the Riemann Hypothesis holds which implies that zeros of $W(p, p^{-s})$ on the right of $\Re(s) = \beta$ can't get cancelled by zeros of the Riemann zeta function. There is a subcase which doesn't require the Riemann Hypothesis. This depends on the second term of the Puiseux power series. In this subcase we show that the current estimates for the number of Riemann zeros off the line $\Re(s) = \frac{1}{2}$ are sufficient to show that there are not enough to cancel local zeros. We shall consider this subcase in Sect. 5.3.

We now prove the following:

Theorem 5.6. *Suppose that $W(X, Y) \neq 1$ and has no unitary factors and is a polynomial of type I, II or III. Then $\Re(s) = \beta$ is a natural boundary for $Z(s)$ (where we assume the Riemann Hypothesis for polynomials of Type III).*

Proof. In all three cases the zeros

$$U = \Omega_i(V^{1/q}) = c_i + c_{i,1}\left(V^{1/q}\right)^{\gamma_i} + \Omega_{i,2}(V^{1/q})$$

lie within the unit circle for V small enough.

Let $\Delta_{\nu,\eta}$ be the region $z = \sigma + \tau i$ with

$$\beta + \frac{1}{\nu + 1} < \sigma \leq \beta + \frac{1}{\nu}, \quad 0 < u < \tau < u + \eta,$$

where ν is a positive integer, $\eta > 0$ and $u > 0$.

We have a lot of candidate zeros of $Z(s)$ produced by the zeros of $\Omega_i(V^{1/q})$, namely:

$$s_{n,p} = \beta - \frac{\log\left(c_i + c_{i,1}p^{-\gamma_i/qj} + \Omega_{i,2}(p^{-1/qj})\right)}{\log p} + \frac{2\pi n i}{\log p}$$

with $n \in \mathbb{Z}$ and p ranging over all primes. Note that our conditions in case 1, 2 and 3 imply that for p sufficiently large, where the term $\Omega_{i,2}(p^{-1/qj})$ becomes negligible, these zeros lie on the right-hand side of $\Re(s) = \beta$. This follows because the modulus of $c_i + c_{i,1}p^{-\gamma_i/qj} + \Omega_{i,2}(p^{-1/qj})$ will be less than 1. For p large enough $c_i + c_{i,1}p^{-\gamma_i/qj} + \Omega_{i,2}(p^{-1/qj})$ will lie within the unit circle. Let $S(\nu, \eta)$ denote the number of zeros $s_{n,p}$ in $\Delta_{\nu,\eta}$.

We start with case 1. Let

$$C_i = |c_i^{-1}| > 1 .$$

For $2\pi/\log p < \eta$ and

$$\frac{1}{\nu+1} < \frac{-\log\left|c_i + c_{i,1}p^{-\gamma_i/qj} + \Omega_{i,2}(p^{-1/qj})\right|}{\log p} \le \frac{1}{\nu} , \tag{5.4}$$

there exists n such that $s_{n,p} \in \Delta_{\nu,\eta}$.

For p large enough, one of the following cases holds

$$\left|1 + \frac{c_{i,1}}{c_i}p^{-\gamma_i/qj} + \frac{\Omega_{i,2}(p^{-1/qj})}{c_i}\right| \ge 1 \text{ for all } p > N , \tag{5.5}$$

$$\left|1 + \frac{c_{i,1}}{c_i}p^{-\gamma_i/qj} + \frac{\Omega_{i,2}(p^{-1/qj})}{c_i}\right| \le 1 \text{ for all } p > N . \tag{5.6}$$

In case (5.5) set $\varepsilon = 1$ and in case (5.6) set $\varepsilon = 0$. We can then choose N large enough to ensure that

$$(-1)^\varepsilon \left|1 + \frac{c_{i,1}}{c_i}p^{-\gamma_i/qj} + \frac{\Omega_{i,2}(p^{-1/qj})}{c_i}\right|^{-\nu-\varepsilon} < \left(\frac{1+(-1)^\varepsilon C_i^{\nu+\varepsilon}}{C_i^{\nu+\varepsilon}}\right) . \tag{5.7}$$

Lemma 5.7. *If $p > N$ and*

$$C_i^\nu + 1 \le p \le C_i^{\nu+1} - 1 , \tag{5.8}$$

then (5.4) holds.

Proof. Condition (5.7) implies that

$$C_i^{\nu+1} - 1 < C_i^{\nu+1}\left|1 + \frac{c_{i,1}}{c_i}p^{-\gamma_i/qj} + \frac{\Omega_{i,2}(p^{-1/qj})}{c_i}\right|^{-\nu-1}$$

and that

$$C_i^\nu \left|1 + \frac{c_{i,1}}{c_i}p^{-\gamma_i/qj} + \frac{\Omega_{i,2}(p^{-1/qj})}{c_i}\right|^{-\nu} < C_i^\nu + 1 \,.$$

Hence (5.8) implies

$$C_i^\nu \left|1 + \frac{c_{i,1}}{c_i}p^{-\gamma_i/qj} + \frac{\Omega_{i,2}(p^{-1/qj})}{c_i}\right|^{-\nu} < p$$

and

$$p < C_i^{\nu+1} \left|1 + \frac{c_{i,1}}{c_i}p^{-\gamma_i/qj} + \frac{\Omega_{i,2}(p^{-1/qj})}{c_i}\right|^{-\nu-1} \,.$$

Now just take logs of both sides. (Note that, unlike C_i, we don't know whether

$$\left|1 + \frac{c_{i,1}}{c_i}p^{-\gamma_i/qj} + \frac{\Omega_{i,2}(p^{-1/qj})}{c_i}\right| \tag{5.9}$$

is greater or less than 1. In Case 2 and 3 when $C_i = 1$ we are going to make an assumption to control the size of (5.9) which we don't need to do in this case since $C_i > 1$.)   □

So for a fixed choice of η, if we take ν such that $2\pi/\log(C_i^\nu + 1) < \eta$ and $N < C_i^\nu + 1$ then for any prime satisfying $C_i^\nu + 1 \le p \le C_i^{\nu+1} - 1$ we will get $\frac{\eta \log p}{2\pi} + \theta$ zeros $s_{n,p}$ in $\Delta_{\nu,\eta}$ where $|\theta| \le 1$. Note that $s_{n,p} \ne s_{n',p'}$ if p and p' are distinct primes. This depends on the fact that $\log p$ and $\log p'$ are algebraically independent. Hence

$$S(\nu, \eta) > \sum_{C_i^\nu + 1 \le p \le C_i^{\nu+1} - 1} \left(\frac{\eta \log p}{2\pi} + \theta\right) \,.$$

Since $\sum_{p \le x} \log p \sim x$ we get $S(\nu, \eta) \sim \frac{\eta C_i^{\nu+1}}{2\pi}$ as ν tends to infinity.

We now want to check that these zeros don't get cancelled by singularities in $\Delta_{\nu,\eta}$ produced by the ζ-factors. We only have to consider $\zeta(ms - n)$ for which $n/m \le \beta$ since $c_{n,m} = 0$ otherwise. If $\rho \in \Delta_{\nu,\eta}$ and $m\rho - n$ is a zero of $\zeta(s)$, then since zeros of the Riemann zeta function have real part less than 1,

$$\Re(\rho) < 1/m + n/m \le 1/m + \beta \,.$$

But $\Re(\rho) > \beta + 1/(\nu+1)$. Hence $m < \nu + 1$. Since $\Im(\rho) < u + \eta$, this implies that singularities ρ must have their source in some zero $m\rho - n$ of $\zeta(s)$ situated below the line $(u+\eta)(\nu+1)$. The number of such zeros according to a classical result is

$$O((\nu + 1)\log(\nu + 1)) \,.$$

Each zero can appear as a singularity of at most $\beta(\nu+1)^2$ zeta functions $\zeta(ms-n)$ since $0 \le m < (\nu+1)$ which in turn implies that $n \le m\beta \le \beta(\nu+1)$. Hence in the region $\Delta_{\nu,\eta}$ we can get at most $O((\nu+1)^3 \log(\nu+1))$ singularities, which is not enough to kill the $S(\nu,\eta) \sim \frac{\eta C_i^{\nu+1}}{2\pi}$ zeros in this region.

This means that there are an infinity of zeros in the region

$$\Delta_\eta = \lim_{\nu\to\infty} \bigcup_{\nu' \le \nu} \Delta_{\nu',\eta} \ .$$

These zeros have at least one limit point which must lie on $\Re(s) = \beta$. This limit point is a singularity of $Z(s)$. The analysis above was valid for any choice of $u > 0$ and any positive η. However the result is still true for $u < 0$ since $\zeta(s)$ takes conjugate values in conjugate points. Hence $\Re(s) = \beta$ is a natural boundary for $Z(s)$. This completes the proof for polynomials of Type I.

For polynomials of Type II there are only finitely many pairs (n,m) with $c_{n,m} > 0$ and $(n+1)/m > \beta$. Then since zeros of $\zeta(s)$ have real part less than 1, choosing ν large enough that

$$\beta + 1/\nu < \min\{\, (n+1)/m > \beta \text{ and } c_{n,m} > 0 \,\} \,,$$

then there are no singularities from the zeta functions $\zeta(ms - n)^{-c_{n,m}}$ in $\Delta_{\nu,\eta}$. Hence the region $\Delta_\eta = \lim_{\nu\to\infty} \bigcup_{\nu' \le \nu} \Delta_{\nu',\eta}$ contains only finitely many singularities coming from the zeta functions $\zeta(ms - n)^{-c_{n,m}}$.

We show now that this region contains infinitely many zeros $s_{n,p}$ of $Z(s)$. Recall

$$s_{n,p} = \beta - \frac{\log\left(c_i + c_{i,1}p^{-\gamma_i/qj} + \Omega_{i,2}(p^{-1/qj})\right)}{\log p} + \frac{2\pi ni}{\log p} \ .$$

Type II assumes that for some choice of i,

$$\left| 1 + (c_{i,1}/c_i)p^{-\gamma_i/qj} + \Omega_{i,2}(p^{-1/qj})/c_i \right| \to 1$$

from below as $p \to \infty$. Hence for p large enough,

$$\log\left| \left(c_i + c_{i,1}p^{-\gamma_i/qj} + \Omega_{i,2}(p^{-1/qj})\right) \right|$$
$$= \log\left| \left(1 + (c_{i,1}/c_i)p^{-\gamma_i/qj} + \Omega_{i,2}(p^{-1/qj})/c_i\right) \right|$$
$$< 0$$

and

$$\log\left| \left(c_i + c_{i,1}p^{-\gamma_i/qj} + \Omega_{i,2}(p^{-1/qj})\right) \right| > -1 \ .$$

For p big enough we also get $2\pi/\log p < \eta$. Hence for some N for each $p > N$, we get a zero $s_{n,p}$ in the region Δ_η, i.e. infinitely many zeros of $Z(s)$

sit inside Δ_η which can't get cancelled by singularities since there are only finitely many possible in this region. Hence the same argument as for Type I polynomials implies that $\Re(s) = \beta$ is a natural boundary for $Z(s)$.

Note that case 2 includes an interesting subcase when there are only finitely many (n, m) with $c_{n,m} \neq 0$ and $(n + 1)/m > \beta$. In this case we only need a finite number of Riemann zeta functions to continue to $\Re(s) = \beta$. Below we shall consider a strategy for proving the natural boundary in this case (which we call *Type V*) regardless of an assumption on the local zeros.

There is particular case where this happens which is worth recording, namely when β is an integer:

Corollary 5.8. *Let $W(X, Y)$ be a polynomial with an infinite cyclotomic expansion $W(X, Y) = \prod_{(n,m)}(1 - X^n Y^m)^{c_{n,m}}$ and suppose that $\widetilde{W}_1(U^{t_1})$, the piece of the ghost corresponding to the first gradient of the Newton polygon, is cyclotomic. Suppose that $\beta = \max\{n/m : c_{n,m} \neq 0\}$ is an integer. Then there are only finitely many (n, m) with $c_{n,m} \neq 0$ and $(n + 1)/m > \beta$.*

Proof. Since $\widetilde{W}_1(U^{t_1})$ is cyclotomic, there are only finitely many pairs (n, m) with $c_{n,m} \neq 0$ and $n/m = \beta$. We need to prove that if $n/m < \beta$ then $(n + 1)/m \leq \beta$. If not, then

$$n < \beta m < n + 1 \, .$$

But since we are assuming that β is an integer, all the terms are integers, hence we get a contradiction. $\qquad \square$

Corollary 5.9. *Let $W(X, Y)$ be a polynomial with an infinite cyclotomic expansion $W(X, Y) = \prod_{(n,m)}(1 - X^n Y^m)^{c_{n,m}}$ and suppose that $\beta = \max\{n/m : c_{n,m} \neq 0\}$ is an integer. Then if there are zeros of $W(p, Y)$ for p large enough with $|Y| < p^\beta$ then $\Re(s) = \beta$ is a natural boundary.*

Proof. If the ghost is not friendly we are done by case 1 above. If the ghost is friendly then we are in case 2 of the above. $\qquad \square$

Note that the same argument as in Corollary 5.8 implies the following:

Corollary 5.10. *Suppose the inverse $\beta_1 = s_1/t_1$ of the first gradient of the Newton polygon of $W(X, Y)$ is an integer and*

$$t_1 = \min\{t : s/t = \beta \text{ and } X^s Y^t \text{ is a monomial in } W(X, Y)\} \, .$$

Then the abscissa of convergence of $Z(s) = \prod_p W(p, p^{-s})$ is $\alpha = (s_1 + 1)/t_1$, which is the same as that of the ghost $\widetilde{Z}(s) = \prod_p \widetilde{W}(p, p^{-s})$.

We return to the proof of Theorem 5.6. Case 3 assumes that there are infinitely many pairs (n, m) with $c_{n,m} > 0$ and $(n + 1)/m > \beta$. Hence there is the possibility that zeros from the corresponding Riemann zeta functions might create singularities that would cancel the local zeros we are trying to use to get a natural boundary.

We state the following Lemma which shows that at least for large enough m the Riemann zeros on $\Re(s) = \frac{1}{2}$ cannot cancel local zeros.

Lemma 5.11. *Suppose that $W(X,Y)$ has degree d in Y and $\beta = c/e$ with e minimal. Suppose $s_{n,p}$ is a local zero of $W(p, p^{-s})$ with $\Re(s) > \beta$. If $m \geq ed^2$ then the following is not true: $\zeta(ms_{n,p} - n) = 0$ and $\Re(ms_{n,p} - n) = \frac{1}{2}$.*

Proof. We suppose that we have a Riemann zeta function $\zeta(ms - n)$ with $(n + \frac{1}{2})/m > \beta \geq n/m$ and an s such that $ms - n = \rho$ is a zero on the line $\Re(\rho) = \frac{1}{2}$. Hence $\Re(s) = (2n + 1)/2m$.

Now $\beta = c/e$ which we suppose is with e minimal. Suppose d is the degree of the original polynomial $W(X,Y)$ in Y. We show now that provided $m \geq ed^2$, s cannot be a local zero.

Suppose that $p^{-s} = \alpha$ where $W(p, \alpha) = 0$ and $W(p, Y) \in \mathbb{Z}[X]$. Then $\alpha \in K$ a field of degree $\leq d$ over $\mathbb{Q}$ and $\overline{\alpha} \in \overline{K}$. Hence $\alpha\overline{\alpha} \in L$ where $[L : \mathbb{Q}] \leq d^2$. Now $\Re(\alpha\overline{\alpha}) = p^{-2\Re(s)} = p^{-(2n+1)/m}$. If we put $(2n+1)/m = a/b$ with b minimal then this implies that $b \leq d^2$.

Recall though that $(n + \frac{1}{2})/m > \beta \geq n/m$. Hence $1/(2m) > a/2b - c/e \geq 1/2be \geq 1/2ed^2$, i.e. $m < ed^2$. This confirms the Lemma. $\qquad\square$

We resume the proof of Theorem 5.6. In view of Lemma 5.11, choose ν large enough that

$$\beta + 1/\nu < \min\{ (n + \tfrac{1}{2})/m : (n + \tfrac{1}{2})/m > \beta, m < ed^2 \text{ and } c_{n,m} > 0 \} .$$

Then under the Riemann Hypothesis we will get no singularities from the zeta functions $\zeta(ms - n)^{-c_{n,m}}$ which can coincide with a local zero $s_{n,p}$ in Δ_η. Hence the same argument as in (2) will suffice to prove that $\Re(s) = \beta$ is a natural boundary for $Z(s)$. This completes the proof of Theorem 5.6. $\qquad\square$

We summarise some of the current state of the conjecture which we can test for various explicit polynomials.

Corollary 5.12. *(a) If the ghost polynomial is not cyclotomic then $Z(s)$ has a natural boundary at $\Re(s) = \beta$.*
(b) If there are zeros of $W(p, Y)$ for p large enough with $|Y| < p^\beta$ then, under the assumption of the Riemann Hypothesis, $Z(s)$ has a natural boundary at $\Re(s) = \beta$.

The strong assumption of part (b) is not needed for all cases. If there are finitely many pairs (n, m) with $c_{n,m} > 0$ and $(n + 1)/m > \beta$ then we don't need the Riemann Hypothesis. We shall also discuss below subcases of case (3) where we can avoid the Riemann Hypothesis.

5.2.1 Practicalities

It is useful to have an easy way to check which side zeros will be on. We give now a criterion provided the implicit function theorem applies which generalises the example at the start of the chapter. We also show how the implicit function theorem relates to the combinatorial procedure to calculate Puiseux power series based on successive Newton polygons.

Recall that we had the following. Let $j \in I$ be as small as possible with the property that $\frac{n_j}{j} = \beta$ and put

$$U = X^\beta Y, \qquad\qquad V = X^{-1/j} . \qquad (5.10)$$

Then setting $J = \{ (k,i) : i = 0, \ldots, n_k \text{ and } k = 1, \ldots, l \}$,

$$W(X,Y) = F(V,U) = 1 + \sum_{(k,i)\in J} a_{i,k} V^{(n_j k - ij)} U^k$$

$$= 1 + \sum_{(k,i)\in K} b_{i,k} V^i U^k .$$

Note that $n_j k - ij \geq 0$.

Let

$$A(U) = F(0,U) = 1 + \sum_{\frac{n_k}{k}=\beta} a_{n_k,k} U^k .$$

Choose a root ω of $A(U)$ with the property that $|\omega| \leq 1$ which always exists since the constant term is 1 and the coefficients are integers.

Hypothesis 1 Suppose that ω is not a root of $A'(U)$, i.e. it is not a multiple root of $A(U)$.

We can then apply the implicit function theorem to $F(V,U)$ around the zero $(0,\omega)$ so that in some neighbourhood of $(0,\omega)$,

$$U = \omega - \frac{B_\gamma(\omega)}{A'(\omega)} V^\gamma + \Omega(V) ,$$

where

$$B_n(U) = \frac{1}{n!} \frac{\partial}{\partial V^n} F(V,U)\bigg|_{V=0} = \sum_{n_j k - ij = n} a_{i,k} U^k$$

and

$$\gamma = \min\{ n : B_n(\omega) \neq 0 \}$$

and $\Omega(V)$ is a power series in V of degree greater than γ.

This gives us then the first approximation to one branch of the Puiseux power series. This zero will then give zeros in the unit circle for p large enough provided the following hypothesis holds:

Hypothesis 2 Either (1) $|\omega| < 1$ for some choice of a zero ω of $A(U)$ or (2) if $|\omega| = 1$ for all zeros ω of $A(U)$ then for all sufficiently large p, we can choose a zero ω of $A(U)$ such that

$$\log \left| \omega - \frac{B_\gamma(\omega)}{A'(\omega)} p^{-\gamma/j} + \Omega(p^{-1/j}) \right| < 0 . \tag{5.11}$$

If this inequality fails for one zero, it need not fail for all zeros.

Hypothesis 2 is equivalent to the condition

$$\Re \left(-\frac{B_\gamma(\omega)}{\omega A'(\omega)} \right) < 0 , \tag{5.12}$$

provided the LHS of 5.12 is nonzero. To see this put $-\frac{B_\gamma(\omega)}{\omega A'(\omega)} = x + iy$ where $x, y \in \mathbb{R}$. Now

$$|1 + (x + iy)V^\gamma + \Omega(V)|^2 = 1 + 2xV^\gamma + \Omega_1(V) ,$$

where $\Omega_1(V)$ is a power series in V starting with $V^{\gamma+1}$ or some higher term. Assuming $x \neq 0$, it is now clear that (5.11) holds for all sufficiently large p if and only if $x < 0$. If $x = 0$, (5.12) is no use to us. We must instead compute the next term of the power series, and if that doesn't help, the next one after that, and so on.

We therefore have the following

Corollary 5.13. *Suppose that the Riemann Hypothesis is true. Suppose Hypothesis 1 and 2 hold. Then $Z(s)$ has a natural boundary at $\Re(s) = \beta$.*

The assumption of Hypothesis 1 can be interpreted as simplifying the construction of the Puiseux power series by successive use of Newton polygons. For details of this procedure see for example Appendix B of [53]. Note for example that it implies that the value of q, namely the rational power of the variable V that the Puiseux power series are defined in, is 1. Our polynomial is of the form $A(U) + VG(V, U)$. The first Newton polygon therefore has a line corresponding to the polynomial $A(U)$ of which we take a root ω as our first approximation. We then substitute $U = \omega + U_1$ and look at the Newton diagram of this. Hypothesis 1 implies that ω is not a repeated root of $A(U)$. Hence we get $A(U) = U_1 A_1(U_1)$ where $A_1(U_1)$ has non-zero constant term $A'(\omega)$. So at the second Newton polygon we are going to pick the unique piece with negative gradient consisting of the point $A'(\omega)U_1$ and a term $b_0 V^\gamma$ (where V corresponds to the vertical and U to the horizontal in contrast to the choice of [53]). Note that since we are assuming that $U - \omega$ is not a factor of $F(V, U)$, there is a term $b_0 V^\gamma$. The value of b_0 is then $B_\gamma(\omega)$. Hence the next approximation from this second Newton polygon gives $B_\gamma(\omega)V^\gamma + A'(\omega)U_1 = 0$, i.e. $U = \omega - \frac{B_\gamma(\omega)}{A'(\omega)}V^\gamma + U_2$. Hence the denominator of the gradient will not introduce any change to our fractional power of V. This process repeats itself to give the Puiseux expansion provided by the implicit function theorem.

5.2.2 Distinguishing Types I, II and III

It is also useful to be able to effectively distinguish between the three types of polynomial we introduced above. Determining whether a polynomial is of Type I is easy: it suffices to check whether the polynomial does *not* have all its roots on the unit circle. However, it is not so easy to distinguish Types II and III. Corollary 5.8 provides a useful sufficient condition for a polynomial to *not* be of Type III, namely if the natural boundary is an integer. Indeed, this case covers the zeta functions of algebraic groups that we consider in the following chapter. We shall see that there are many local zeta functions of nilpotent groups and Lie rings that do not have integral natural boundaries. It is therefore worthwhile to prove a few propositions which can help us distinguish Types II and III.

Proposition 5.14. *Suppose $A(U)$ is squarefree. Then the cyclotomic expansion of $W(X, Y)$ has only finitely many $(n, m) \in \mathbb{N}^2$ such that $c_{n,m} \neq 0$ and $(n + \gamma/j)/m > \beta$, and infinitely many such that $(n + \gamma/j)/m = \beta$.*

Proof. Set $U = X^\beta Y$, $V = X^{-1/j}$ as per (5.10), and put $F(V, U) = W(X, Y)$. Note that $(n + \gamma/j)/m > \beta$ iff $n_j m - jn < \gamma$.

We may express $F(V, U)$ in the form

$$F(V, U) = A(U) + VG(V, U) + V^\gamma B_\gamma(U) + V^{\gamma+1}H(V, U) \qquad (5.13)$$

for some bivariate polynomials $G(V, U)$ and $H(V, U)$, with $VG(V, U)$ of degree less than $\gamma - 1$ in V and $G(V, \omega) = 0$ for all roots ω of $A(U)$. Since $A(U)$ is squarefree, $G(V, U)$ must be divisible by $A(U)$. Hence

$$F(V, U) = A(U)(1 + V\tilde{G}(V, U)) + V^\gamma B_\gamma(U) + V^{\gamma+1}H(V, U)$$

for some polynomial $\tilde{G}(V, U)$. It can now be seen that $F(V, U)$ has a cyclotomic expansion with only finitely many factors $(1 - U^m V^{n_j m - jn})^{\pm 1}$ for $n_j m - jn < \gamma$.

For a contradiction, suppose the cyclotomic expansion of $F(V, U)$ has only finitely many such factors with $n_j m - jn = \gamma$. We may then write

$$F(V, U) \equiv A(U) \prod_{i=1}^{r} (1 - U^{m_i} V^{n_i})^{\varepsilon_i} \pmod{V^{\gamma+1}}$$

for suitable $n_i \in \mathbb{N}$, $m_i \in \mathbb{N}_{>0}$, $\varepsilon_i \in \mathbb{Z} \setminus \{0\}$. In particular, this implies $F(V, \omega) \equiv 0 \pmod{V^{\gamma+1}}$. But since $B_\gamma(\omega) \neq 0$,

$$F(V, \omega) \equiv V^{\gamma} B_{\gamma}(\omega) \pmod{V^{\gamma+1}}$$

$$\not\equiv 0 \pmod{V^{\gamma+1}}. \tag{5.14}$$

This is the contradiction we sought.   □

Recall from above Corollary 5.8, which asserts that if $\beta \in \mathbb{N}$, $W(X, Y)$ is not of Type III. The following corollary is in a similar vein:

Corollary 5.15. *Suppose $W(X, Y)$ satisfies Hypotheses 1 and 2 and is not of Type I. If $\gamma \geq j$, $W(X, Y)$ is of Type II.*

Proof. Clear.   □

To demonstrate that a polynomial is of Type III, we need to find an infinite number of pairs (n, m) with $c_{n,m} > 0$ and $(n + 1)/m > \beta$. The following proposition and its corollaries provide one way of doing this:

Proposition 5.16. *Let*

$$W(X, Y) = 1 + \sum_{(n,m) \in S} a_{n,m} X^n Y^m,$$

where $S = \{(n_1, m_1), \ldots, (n_r, m_r)\}$ is a finite nonempty subset of $\mathbb{N} \times \mathbb{N}_{>0}$ with $n_i/m_i \geq n_{i+1}/m_{i+1}$ for $1 \leq i \leq r - 1$ and $a_{n,m} \in \mathbb{Z} \setminus \{0\}$. Suppose $W(X, Y)$ is not of Type I. Let $\beta = n_1/m_1$, suppose $n_r/m_r < \beta$ and let $d = \min\{i : n_i/m_i < \beta\}$. We are interested in the solutions $(\lambda_d, \ldots, \lambda_r)$ of the inequality

$$\frac{(\sum_{i=d}^{r} \lambda_i n_i) + 1}{\sum_{i=d}^{r} \lambda_i m_i} > \frac{n_1}{m_1}. \tag{5.15}$$

1. *Equation (5.15) has only finitely many solutions.*
2. *Suppose that $d = 2$, so that $a_{n_1, m_1} = \pm 1$. Suppose also that there exists $\ell \in \{2, \ldots, r\}$ such that the only simultaneous solution of (5.15) and the congruence*

$$\sum_{i=2}^{r} \lambda_i m_i \equiv m_\ell \pmod{m_1} \tag{5.16}$$

is $\lambda_\ell = 1$, $\lambda_i = 0$ for $2 \leq i \leq r$, $i \neq \ell$. Then, for each $N \in \mathbb{N}$, the cyclotomic expansion of $W(X, Y)$ contains the factor

$$(1 + (-\varepsilon_1)^N \varepsilon_\ell X^{n_\ell + N n_1} Y^{m_\ell + N m_1})^{b_\ell}, \tag{5.17}$$

where $a_{n_1, m_1} = \varepsilon_1$ and $a_{n_\ell, m_\ell} = \varepsilon_\ell b_\ell$ where $\varepsilon_\ell = \pm 1$ and $b_\ell \in \mathbb{N}_{>0}$.

Proof. 1. Clear, once one rearranges (5.15) to

$$m_1 > \sum_{i=d}^{r} \lambda_i(n_1 m_i - n_i m_1)$$

and notes that $n_1/m_1 > n_i/m_i$ for $d \le i \le r$.

2. By induction on N. Since n_1/m_1 is maximal, the only way we can 'clear' the term $a_{n_1,m_1} X^{n_1} Y^{m_1}$ is to introduce the factor $(1 + \varepsilon_1 X^{n_1} Y^{m_1})$. Since the congruence (5.16) has a unique solution, the only way to clear the term $\varepsilon_\ell b_\ell X^{n_\ell} Y^{m_\ell}$ is to multiply in the factor $(1 - \varepsilon_\ell X^{n_\ell} Y^{m_\ell})^{b_\ell}$. Now, suppose that in our cyclotomic expansion we have the factors

$$(1 + \varepsilon_1 X^{n_1} Y^{m_1}), \tag{5.18}$$

$$(1 + (-\varepsilon_1)^N \varepsilon_\ell X^{n_\ell + Nn_1} Y^{m_\ell + Nm_1})^{b_\ell}. \tag{5.19}$$

Multiplying the second terms of (5.18) and (5.19) contributes a term

$$(-\varepsilon_1)^N \varepsilon_1 \varepsilon_\ell b_\ell X^{n_\ell + (N+1)n_1} Y^{m_\ell + (N+1)m_1}. \tag{5.20}$$

Now

$$\frac{n_\ell + (N+1)n_1 + 1}{m_\ell + (N+1)m_1} > \beta$$

since $(n_\ell + 1)/m_\ell > \beta$. By our congruence assumption, (5.20) cannot coincide with another term in $W(X, Y)$, nor can it be affected by multiplying in factors for clearing other terms. Hence to clear this term we must introduce a factor

$$(1 + (-\varepsilon_1)^{N+1} \varepsilon_\ell X^{n_\ell + (N+1)n_1} Y^{m_\ell + (N+1)m_1})^{b_\ell}.$$

This establishes the induction.   □

Corollary 5.17. *Suppose additionally that $\varepsilon_1 = 1$ or $\varepsilon_1 = \varepsilon_\ell = -1$. Then $W(X, Y)$ is of Type III.*

Proof. It suffices to show that $(-\varepsilon_1)^N \varepsilon_\ell = -1$ for infinitely many $N \in \mathbb{N}$. If $\varepsilon_1 = 1$ then $(-\varepsilon_1)^N \varepsilon_\ell = (-1)^N \varepsilon_\ell$, and if $\varepsilon_1 = \varepsilon_\ell = -1$, then $(-\varepsilon_1)^N \varepsilon_\ell = -1$.   □

Corollary 5.18. *In (5.17), suppose that $\varepsilon_1 = -1$, $\varepsilon_\ell = 1$, $\gamma < \frac{1}{2}j$ and $(n_\ell + \gamma/j)/m_\ell = \beta$. Then $W(X, Y)$ is of Type III.*

Proof. By Proposition 5.16, the cyclotomic expansion of $W(X, Y)$ contains the factor $(1 - X^{n_1} Y^{m_1})$ and the factors

$$(1 + X^{n_\ell + \mu n_1} Y^{m_\ell + \mu m_1}) \tag{5.21}$$

for all $\mu \in \mathbb{N}$. In particular, these generate terms

$$\left\lfloor \tfrac{1}{2}(\mu - 1) \right\rfloor X^{2n_\ell + \mu n_1} Y^{2m_\ell + \mu m_1} , \tag{5.22}$$

and by our supposition, $(2n_\ell + \mu n_1 + 1)/(2m_\ell + \mu m_1) > \beta$. The only way to cancel such terms without introducing a factor of the form

$$(1 - X^{2n_\ell + \mu n_1} Y^{2m_\ell + \mu m_1}) \tag{5.23}$$

is if the polynomial $W(X, Y)$ contains (finitely many) terms of the form $e_i X^{2n_\ell + \mu_i n_1} Y^{2n_\ell + \mu_i m_1}$ for $e_i \in \mathbb{N}_{>0}$, $\mu_i \in \mathbb{N}$. Each such term contributes an infinite number of terms of the form

$$e_i X^{2n_\ell + \mu n_1} Y^{2m_\ell + \mu m_1} \tag{5.24}$$

for $\mu \in \mathbb{N}$. However, the coefficients e_i in (5.24) are all constant, whereas the coefficient in (5.22) grows linearly with μ. Hence we must introduce infinitely many factors of the form (5.23), so $W(X, Y)$ is of Type III.   □

5.3 Avoiding the Riemann Hypothesis

There are two subcases of polynomials of Type III which avoid the Riemann Hypothesis.

Case (a) Suppose that we are in case (3) and $\gamma_i/qj < 1/3$ for some choice of Puiseux branch $\Omega_i(V^{1/q})$ corresponding to the first gradient of the Newton polygon, also satisfying our condition (5.3). We shall see below that this isn't far from being forced on us by our assumption that there are infinitely many pairs (n, m) with $c_{n,m} > 0$ and $(n + 1)/m > \beta$.

We are going to choose a different box to estimate the number of zeros of the local factors corresponding to this branch of the Puiseux power series. Instead of a box of width $[\beta + 1/(\nu + 1), \beta + 1/\nu]$ we are going to take a box with $\Re(s) \in [\beta + \delta, \beta + 2\delta]$.

By estimating the size of

$$\left| c_i + c_{i,1} p^{-\gamma_i/qj} + \Omega_{i,2}(p^{-1/qj}) \right| = 1 - p^{-\gamma_i/qj} |c_{i,1}| \sin \theta + \Omega$$

(which by assumption (5.3) is within the unit circle for p large enough) we see that the real part of a zero $s_{n,p}$ is

$$\Re(s_{n,p}) = \beta + p^{-\nu} c + \Omega ,$$

where $\nu = \gamma_i/qj$, c is constant and Ω is something small compared to $p^{-\nu}$. So we need to estimate how many primes p there are in the interval $[(2\delta)^{-1/\nu}, \delta^{-1/\nu}]$. With the assumption on ν, this interval contains

$[(\frac{1}{2})\delta^{-1/\nu}, \delta^{-1/\nu}]$. So, provided δ is sufficiently small, with the help of the Prime Number Theorem we can deduce that there are approximately

$$C\frac{\delta^{-1/\nu}}{\log \delta^{-1/\nu}}$$

primes in the interval $\left[(\frac{1}{2})\delta^{-1/\nu}, \delta^{-1/\nu}\right]$, for some constant C. So in the box

$$\{\, s \in \mathbb{C} : \Re(s) \in [\beta + \delta, \beta + 2\delta], \Im(s) \in [u, u + \eta] \,\}\,,$$

where u and η are fixed, we get approximately

$$C\frac{\delta^{-1/\nu}}{\log \delta^{-1/\nu}} \eta \log p = C'\frac{\delta^{-1/\nu}}{\log \delta^{-1/\nu}} \log(\delta^{-1})$$

zeros since $p \in \left[(2\delta)^{-1/\nu}, \delta^{-1/\nu}\right]$.

The estimate of the number of Riemann zeros in this box which we derived in Case 1 would give

$$K\delta^{-3}\log(\delta^{-1}).$$

Hence a bound of $\nu < 1/3$ would suffice to show that local zeros dominate Riemann zeros in case (3). We shall call such polynomials of *Type IIIa*.

Case (b) We can do a little better if we assume (i) that of the infinitely many pairs (n, m) with $c_{n,m} > 0$ and $(n+1)/m > \beta$, only finitely many pairs have $(n + \frac{1}{2})/m > \beta$. In this case the Riemann zeros that appear must be increasingly far from $\Re(s) = \frac{1}{2}$ that we can use stronger estimates for the number of zeros off the line to weaken the condition on $\nu = \gamma_i/qj$.

Suppose (ii) $\nu < \frac{1}{2}$ for some choice of Puiseux branch $\Omega_i(V^{1/q})$ corresponding to the first gradient of the Newton polygon, also satisfying our condition (5.3). Type III polynomials satisfying (i) and (ii) will be called polynomials of *Type IIIb*.

We shall show for Type IIIb polynomials that local zeros dominate Riemann zeros without an assumption of the Riemann Hypothesis.

If a local zero $s_{n,p}$ meets a Riemann zero ρ of $\zeta(ms-n)$ where $(n+\frac{1}{2})/m \le \beta$ then $ms_{n,p} - n = \rho$, so

$$\Re(\rho) \approx m(\beta + p^{-\nu}) - n$$
$$= (m\beta - n) + mp^{-\nu}$$
$$\ge \tfrac{1}{2} + mp^{-\nu}$$
$$\Im(\rho) \approx mu$$

So the zero will have to be further and further from $\frac{1}{2}$ as m gets bigger. Note that since $\Re(\rho) < 1$ we get that m is bounded by $m \le 1/(2\delta)$ since $p^{-\nu} \ge \delta$. We are looking at getting zeros in the box

$$u \leq \Im(s) \leq u + \eta, \quad \beta + \delta \leq \Re(s) \leq \beta + 2\delta .$$

Define

$$N(\sigma, T) := \text{card}\{ \rho : \zeta(\rho) = 0, |\Im(\rho)| \leq T, \Re(\rho) \geq \sigma \} ,$$

then

$$N(\sigma, T) \ll T^{1-(\sigma-\frac{1}{2})/4} \log T .$$

We use this then to estimate how many zeros there are for values $m \in [M, 2M]$ with $\Im(\rho) \leq uM$ and $\Re(\rho) \geq \frac{1}{2} + M\delta$:

$$(uM)^{1-M\delta/4} \log uM = cM^{1-M\delta/4} \log M .$$

Summing over values $m \in [M, 2M]$ we get approximately

$$\sum_{M=2^k \leq 1/(2\delta)} M^{1-M\delta/4} \log M \ll \log^2 N \max_{M \leq N/2} M^{1-M/4N}$$

Riemann zeros where $N = 1/\delta$. Now summing over the subdivision of $m \leq 1/(2\delta)$ we get at most

$$\sum_{M \leq 1/(2\delta)} M \log M \text{ zeros for each } M$$

$$\approx 1/\delta^2 \log 1/\delta .$$

We now compare this against our estimate of $(1/\delta)^{1/\nu}$ local zeros. So provided $\nu < \frac{1}{2}$, the local zeros will dominate.

What can we say in general about the value of ν, the degree of the second term in our Puiseux power series expansion?

Note that we are assuming that there are infinitely many pairs (n, m) with $c_{n,m} > 0$ and $(n + 1)/m > \beta$ but that $\widetilde{W}_1(U^{t_1})$ is cyclotomic (hence there are finitely many (n, m) with $c_{n,m} > 0$ and $n/m = \beta$). This means that when we draw the lattice points representing $W(X, Y)$ we have at least one lattice point (n, m) with $(n+1)/m > \beta$. (The infinitely many other cases come when we introduce this into our cyclotomic expression and correct the error term. This produces things of the form $(n_1 + n_2, m_1 + m_2)$ where $n_1/m_1 = \beta$ and $(n_2 + 1)/m_2 > \beta$. Hence $(n_1 + n_2 + 1)/(m_1 + m_2) > \beta$.)

To understand where the ν is coming from we have to understand how to build the second Newton polygon after the first approximation of $U = c_1$. The Newton polygon corresponding to our transformed polynomial in U and V has a lattice point (m', n') (where we change the ordering to make it consistent with [53] for the moment) with the property that with $n' - j < 0$. (Note that the first piece of the Newton polygon is now a horizontal line representing a polynomial in U.) Now substitute $c_1 + U_1$ into this polynomial to get a new polynomial in U_1 and V. We are interested now in the pieces of negative gradient. Note that substituting $c_1 + U_1$ into the polynomial corresponding to the first piece of the lower convex hull produces a polynomial in U_1 with zero constant term since c_1 is a root of this polynomial. Let U_1^d be the first term. (Note that we considered above the situation where the Puiseux power series analysis reduces to the implicit function theorem. There we needed an assumption that c_1 is not a repeated root of $A(U)$ which implies $d = 1$.)

Substituting $c_1 + U_1$ into the terms of the polynomial corresponding to the terms on a horizontal line through a (m', n') with $n' - j < 0$, either produces a term on the V-axis or we get that c_1 is a root of the polynomial corresponding to this line. If we have any choice of c_1 and any choice of n' with $n' - j < 0$ which produces a non-zero term on the V-axis, then the second term of the Puiseux power series is of the form

$$U = c_i + c_{i,1} V^{\gamma_i/q} + \Omega_{i,2} ,$$

where $-\gamma_i/q$ is a gradient of a line in this second Newton polygon in the section starting at some $cV^{n'}$ and ending at U_1^d. The gradient can therefore be chosen to have slope at most n'/d. Hence $\nu = \gamma_i/qj \le n'/dj < 1/d$ by our assumption that $n' - j < 0$. So we can see that our assumption on ν which helps us avoid the Riemann Hypothesis is not so far away from what might happen in this setting anyway. The only problem is that every choice of root c_1 of $A(U)$ might make the polynomials $V^{n'} P_{n'}(U)$ vanish where these polynomials are defined by the horizontal lines through each (m', n') with $n' - j < 0$. This doesn't seem so far from being related to the fact that we might actually only have finitely many terms $(1 + X^n Y^m)$ with $(n + 1)/m > \beta$, contradicting our assumption for polynomials of Type III. For example if $V^{n'} P_{n'}(U) = V^{n'} A(U) U^r$, then this certainly vanishes on all roots c_1 of $A(U)$. However it also means that we can get rid of all these terms without introducing more error terms in the next approximation of the cyclotomic expansion of $W(U, V)$:

$$W(U, V) = A(U) + V^{n'} A(U) U^r + V B(U, V)$$
$$= A(U)(1 + V^{n'} U^r) + V B(U, V) .$$

So if we could do this for all n' with $n' - j < 0$ we would not get infinitely many terms $(1 + X^n Y^m)$ with $(n + 1)/m > \beta$ in the cyclotomic expansion.

Although inconclusive, the discussion above hints that in any particular case there is some concrete analysis which can be executed with the hope that the Riemann Hypothesis can be avoided.

5.4 All Local Zeros on or to the Left of $\Re(s) = \beta$

We discuss here some tactics that might yield a natural boundary if we don't have local zeros available to us on the right-hand side of $\Re(s) = \beta$. Before we do so, however, it is instructive to consider what polynomials don't have local zeros to the right of $\Re(s) = \beta$.

Lemma 5.19. *Let $W(X, Y)$ be a polynomial for which all the local zeros of $W(p, p^{-s})$ lie in the closed half-plane $\Re(s) \leq \beta$ for sufficiently large p. Then, for all non-repeated roots ω of $A(U)$, $\gcd(\gamma, j) \in \{\frac{1}{2}j, j\}$.*

Proof. We may suppose that our polynomial $W(X, Y)$ is not of Type I, so the polynomial $A(U)$ has all its roots on the unit circle, and is thus cyclotomic. Let $\beta = a/b$ where we may assume $\gcd(a, b) = 1$. Set $U = X^{n_j/j}Y$, $V = X^{-1/j}$, $F(V, U) = W(X, Y)$, $A(U) = F(0, U)$. Let ω be any non-repeated root of $A(U)$. We can write

$$F(V, U) = A(U) + VG(V, U) + V^\gamma B_\gamma(U) + V^{\gamma+1}H(V, U)$$

for some bivariate polynomials $G(V, U)$ and $H(V, U)$, with $VG(V, U)$ of degree less than γ in V and $G(V, \omega) = 0$. Furthermore, $X^n Y^m = U^m V^{n_j m - jn}$, so $(n+1)/m > \beta$ if and only if $j > n_j m - jn$.

It is clear that $A(U) = f(U^b)$ and $UA'(U) = g(U^b)$ for some polynomials $f(U)$ and $g(U)$. Let $d = \gcd(n_j, j)$, so that $n_j = da$, $j = db$. It is also clear that $d \mid \gamma$. There exists a unique λ such that $0 \leq \lambda < b$ and $\lambda a \equiv \gamma/d \pmod{b}$, so $B_\gamma(U) = U^\lambda h(U^b)$ for some polynomial $h(U)$. Hence

$$-\frac{B_\gamma(\omega)}{\omega A'(\omega)} = -\frac{\omega^\lambda h(\omega^b)}{g(\omega^b)} . \tag{5.25}$$

Let ω be a root of $A(U)$, so that ω^b is then a root of $f(U)$. For $0 \leq n < b$, put $\xi_n = e^{2\pi i n/b}\omega$. Then $\xi_n^b = \omega^b$ for $0 \leq n < b$, so ξ_n is a root of $A(U)$ for all such n. Furthermore, $B_\gamma(\xi_n)$ and $\xi_n A'(\xi_n)$ remain constant as n varies.

$\xi_n^\lambda = e^{2\pi i \lambda n/b}$ and as n varies this takes all Nth roots of unity, where $N = b/\gcd(\lambda, b)$. $h(\xi_n^b) \neq 0$ by definition of $B_\gamma(U)$. If $N \geq 3$, it is clear that

$$\Re\left(-\frac{\xi_n^\lambda h(\xi_n^b)}{g(\xi_n^b)}\right) < 0$$

for some ξ_n, and thus we have a root of $W(X, Y)$ to the right of $\Re(s) = \beta$. Thus, if $\Re(-\xi_n^\lambda h(\xi_n^b)/g(\xi_n^b)) \geq 0$ for all n, we must have $N \leq 2$. Furthermore, if $N = 2$, $B_\gamma(\omega)/(\omega A'(\omega))$ must be purely imaginary. Hence $\gcd(\lambda, b) \in \{\frac{1}{2}b, b\}$, and since $\gcd(\lambda, b) = \gcd(\gamma/d, b)$, $\gcd(\gamma, j) = d\gcd(\gamma/d, b) \in \{\frac{1}{2}j, j\}$. $\qquad\square$

5.4.1 Using Riemann Zeros

Case 4: Assume that there are an infinite number of pairs (n, m) with $c_{n,m} \neq 0$ and $(n + \frac{1}{2})/m > \beta$, but that all local zeros (for large enough primes) are on or to the left of $\Re(s) = \beta$. We call these polynomials of *Type IV*. In this case we can estimate a lot of singularities or zeros coming from $\zeta(ms - n)^{-c_{n,m}}$

in the region $\Delta(\nu, \eta)$. However in order to make sure that they don't cancel each other we need to make the extra assumption that there is no rational dependence between the values of the imaginary parts of nontrivial zeros of the Riemann zeta function. We have in the region $\Delta(\nu, \eta)$ for M large enough:

$$Z(s) = \prod_{\substack{(n,m) \in \mathbb{N}^2 \\ m \le M}} \zeta(ms - n)^{-c_{n,m}} W_M(s) .$$

Let $N_0(T)$ denote the number of distinct zeros of the Riemann zeta function on the critical line $\Re(s) = \frac{1}{2}$. Hardy and Littlewood [33] proved the following:

Theorem 5.20. *Let $U = T^a$ where $a > \frac{1}{2}$. Then, there exists $K(a)$ and $T_0(a)$ such that*

$$N_0(T + U) - N_0(T) > K(a)U$$

for all $T > T_0(a)$.

We show that there an infinite number of singularities or zeros of $Z(s)$ inside the rectangle $\widetilde{\Delta}(\nu, \eta u) = \bigcup_{\nu \le \nu'} \Delta_{\nu', \eta u}$.

The translate $\zeta(ms - n)$ of $\zeta(s)$ has its 'critical line' at $\Re(s) = (n + \frac{1}{2})/m$. Hence, for each pair (n, m) with $c_{n,m} \ne 0$ and $(n + \frac{1}{2})/m > \beta$, $\zeta(ms - n)$ has infinitely many zeros to the right of $\Re(s) = \beta$. Now $n/m \le \beta$ so $(n + \frac{1}{2})/m \le \beta + 1/(2m)$. Thus, for $m > \frac{1}{2}\nu$, the zeros of $\zeta(ms - n)$ on its critical line are in the strip $\beta < \Re(s) < \beta + 1/\nu$.

If $mu > \eta^{-3}$, then $m\eta u > (mu)^{2/3}$. If also $mu > T_0(\frac{2}{3})$, Theorem 5.20 then implies that there exist at least $K(\frac{2}{3})(m\eta u)^{2/3}$ zeros of $\zeta(ms - n)$ inside the box $\widetilde{\Delta}(\nu, \eta u)$.

By supposition, there exist infinitely many pairs (n, m) such that $c_{n,m} \ne 0$ and $(n + \frac{1}{2})/m > \beta$. For fixed η, u, ν, it is clear that infinitely many m will satisfy $mu > \max(T_0(\frac{2}{3}), \eta^{-3})$ and $m > \frac{1}{2}\nu$. Hence $\widetilde{\Delta}(\nu, \eta u)$ contains infinitely many singularities or zeros. These singularities or zeros come from zeros of $\zeta(s)$ on the line $\Re(s) = \frac{1}{2}$ between $\Im(s) = mu$ and $\Im(s) = mu + m\eta u$. Our assumption of rational independence of nontrivial zeros of $\zeta(s)$ implies that none of these can coincide with zeros or singularities coming from another pair (n', m'), otherwise there is a point $s \in \widetilde{\Delta}(\nu, \eta u)$ which can be written as $s = m\rho - n = m'\rho' - n'$ where $\rho = \sigma + i\tau$ and $\rho' = \sigma' + i\tau'$ are zeros of the Riemann zeta function, i.e. $\sigma = (m'/m)\sigma'$.

So by using the zeros of the Riemann zeta function we can realise $\Re(s) = \beta$ as a natural boundary under the strong assumption of rational independence of nontrivial zeros $\zeta(s)$ on $\Re(s) = \frac{1}{2}$.

Theorem 5.21. *Suppose that $W(X, Y) \ne 1$ and has no unitary factors. Suppose that there are an infinite number of pairs (n, m) with $c_{n,m} \ne 0$ and $(n + \frac{1}{2})/m > \beta$. Under the assumption that the nontrivial Riemann zeros are* *rationally independent (i.e. if $\rho = \tau + \sigma\mathrm{i}$ and $\rho' = \tau' + \sigma'\mathrm{i}$ are nontrivial zeros of $\zeta(s)$ then $\sigma/\sigma' \notin \mathbb{Q}$) then $\Re(s) = \beta$ is a natural boundary for $Z(s)$.*

If one preferred to make this assumption about rational independence of zeros against an assumption of the Riemann Hypothesis then we can use a similar argument for Type III polynomials satisfying: there are an infinite number of pairs (n, m) with $c_{n,m} \neq 0$ and $(n + \frac{1}{2})/m > \beta$. We call these polynomials *Type III-IV*. Note that we would need to add that local zeros can't kill Riemann zeros sitting on $\Re(s) = \frac{1}{2}$ as we proved above. This is the reason we didn't need to include any assumption on the local zeros in the statement of Theorem 5.21.

This argument using the Riemann zeros to get a natural boundary also appears in the paper [4] of Dahlquist where he generalises Estermann's result from a polynomial to a general analytic function in one variable. Dahlquist however is able to use some combinatorial arguments on the cyclotomic representation of the analytic function (which is in one variable unlike our case) to avoid any strong condition on the Riemann zeros. This combinatorial argument shows that one can always find Riemann zeros not cancelled by other Riemann zeros.

5.4.2 Avoiding Rational Independence of Riemann Zeros

It is possible to perform some analysis to ascertain whether one really needs this condition on rational independence of Riemann zeros. This again should yield in any individual case whether such an assumption is really necessary.

We have our cyclotomic expression $W(X, Y) = \prod_{(n,m)} (1 - X^n Y^m)^{c_{n,m}}$. This gives rise then to consideration of an infinite product of Riemann zeta functions $\prod_{(n,m)} \zeta(ms - n)^{-c_{n,m}}$. Take a pair (n_0, m_0) with $c_{n_0 m_0} \neq 0$ and $(2n_0 + 1)/2m_0 = a/b > \beta$. We want to show that we get some zeros on $\Re(s) = a/b$ with $\Im(s) \in [u, u + \eta]$. Namely we want to know that we can't get cancelling out from zeros of other $\zeta(ms - n)^{-c_{n,m}}$. Note that if such a term cancels zeros on $\Re(s) = a/b$ then (assuming the Riemann Hypothesis for the moment) $2m = kb$ and $2n + 1 = ka$ so this puts some restrictions on the possible pairs we can take. Let $I_{a/b} = \{ (n, m) : 2m = kb, 2n + 1 = ka \}$.

We want to prove that the multiplicity of $\prod_{(n,m)} \zeta(ms - n)^{-c_{n,m}}$ at $s = a/b + \mathrm{i}\tau$ is non-zero for some $\tau \in [u, u + \eta]$. Let $\nu(\zeta : \frac{1}{2} + \mathrm{i}\tau)$ denote the multiplicity of a zero $\frac{1}{2} + \mathrm{i}\tau$ of the Riemann zeta function $\zeta(s)$. We are required to prove therefore that

$$\sum_{(n,m) \in I_{a/b}} \nu(\zeta : \tfrac{1}{2} + \mathrm{i}m\tau) c_{n,m} = 0$$

for all $\tau \in [u, u + \eta]$.

Now the number of zeros with imaginary part between $m(u + \eta)$ and mu we estimated as

$$(um/2\pi)(\log(u + \eta) - \log u) + (\eta m/2\pi)\log((u + \eta)m/2\pi)$$
$$- \eta m/2\pi + O(\log(u + \eta)m)$$
$$= cm\eta \log m + O(m) .$$

So this implies that

$$\sum_{(n,m)\in I_{a/b}} c_{n,m} m \log m = O(m) . \tag{5.26}$$

There is a geometric picture that one can begin to build up to determine the values of $c_{n,m}$ which appear in our cyclotomic expression. Consider the lattice of points (n, m) with $c_{n,m}$ non-zero. Let $l(a/b)$ be the line through 0 with gradient a/b. The best thing would be to show that there exist infinitely many lines $l(a/b)$ with $a/b > \beta$ containing one and only one point $(n + \frac{1}{2}, m)$ with $c_{n,m}$ non-zero. (We know by assumption of case 4 that there are infinitely many lines with at least one such point.)

Starting from the diagram of finite number of lattice points defining the original polynomial $W(X, Y)$ where each lattice point has the corresponding coefficient $a_{n,m}$ attached to it, can we build up a picture of what the lattice points corresponding to the cyclotomic expression looks like? Somehow, once we have cleared all the finite number of terms of the polynomial, we should get some nice recurrence relations generating all the lattice points.

We can forget about anything which sits too far to the right of $l(\beta)$, i.e. with $(n + \frac{1}{2}, m)$ also sitting to the right of $l(\beta)$.

We first write

$$W(X, Y) = \prod_{(n,m)\in J(\beta)} (1 - X^n Y^m)^{c_{n,m}} + \sum_{n/m<\beta} a_{n,m} X^n Y^m$$
$$= \widetilde{W}_1(X, Y) + \sum_{n/m<\beta} a_{n,m} X^n Y^m ,$$

where $J(\beta)$ consists of pairs with $n/m = \beta$. $J(\beta)$ is finite since we are assuming that the polynomial corresponding to the first gradient of the Newton polygon is cyclotomic. So this precisely clears the terms of $W(X, Y)$ which sit on $l(\beta)$ and we can do this in a finite number of steps without introducing extra stuff which needs to be cleared.

Now take a lattice point with $(n + \frac{1}{2}, m)$ to the right of $l(\beta)$. Lets take it so it lies on $l(a/b)$ with a/b minimal. (There may be some other points also on this line but lets just see what happens to this one.) Lets just assume that the coefficient $a_{n,m} = -1$. Then we introduce a term $(1 - X^n Y^m)$ into the cyclotomic expression, but we introduce a load of extra terms that will need to be cleared, namely $(\widetilde{W}_1(X, Y) - 1)X^n Y^m$. Geometrically, this is then a finite set of lattice points (n', m') such that the lattice points $(n' + \frac{1}{2}, m')$ all sit on the line $l(\beta, (n + \frac{1}{2}, m))$ of gradient β passing through the lattice point $(n + \frac{1}{2}, m)$. When we try to clear these points, we will introduce more points on this line coming from multiplication by $(\widetilde{W}_1(X,Y) - 1)$. Note that all of these new points sit on distinct lines $l(a'/b')$ of larger and larger gradient (bounded by β though). If we had another lattice point $(n' + \frac{1}{2}, m')$ on this original line $l(a/b)$ sitting further out, then we should get the same picture of new lattice points sitting on the line $l(\beta, (n' + \frac{1}{2}, m'))$ but now we'll be able to get points on $l(\beta, (n' + \frac{1}{2}, m'))$ such that the line $l(a_1/b_1)$ through this point doesn't pick up a point on the line $l(\beta, (n + \frac{1}{2}, m))$.

With a small example like $1 + Y + XY^2$ one can see why we get lines $l(a/b)$ with only one lattice point in the cyclotomic expression sitting on it. The trouble is when there are lots of terms it's hard to keep track of whether you can't always get several points on lines as a/b gets bigger. My feeling is that one starts with a finite number of lattice points of the form $(n + \frac{1}{2}, m)$ coming from the original polynomial. Then one generates a finite number of lines of gradient β emanating from these points, with an infinite number of lattice points sitting on them regularly distributed up the lines according somehow to what $\widetilde{W}_1(X,Y)$ looks like, and then one has to show that sufficiently high up these lines, you can get $l(a/b)$ which picks up exactly one point. I'm not sure if this is true, but certainly the possible relations between the $c_{n,m}$ that we would get come from the points that we pick up on these lines $l(a/b)$.

I have ignored in this an extra complication which I hope won't be too much of a problem. For example when we have $a_{n,m} \neq -1$, then we are going to get extra error terms from $(1 - X^n Y^m)^{a_{n,m}}$ which could have $(\lambda n + \frac{1}{2}, \lambda m)$ to the right of $l(\beta)$. However this will only happen finitely often since as λ grows this will fall to the left of $l(\beta)$ since $n/m < \beta$.

This geometric picture could be helpful, but things are still quite complicated.

Note that we at least have a condition (5.26) on $c_{n,m}$ which for any particular polynomial one can check is satisfied infinitely often and hence avoid invoking anything as strong as the rational independence.

There was some hope that the assumption about the rational independence of zeros of the Riemann zeta function could be avoided by using Von Mangoldt's estimate for the number of zeros below $\Im(s) = T$ and the Riemann Hypothesis. Von Mangoldt's estimate for the number of zeros below $\Im(s) = T$ is (see [63]):

$$(T/2\pi) \log(T/2\pi) - T/2\pi + O(\log T) .$$

We get singularities in $\widetilde{\Delta}(\nu, \eta) = \bigcup_{\nu' \leq \nu} \Delta_{\nu, \eta}$ for each pair (n, m) with $c_{n,m} > 0$ and $\left(\frac{1}{2} + n\right)/m > \beta + 1/(\nu + 1)$ and zeros possibly cancelling these singularities from each pair (n', m') with $c_{n'm'} < 0$ and $(\frac{1}{2} + 2n')/2m' > \beta + 1/(\nu + 1)$. The imaginary part of these zeros of $\zeta(s)$ then lie between $(u + \eta)m$ and um. We therefore choose

$$m_0 = \max\{ m : c_{n,m} > 0, (\tfrac{1}{2} + n)/m > \beta + 1/(\nu + 1) \} .$$

If there is a zero which would cancel these singularities coming from a $\zeta(m_0 s - n_0)^{-c_{n_0, m_0}}$, the claim is that it must come from a pair $(n', m') \neq$

(n_0, m_0) with $m' < m_0$. Since we are assuming the Riemann Hypothesis, we know the precise location of the real part of the zeros. Hence if $m' = m_0$, then the real part of the singularity is $(\frac{1}{2} + n')/m' = (\frac{1}{2} + n_0)/m_0$ which implies that $n_0 = n'$. Therefore $m' < m_0$.

The number of singularities on $\Re(s) = (\frac{1}{2} + n_0)/m_0$ is then the number of zeros of the Riemann zeta function between $(u + \eta)m_0$ and um_0, which is:

$$((u + \eta)m_0/2\pi) \log((u + \eta)m_0/2\pi) - (u + \eta)m_0/2\pi + O(\log(u + \eta)m_0)$$
$$- ((um_0/2\pi) \log(um_0/2\pi) - um_0/2\pi + O(\log um_0)) \qquad (5.27)$$
$$= (um_0/2\pi)(\log(u + \eta) - \log u) + (\eta m_0/2\pi) \log((u + \eta)m_0/2\pi)$$
$$- \eta m_0/2\pi + O(\log((u + \eta)m_0)) \ .$$

The number of singularities on $(\frac{1}{2} + n')/m' = (\frac{1}{2} + n_0)/m_0$ between $(u + \eta)m'$ and um' is then

$$(um'/2\pi)(\log(u + \eta) - \log u) + (\eta m'/2\pi) \log((u + \eta)m'/2\pi) \qquad (5.28)$$
$$- \eta m'/2\pi + O(\log((u + \eta)m')) \ .$$

There could be at worst singularities for every $m' < m_0$ killing those from m_0. So we need to consider:

$$\sum_{m'<m_0} \frac{um'}{2\pi}(\log(u + \eta) - \log u) + \frac{\eta m'}{2\pi} \log\left(\frac{(u + \eta)m'}{2\pi}\right)$$
$$- \frac{\eta m'}{2\pi} + O(\log((u + \eta)m'))$$
$$= \frac{um_0(m_0 - 1)(\log(u + \eta) - \log u)}{4\pi} + \left(\sum_{m'<m_0} \left(\frac{\eta m'}{2\pi}\right) \log\left(\frac{(u + \eta)m'}{2\pi}\right)\right)$$
$$- \frac{\eta m_0(m_0 - 1)}{4\pi} + O(\log((u + \eta)(m_0 - 1)!)) \ .$$

This looks pretty deadly against (5.27). Even consider the difference for a fixed m'.

We have to prove that the difference of (5.27) and (5.28) is positive:

$$(5.27) - (5.28)$$
$$= (m_0 - m')u/2\pi (\log(u + \eta) - \log u) \qquad (5.29)$$
$$+ (\eta m_0/2\pi) \log((u + \eta)m_0/2\pi) - (\eta m'/2\pi) \log((u + \eta)m'/2\pi) \qquad (5.30)$$
$$- (m_0 - m')\eta/2\pi \qquad (5.31)$$
$$+ O(\log((u + \eta)m_0)) \ . \qquad (5.32)$$

The trouble is that m' could be close to m_0, e.g. $m' = m_0 - 1$. Then $O(\log((u + \eta)m_0))$ is getting bigger whilst

$$(m_0 - m')u/2\pi (\log(u + \eta) - \log u) = u/2\pi (\log(u + \eta) - \log u)$$

is not. Consider the second line of this expression (5.30):

$$(\eta m_0/2\pi) \log((u+\eta)m_0/2\pi) - (\eta(m_0 - 1)/2\pi) \log((u+\eta)(m_0-1)/2\pi)$$
$$= (\eta m_0/2\pi) \log((u+\eta)m_0/2\pi)$$
$$\qquad - (\eta m_0/2\pi - \eta/2\pi) \left(\log((u+\eta)m_0/2\pi) + \log(1 - 1/m_0) \right)$$
$$= \eta/2\pi \log((u+\eta)(m_0-1)/2\pi) - \eta m_0/2\pi \log(1 - 1/m_0) \ .$$

Again this is deadly since $\eta/2\pi \log((u+\eta)(m_0 - 1)/2\pi)$ compared to $O(\log(u+\eta)m_0)$ with η small doesn't look good.

Despite all this analysis we have not been able to construct a polynomial of Type IV. For a polynomial to be of this type, Proposition 5.14 and Lemma 5.19 together imply that the polynomial $A(U)$ cannot be squarefree. Non-squarefree polynomials have repeated roots and these provide another difficulty to overcome. Frequently there are multiple Puiseux branches at these repeated roots, and we must somehow force the zeros on *all* the branches to lie outside the unit circle.

We return now to another possible strategy for polynomials that only involve a finite number of Riemann zeta functions to continue to the candidate natural boundary and local zeros to the left of this boundary.

5.4.3 Continuation with Finitely Many Riemann Zeta Functions

Case 5: where $\widetilde{W}_1(U^{t_1})$ is cyclotomic but the zeros of $W(p, p^{-s})$ all lie on or to the left of $\Re(s) = \beta$ for p large enough but there exist only finitely many (n, m) with $c_{n,m} \neq 0$ and $(n+1)/m > \beta$. We call these polynomials of *Type V*. It is clear that in this case we have $\gamma \geq j$, so the finitely many (n, m) with $c_{n,m} \neq 0$ and $(n+1)/m > \beta$ are those that form the cyclotomic expansion of $\widetilde{W}_1(U^{t_1})$.

The strategy here would be to demonstrate that there is a dense set of points on $\Re(s) = \beta$ for which the function blows up as we approach it along a line of fixed imaginary part.

The meromorphic function on $\Re(s) > \beta$ in this case looks like, for some M

$$Z(s) = \prod_{\substack{(n,m)\in\mathbb{N}^2 \\ m \leq M}} \zeta(ms - n)^{-c_{n,m}} \prod_p \left(1 + \frac{\sum_{m>M} e_{n,m} p^{n-ms}}{\prod_{m \leq M}(1 - p^{n-ms})^{c_{n,m}}} \right)$$

$$= \prod_{\substack{(n,m)\in\mathbb{N}^2 \\ m \leq M}} \zeta(ms - n)^{-c_{n,m}} W_M(s)$$

$$= W(q, q^{-s}) \prod_{\substack{(n,m)\in\mathbb{N}^2 \\ m \leq M}} \zeta_q(ms - n)^{c_{n,m}} \prod_{\substack{(n,m)\in\mathbb{N}^2 \\ m \leq M}} \zeta(ms - n)^{-c_{n,m}} W_{M,q}(s) \ ,$$

where

$$W_{M,q}(s) = \prod_{p \neq q} \left(1 + \frac{\sum_{m>M} e_{n,m} p^{n-ms}}{\prod_{m \leq M}(1 - p^{n-ms})^{c_{n,m}}} \right).$$

The task is to show that we get $Z(s)$ blowing up for $s = \beta + \varepsilon + ri\pi/\log q$ for fixed prime q and integer r and $\varepsilon \to 0$. Let $I = \{(n,m) : n/m = \beta, c_{n,m} \neq 0\}$. Since $\widetilde{W}_1(U^{t_1})$ is cyclotomic, this set is finite and non-empty. If one recalls the way $c_{n,m}$ are defined, if there is a $c_{n,m} < 0$ then this has come from a term $(1 + X^n Y^m) = (1 - X^n Y^m)^{-1}(1 - X^{2n} Y^{2m})$. In fact it is simpler to rewrite $\prod_{(n,m) \in I}(1 - p^{n-ms})^{c_{n,m}} = \prod_{(n,m) \in I'}(1 + (-1)^{\delta_{n,m}} p^{n-ms})^{c'_{n,m}}$ where $c'_{n,m} > 0$ and I' is non-empty and $\delta_{n,m} \in \{0,1\}$. Now certainly

$$\left| 1 + \frac{\sum_{m>M} e_{n,m} q^{n-ms}}{\prod_{(n,m) \in I'}(1 + (-1)^{\delta_{n,m}} p^{n-ms})^{c'_{n,m}} \prod_{m \leq M, n/m \neq \beta}(1 - q^{n-ms})^{c_{n,m}}} \right|$$

$$= \left| W(q, q^{-s}) \prod_{\substack{(n,m) \in \mathbb{N}^2 \\ m \leq M}} \zeta_q(ms - n)^{c_{n,m}} \prod_{(n,m) \in I'}(1 + (-1)^{\delta_{n,m}} p^{n-ms})^{-c'_{n,m}} \right|$$

$$\to \infty$$

for $s = \beta + \varepsilon + ri\pi/\log q$ as $\varepsilon \to 0$ for r odd if there exists $\delta_{n,m} = 0$ or r even if there exists $\delta_{n,m} = 1$. Note that because $\widetilde{W}_1(U^{t_1})$ is not a factor of W, the zeros of $W(q, q^{-s})$ are bounded away from $\Re(s) = \beta$. So the task is to show that

$$\left| \prod_{\substack{(n,m) \in \mathbb{N}^2 \\ m \leq M}} \zeta(ms - n)^{-c_{n,m}} W_{M,q}(s) \right|$$

does not tend to zero along $\Im(s) = 2\pi r/\log q$.

There is a subcase of polynomials of Type V for which it may be possible to see the elusive zeros to the left of $\Re(s) = \beta$.

Case 5(a): where $\widetilde{W}_1(U^{t_1})$ is cyclotomic but the zeros of $W(p, p^{-s})$ all lie to the left of $\Re(s) = \beta$ for p large enough but there exist only finitely many (n,m) with $c_{n,m} \neq 0$ and $(n+1)/m \geq \beta$. In this case it may be possible to show that any meromorphic continuation beyond $\Re(s) = \beta$ would have to pick up the zeros of $W(p, p^{-s})$ on the left of $\Re(s) = \beta$. Call polynomials in this case of *Type Va*.

5.4.4 Infinite Products of Riemann Zeta Functions

The only case which is missing from the analysis above is the following.

Case 6: This is the case where $W(X,Y) = \prod_{(n,m)}(1 - X^nY^m)^{c_{n,m}}$ and

(1) There are finitely many pairs (n, m) with $c_{n,m} \neq 0$ and $n/m = \beta$ (in which case the part of the ghost corresponding to the first gradient is cyclotomic and all zeros of the local factors are clustered round the unit circle).
(2) Only finitely many zeros of $W(p, Y)$ for all p lie within the unit circle (this of course implies condition (1)).
(3) There are infinitely many pairs (n, m) with $c_{n,m} \neq 0$ and $(n + 1)/m > \beta$ (which means we need an infinite number of Riemann zeta functions to meromorphically continue $Z(s)$) but none with $(n + \frac{1}{2})/m > \beta$ (which means we can't get enough zeros of these infinite number of Riemann zeta functions to cause trouble in the region into which we have meromorphically continued).

Polynomials in this case will be called of *Type VI*.
    There are two subcases which are probably relevant to this case:

(a) There are infinitely many pairs (n, m) with $c_{n,m} \neq 0$ and $(n + \frac{1}{2})/m = \beta$.
(b) There are only finitely many pairs (n, m) with $c_{n,m} \neq 0$ and $(n+\frac{1}{2})/m = \beta$.

    An example of case (a) is $W(X,Y) = 1 + Y + XY^2$, an embarrassingly innocuous looking polynomial. We can apply the quadratic formula to see that the zeros are on the candidate natural boundary at $\Re(s) = \beta = \frac{1}{2}$. We take

$$U = X^{1/2}Y, V = X^{-1/2} ,$$
$$F(V,U) = 1 + VU + U^2 .$$

It is then clear that $A(U) = 1 + U^2$, $A'(U) = 2U$ and $B_1(U) = U$. It is elementary to see that the zeros of $F(V, U)$ are on the unit circle. As expected, taking either zero $\omega = \pm i$ of $A(U) = 1 + U^2$ gives us

$$\Re\left(-\frac{B_\gamma(\omega)}{\omega A'(\omega)}\right) = 0 .$$

    We get the following cyclotomic expansion of $W(X,Y)$:

$$1 + Y + XY^2$$
$$= (1 + Y)(1 + XY^2)\prod_{n\geq 1}(1 + (-1)^n X^nY^{2n+1}) \prod_{(n,m)\in I}(1 - X^nY^m)^{c_{n,m}}$$
$$= (1 - Y)^{-1}(1 - Y^2)(1 - XY^2)^{-1}(1 - X^2Y^4)\prod_{(n,m)\in I}(1 - X^nY^m)^{c_{n,m}}$$
$$\times \prod_{m\geq 0}(1 - X^{2m+1}Y^{4m+3})\prod_{m\geq 1}\left(1 - X^{2m}Y^{4m+1}\right)^{-1}\left(1 - X^{4m}Y^{8m+2}\right),$$

where $(n, m) \in I$ if and only if $(n + 1)/m < \beta = \frac{1}{2}$, i.e. I consists of all the terms which will not contribute anything critical.

This requires then an infinite number of Riemann zeta functions to continue to $\Re(s) = \beta = \frac{1}{2}$ since $(n+1)/(2n+1) = \frac{1}{2} + 1/(4n+2)$. For any fixed N we get a meromorphic continuation to $\Re(s) > \frac{1}{2} + 1/(4N+2)$ where the function $Z(s)$ is defined by

$$
Z(s) = \prod_{\substack{(n,m)\in\mathbb{N}^2 \\ m\leq 2N+1}} \zeta(ms-n)^{-c_{n,m}} \prod_p \left(1 + \frac{\sum_{m>2N+1} e_{n,m} p^{n-ms}}{\prod_{m\leq 2N+1}(1-p^{n-ms})^{c_{n,m}}}\right)
$$

$$
= W(q, q^{-s}) \prod_{\substack{(n,m)\in\mathbb{N}^2 \\ m\leq 2N+1}} \zeta_q(ms-n)^{c_{n,m}} \prod_{\substack{(n,m)\in\mathbb{N}^2 \\ m\leq 2N+1}} \zeta(ms-n)^{-c_{n,m}} W_{M,q}(s) ,
$$

where

$$
W_{M,q}(s) = \prod_{p\neq q} \left(1 + \frac{\sum_{m>2N+1} e_{n,m} p^{n-ms}}{\prod_{m\leq 2N+1}(1-p^{n-ms})^{c_{n,m}}}\right).
$$

Hence we could try to play the same trick as in case 4 above by looking to prove that the function blows up as we tend to $\frac{1}{2} + (2r+1)\pi i/\log q$ along the line $\Im(s) = (2r+1)\pi i/\log q$, by exploiting the blowing up of the local factor

$$
\left(1 + \frac{\sum_{m>2N+1} e_{n,m} q^{n-ms}}{\prod_{m\leq 2N+1}(1-q^{n-ms})^{c_{n,m}}}\right) = W(q, q^{-s}) \prod_{\substack{(n,m)\in\mathbb{N}^2 \\ m\leq 2N+1}} \zeta_q(ms-n)^{c_{n,m}}
$$

caused by $(1+q^{2s-1})^{-1}$.

However now we run up against the difficult problem of the behaviour of an infinite product of Riemann zeta functions on $\Re(s) = \frac{1}{2}$. We need to prove that

$$
\prod_{m\geq 0} \zeta\left((4m+3)s - (2m+1)\right)^{-1} \prod_{m\geq 1} \frac{\zeta\left((4m+1)s - 2m\right)}{\zeta\left((8m+2)s - 4m\right)}
$$

does not tend to zero as s tends to $\frac{1}{2} + (2r+1)\pi i/\log q$ along the line $\Im(s) = (2r+1)\pi i/\log q$. Essentially we need to understand the behaviour of

$$
\prod_{m\geq 0} \zeta(\tfrac{1}{2} + (2m+1)\pi i/\log q)
$$

which does not appear to be known.

The other approach is to use the fact that on the candidate natural boundary we have a lot of potential poles coming from all the zeros of the Riemann zeta function. This is a case where there are an infinite number of pairs (n, m) with $c_{n,m} > 0$ and $(n+\frac{1}{2})/m = \beta$, but only finitely many with $(n+\frac{1}{2})/m > \beta$, so we can't quite see the zeros of the Riemann zeta function because they don't lie in the region of continuation, but they surely will cause troubles on $\Re(s) = \beta$.

However this runs into the same problems. We take some $\zeta((4m_0 + 3)s - (2m_0 + 1))^{-1}$ and a zero of the Riemann zeta function ρ. Then we have to show why

$$\prod_{\substack{m \geq 0 \\ m \neq m_0}} \zeta((4m + 3)s - (2m + 1))^{-1} \prod_{m \geq 1} \frac{\zeta((4m + 1)s - 2m)}{\zeta((8m + 2)s - 4m)}$$

does not tend to zero as we approach ρ from the right along a horizontal line. Certainly we'll want to know that we aren't picking up a zero of $\zeta((4m+1)s - 2m)$ for some other m. But there is the same problem as above that although each individual zeta function doesn't tend to zero, the infinite product of Riemann zeta functions might tend to zero.

Nonetheless it is at least worth remarking that the point $\frac{1}{2}$ must lie outside the region of analytic continuation. The factor $\zeta((4m + 3)s - (2m + 1))^{-1}$ has a zero at $s = \frac{1}{2} + (8m + 6)^{-1}$ arising from the singularity of $\zeta(s)$ at $s = 1$, and hence $\frac{1}{2}$ is a limit point of zeros. It does raise the interesting question of what shape region a Dirichlet series can be analytically continued into, if it can't be meromorphically continued to $\mathbb{C}$. Is it always a right half-plane?

Polynomials that fall into case (b) above however will not have the luxury of this second approach. However, we have been unable to come up with a polynomial that will have all zeros of $W(p, Y)$ outside the unit circle but satisfy the conditions for case (b). We run into the same difficulties we encountered with Type IV.

6

Natural Boundaries II: Algebraic Groups

6.1 Introduction

In this chapter, we use the analysis of the previous section to prove that the zeta functions of the classical groups GO_{2l+1}, GSp_{2l} or GO_{2l}^+ of types B_l for $l \geq 2$, C_l for $l \geq 3$ and D_l for $l \geq 4$ have natural boundaries. These results were announced in [18]. We recall the definition of the local factors and the formula in terms of the root system established in [36] and [21].

Let G be one of the classical reductive groups $GL_{l+1}, GO_{2l+1}, GSp_{2l}$ or GO_{2l}^+. For any field K, $G(K)$ will denote the appropriate subgroup of $GL_n(K)$. Hey [35] and Tamagawa [56] proved that when $G = GL_{l+1}$, the zeta function of G is something very classical, namely $Z_G(s) = \zeta(s) \ldots \zeta(s-l)$, and hence has meromorphic continuation to the whole complex plane. Emboldened by the case of GL_{l+1}, the following definition of the zeta function of the classical group G had been proposed:

Definition 6.1. *1. For each prime p, let μ_G denote the Haar measure on $G(\mathbb{Q}_p)$ normalised such that $\mu_G(G(\mathbb{Z}_p)) = 1$. Define the local or p-adic zeta function of G to be*

$$Z_{G,p}(s) = \int_{G_p^+} |\det(g)|_p^s \, \mu_G(g) \,,$$

where $G_p^+ = G(\mathbb{Q}_p) \cap M_n(\mathbb{Z}_p)$, the set of matrices whose entries are all p-adic integers, and $|\cdot|_p$ denotes the p-adic valuation.
2. Define the global zeta function of G to be

$$Z_G(s) = \prod_{p \text{ prime}} Z_{G,p}(s) \,.$$

Given any algebraic group G defined over a number field K and some K-rational representation $\rho : G \to GL_n$ we can define in a similar manner an associated zeta function. In this paper we restrict ourselves to the above case of the classical groups, i.e. $\mathbb{Q}$-split reductive algebraic groups of type A_l, B_l, C_l, and D_l and their natural representations. In [18] we consider the exceptional types and the effect of changing the representation.

We describe now the formula in terms of the root system for the local zeta functions.

Let T denote the diagonal matrices of $G(\mathbb{Q}_p)$, namely a maximal split torus for $G(\mathbb{Q}_p)$. Let $\Pi = \{\alpha_1, \ldots, \alpha_l\}$ be a basis for the root system $\Phi \subset \mathrm{Hom}(T, \mathbb{Q}_p)$ of $G(\mathbb{Q}_p)$ and let ϖ be the dominant weight of the contragredient (irreducible) representation $\rho^* = {}^T\rho^{-1}$ of the natural representation ρ that we are taking for $G(\mathbb{Q}_p)$. Let m denote the order of the centre of the derived group $[G(\mathbb{C}), G(\mathbb{C})]$. Note that in particular m divides n. Let $\alpha_0 = \det^{n/m}\big|_T$. Then there exist integers $c_i > 0$ for $1 \leq i \leq l$ such that

$$\varpi^m = \alpha_0^{-1} \cdot \prod_{i=1}^{l} \alpha_i^{c_i} .$$

The second set of numerical data we need for our formula are the positive integers $b_1, \ldots, b_l$ which express the sum of the positive roots in terms of the primitive roots:

$$\prod_{\alpha \in \Phi^+} \alpha = \prod_{i=1}^{l} \alpha_i^{b_i} .$$

We can now write down our formula for the zeta function. Let W denote the finite Weyl group of Φ and $\lambda(w)$ the length of an element w of the Weyl group in terms of the fundamental reflections in the hyperplanes defined by the primitive roots.

Define two polynomials $P_G(X, Y)$, $Q_G(X, Y) \in \mathbb{Z}[X, Y]$ by

$$P_G(X, Y) = \sum_{w \in W} X^{-\lambda(w)} \prod_{\alpha_j \in w(\Phi^-)} X^{b_j} Y^{c_j} ,$$

$$Q_G(X, Y) = (1 - Y^m) \prod_{j=1}^{l} \left(1 - X^{b_j} Y^{c_j}\right) .$$

Then for each prime p,

$$Z_{G,p}(s) = \frac{P_G(p, p^{-(n/m)s})}{Q_G(p, p^{-(n/m)s})} .$$

It was proved in [36] and [21] that these polynomials satisfy a functional equation

$$\frac{P_G(X^{-1}, Y^{-1})}{Q_G(X^{-1}, Y^{-1})} = (-1)^{l+1} X^{\mathrm{card}(\Phi^+)} Y^m \frac{P_G(X, Y)}{Q_G(X, Y)} .$$

We are interested in the global behaviour of the zeta function defined as an Euler product of all these local factors. The denominator is always well behaved since it is just built out of the Riemann zeta function $\zeta_p(s)$. The interest lies in the numerator.

We record now the results of our analysis of the polynomial $P_G(X, Y)$ for the classical groups and in particular that for large enough l the polynomials for $G = \mathrm{GO}_{2l+1}, \mathrm{GSp}_{2l}$ and GO_{2l}^+ satisfy, after some factorisation, the conditions of Corollary 5.9 in the previous chapter and hence have a natural boundary. We tabulate first the combinatorial data for the four examples (Table 6.1):

Table 6.1. Combinatorial data for algebraic groups

		m	b_i		c_i	
GL_{l+1}	A_l	$l+1$	$i(l-i+1)$		$l-i+1$	
GO_{2l+1}	B_l	1	$i(2l-i)$		1	
GSp_{2l}	C_l	2	$\begin{cases} i(2l-i+1) & \text{if } i < l \\ l(l+1)/2 & \text{if } i = l \end{cases}$		$\begin{cases} 2 & \text{if } i < l \\ 1 & \text{if } i = l \end{cases}$	
GO_{2l}^+	D_l	2	$\begin{cases} i(2l-i-1) & \text{if } i < l-1 \\ l(l-1)/2 & \text{if } i \geq l-1 \end{cases}$		$\begin{cases} 2 & \text{if } i < l-1 \\ 1 & \text{if } i \geq l-1 \end{cases}$	

Let $P_G(s) = \prod P_G(p, p^{-s})$ and α_{P_G} be the abscissa of convergence of $P_G(s)$.

To satisfy the conditions of Corollary 5.10 it suffices to know what the ghosts of $P_G(X, Y)$ look like. The following descriptions were announced in [16] and proved in [18]. For convenience, we set $b_0 = 0$.

Proposition 6.2. *1. The ghost polynomial $\widetilde{P}_G(X, Y)$ associated to $G = \mathrm{GO}_{2l+1}$ is*

$$\prod_{i=0}^{l-1} (1 + X^{b_i} Y) \,.$$

Hence $Z_{\mathrm{GO}_{2l+1}}(s)$ has a friendly ghost.

2. The ghost polynomial $\widetilde{P}_G(X, Y)$ associated to GSp_{2l} is

$$\prod_{i=0}^{l-1} (1 + X^{b_i/2} Y) \prod_{i=0}^{l-2} (1 + X^{b_i/2+1} Y) \,.$$

Hence $Z_{\mathrm{GSp}_{2l}}(s)$ has a friendly ghost.

3. The ghost polynomial $\widetilde{P}_G(X, Y)$ associated to GO_{2l}^+ or D_l and its natural representation is

$$\prod_{i=0}^{l-2}(1 + X^{b_i/2}Y)^2 \ .$$

Hence $Z_{\mathrm{GO}_{2l}^+}(s)$ has a friendly ghost.

Corollary 6.3. *If $G = \mathrm{GO}_{2l+1}$, GSp_{2l} or GO_{2l}^+ then the inverse of the gradients of the Newton polygon of $P_G(X, Y)$ are all integers.*

Proof. This follows since the gradients are the same as the gradients of the Newton polygon of the ghost.   □

Corollary 6.4. *The abscissa of convergence α_{P_G} of $P_G(s) = \prod P_G(p, p^{-s})$ for $G = \mathrm{GO}_{2l+1}$, GSp_{2l} or GO_{2l}^+ is b_l.*

Proof. 1. If $G = \mathrm{GO}_{2l+1}$ then b_{l-1} is the maximal inverse gradient in the Newton polygon. Hence $\alpha_{P_G} = b_{l-1} + 1 = b_l$.
  2. If $G = \mathrm{GSp}_{2l}$ then $b_{l-1}/2$ is the maximal inverse gradient in the Newton polygon. Hence $\alpha_{P_G} = b_{l-1}/2 + 1 = b_l$.
  3. If $G = \mathrm{GO}_{2l}^+$ then $b_{l-2}/2$ is the maximal inverse gradient in the Newton polygon. Hence $\alpha_{P_G} = b_{l-2}/2 + 1 = b_l$.   □

Note that in each case there is a term $X^{b_l-1}Y$ appearing in both $P_G(X, Y)$ and its ghost. In fact there is another way to see why b_l is the abscissa of convergence without passing to the ghost although the analysis below was essential in determining the ghost.

We know that $\alpha_{P_G} = \max\{\frac{1+n_k}{k} : k = 1, \ldots, r\}$. We shall need to analyse the root system and the combinatorial data to ascertain the value of α_{P_G}.

Choose a subset of simple roots $\Pi_0 \subseteq \Pi$. Let Φ_0 be the sub-root system that Π_0 generates. Notice that in the expression for $P_G(X, Y)$ we can realise the monomial term $X^{-\lambda(w)} \prod_{\alpha_j \in \Pi_0} X^{b_j} Y^{c_j}$ where w is a Weyl element such that $\Pi_0 = \{w^{-1}\alpha_j\} \subset \Phi^-$. For each choice of Π_0, such elements w exist since we can take $w = w_0$ to be the unique element of W_0, the Weyl group of Φ_0, that sends all positive roots Φ_0^+ to negative roots Φ_0^-. To calculate the abscissa of convergence α_{P_G} we are going to be interested in choosing a w which is of minimal length since

$$\alpha_{P_G} = \max\left\{ \frac{1 - \lambda(w) + \sum_{\alpha_j \in \Pi_0} b_j}{\sum_{\alpha_j \in \Pi_0} c_j} : \Pi_0 \subseteq \Pi, w \in W \text{ s.t. } w^{-1}\Pi_0 \subset \Phi^- \right\}$$

The following lemma tells us that for any choice of a subset of simple roots Π_0, w_0 is the most efficient way to realise the corresponding monomial term:

Lemma 6.5. *Let Π_0 be a subset of the simple roots Π and let Φ_0 be the sub-root system of Φ generated by Π_0. Then the length of the shortest element $w \in W$ with the property that $w(\Pi_0) \subset \Phi^-$ but $w(\Pi \setminus \Pi_0) \subset \Phi^+$ is $\mathrm{card}(\Phi_0^+) = \lambda(w_0)$.*

Proof. Let w_0 be the element which maps Φ_0^+ to Φ_0^-. Certainly then $w_0(\Pi_0) \subset \Phi^-$. The length of this element is card(Φ_0^+) in terms of the natural generators w_α where $\alpha \in \Pi_0$. We can't get any shorter than this using all generators w_α where $\alpha \in \Pi$ since the length is still the number of positive roots sent to negative roots in Φ which is at least card(Φ_0^+). But notice that we have now shown that it is exactly that number hence $\Pi \setminus \Pi_0$ must be sent to positive roots since $(\Pi \setminus \Pi_0) \cap \Phi_0 = \varnothing$. But now the length of any element sending Π_0 to negative roots must be at least card(Φ_0^+) since the length is the number of positive roots in Φ^+ sent to negative roots and if Π_0 gets sent to negative roots then so does Φ_0^+. This completes the proof of the lemma. $\qquad\square$

Lemma 6.6. $\alpha_{P_G} = b_l$.

Proof. First note that $\alpha_{P_G} \geq b_l$ since we can take $\Pi_0 = \{\alpha_l\}$ and $w_0 = w_l$ the reflection in α_l which is a word of length 1. Next note that card(Φ_0^+) $\geq$ card(Π_0). An analysis of the combinatorial data will confirm that $b_l - 1 = (b_l - 1)/c_l = \max\{b_i - 1/c_i : i = 1, \ldots, l\}$. The easiest way to check this is to note that for example in the case C_l we have $(2l - i + 1)/2 = \sum_{j=l-i}^{l-1} j$. Then we can use the fact that for any positive integers $x_1, \ldots, x_r, y_1, \ldots, y_r$ we have $\frac{x_1 + \ldots + x_r}{y_1 + \ldots + y_r} \leq \max \frac{x_i}{y_i}$ to deduce that for $w \in W$ such that $w^{-1}\Pi_0 \subset \Phi^-$,

$$\frac{1 - \lambda(w) + \sum_{\alpha_j \in \Pi_0} b_j}{\sum_{\alpha_j \in \Pi_0} c_j} \leq \frac{1 + \sum_{\alpha_j \in \Pi_0}(b_j - 1)}{\sum_{\alpha_j \in \Pi_0} c_j}$$

$$\leq \left(\sum_{\alpha_j \in \Pi_0} c_j\right)^{-1} + \max\{b_i - 1/c_i : \alpha_i \in \Pi_0\}.$$

Therefore $\alpha_{P_G} \leq 1 + (b_l - 1) = b_l$. This completes the lemma. $\qquad\square$

We now put $\beta_P = b_l - 1 = \max\{\frac{n_k}{k} : k \in I\}$ where $I = \{k : \frac{1 + n_k}{k} = \alpha_P\}$. The three examples B_l, C_l and D_l are perfect to illustrate the application of Hypotheses 1 and 2 (p. 134) of the previous chapter. For B_l with $l \geq 2$, we will find that the two hypotheses are satisfied and that β_P is a natural boundary. For C_l with $l \geq 3$, we will find that Hypothesis 2 actually fails, but because $P(X, Y)$ has a factor of the form $(1 + X^{\beta_P}Y)$ and hence the first candidate natural boundary can be passed. We then show that if $P(X, Y) = (1 + X^{\beta_P}Y)P_1(X, Y)$ then $P_1(X, Y)$ will give us a natural boundary. For D_l with $l \geq 4$, we will find that Hypothesis 1 fails. Again this is due to a factor of the form $(1 + X^{\beta_P}Y)$. Once this is removed we find that both Hypotheses 1 and 2 are satisfied and β_P is in fact a natural boundary.

6.2 $G = GO_{2l+1}$ of Type B_l

Proposition 6.7. *If $G = GO_{2l+1}$ of type B_l and $l \geq 2$, then $P_G(s)$ has a natural boundary at $\beta_P = b_{l-1} = b_l - 1 = l^2 - 1$.*

Proof. We make the change of variable $U = X^{\beta_P}Y$ and $V = X^{-1}$ so that $P(X,Y) = F(U,V)$. Then $A(U) = F(0,U) = 1 + U$. This follows because for all $\Pi_0 \subseteq \Pi, w \in W$ such that $w^{-1}\Pi_0 \subset \Phi^-$ except for the case $\Pi_0 = \{\alpha_l\}$ and $w_0 = w_l$ we have:

$$\frac{-\lambda(w) + \sum_{\alpha_j \in \Pi_0} b_j}{\sum_{\alpha_j \in \Pi_0} c_j} < \beta_P = \frac{b_l - 1}{c_l}$$

We set $\omega = -1$, the unique root of $A(U)$. Clearly Hypothesis 1 is satisfied since $A(U)$ does not have a multiple root at ω.

To check Hypothesis 2 we need to determine

$$B_1(U) = \frac{\partial}{\partial V}F(V,U)\bigg|_{V=0} = \sum_{\beta_P k - i = 1} a_{i,k}U^k .$$

We claim that for $i = \beta_P k - 1$, $a_{i,k} \neq 0$ if and only if $k = 1$. For $k = 1$ we are required to show there is a monomial of the form $X^{b_l-1}-1Y$. This can be realised by taking $\Pi_0 = \{\alpha_{l-1}\}$ and $w_0 = w_{l-1}$ the reflection defined by the root α_{l-1}. For $k > 1$, for each choice of Π_0 with k elements and a corresponding w such that $w^{-1}\Pi_0 \subset \Phi^-$,

$$-\lambda(w) + \sum_{\alpha_j \in \Pi_0} b_j \leq \sum_{\alpha_j \in \Pi_0} (b_j - 1)$$
$$< (k-1)(b_{l-1} - 1) + b_l - 1 \text{ if } \Pi_0 \neq \{\alpha_{l-1}, \alpha_l\}$$
$$\leq (k-1)(\beta_P - 1) + \beta_P \leq i$$

since the b_i are a strictly increasing sequence. For $\Pi_0 = \{\alpha_{l-1}, \alpha_l\}$ we just have to use the stronger inequality that if $w^{-1}\Pi_0 \subset \Phi^-$ then $\lambda(w) \geq 3$. Hence we have shown that for each $k > 1$, there are no monomials of the form $X^{\beta_P k-1}Y^k$. Hence $B_1(U) = a_{\beta_P-1,1}U$ and

$$-\frac{B_1(\omega)}{\omega A'(\omega)} = -a_{\beta_P-1,1}.$$

Since $a_{\beta_P-1,1} > 0$, Hypothesis 2 is satisfied. Therefore we can apply Theorem 5.13 to deduce that $P_G(s)$ has a natural boundary at $\beta_P = b_{l-1} = b_l - 1 = l^2 - 1$. □

Corollary 6.8. *If* $G = GO_{2l+1}$ *of type* B_l *and* $l \geq 2$ *then* $Z_G(s)$ *has abscissa of convergence at* $\alpha_G = b_l + 1$ *and a natural boundary at* $\beta_P = b_{l-1} = b_l - 1 = l^2 - 1$.

Proof. We just have to add that $Q_G(s)^{-1} = \prod Q_G(p, p^{-s})^{-1}$ is a meromorphic function with abscissa of convergence at $\alpha_G = b_l + 1$. □

6.3 $G = \mathrm{GSp}_{2l}$ of Type C_l or $G = \mathrm{GO}_{2l}^+$ of Type D_l

In these two examples, there is an initial problem with performing the analysis of the previous example because $(1 + X^{\beta_P}Y)$ is a factor of $P_G(X,Y)$. This means that after the substitution $U = X^{\beta_P}Y$ and $V = X^{-1}$, $P(X,Y) = F(U,V) = (1+U)F_1(U,V)$ and hence for all n

$$B_n(-1) = \frac{1}{n!} \frac{\partial}{\partial V^n} F(V,U)\bigg|_{V=0,U=-1} = 0 .$$

Hence Hypothesis 2 is never satisfied. This is what we would expect since if $(1 + X^{\beta_P}Y)$ is a factor then the potential natural boundary it might cause at $s = \beta_P$ can be passed by multiplying by the meromorphic function $\prod(1 + p^{\beta_P}p^{-s})^{-1}$. Note that in the case of $G = \mathrm{GO}_{2l}^+$ of type D_l, even Hypothesis 1 fails since $A(U) = F(0,U) = 1 + 2U + U^2$. In this case once the factor $(1 + X^{\beta_P}Y)$ is removed the remaining term $F_1(U,V)$ still has the property that $U = -1$ is a zero of $F_1(U,0)$. We will find that $s = \beta_P$ will now produce a natural boundary. In the case $G = \mathrm{GSp}_{2l}$ of type C_l we will have to move a little further to the left to find our natural boundary.

The polynomial $P_G(X,Y)$ actually has a number of other natural factors, not only $(1 + X^{\beta_P}Y)$. This fact was announced in [16]. Its proof is technical and has been consigned to Appendix B:

Theorem 6.9. *If* $G = \mathrm{GSp}_{2l}$ *of type* C_l *or* $G = \mathrm{GO}_{2l}^+$ *of type* D_l *then* $P_G(X,Y)$ *has a factor of the form*

$$(1+Y)\prod_{i=1}^{r}(1 + X^{b_i/2}Y) ,$$

where $r = l - 1$ *for* $G = \mathrm{GSp}_{2l}$ *and* $r = l - 2$ *for* $G = \mathrm{GO}_{2l}^+$.

Corollary 6.10. *1. If* $G = \mathrm{GSp}_{2l}$ *then*

$$P_G(X,Y) = (1+Y)\prod_{i=1}^{l-1}(1 + X^{b_i/2}Y)R_G(X,Y) ,$$

where $R_G(X,Y)$ *has ghost polynomial*

$$\widetilde{R_G}(X,Y) = (1 + XY)\prod_{i=1}^{l-2}(1 + X^{b_i/2+1}Y) .$$

2. If $G = \mathrm{GO}_{2l}^+$ *then*

$$P_G(X,Y) = (1+Y)\prod_{i=1}^{l-2}(1 + X^{b_i/2}Y)R_G(X,Y) ,$$

where $R_G(X, Y)$ has ghost polynomial

$$\widetilde{R_G}(X, Y) = (1 + Y) \prod_{i=1}^{l-2} (1 + X^{b_i/2} Y) .$$

In Appendix B we give a description of the polynomials $R_G(X, Y)$ in terms of the root system.

6.3.1 $G = \mathbf{GSp}_{2l}$ of Type C_l

Let us recall the structure of the root system C_l and its corresponding Weyl group. Let $\mathbf{e}_i$ be the standard basis for the l-dimensional vector space $\mathbb{R}^l$, where we assume $l \geq 3$.

$C_l^+ = \{\, 2\mathbf{e}_i, \mathbf{e}_i \pm \mathbf{e}_j : 1 \leq i < j \leq l \,\}$ with simple roots $\alpha_1 = \mathbf{e}_1 - \mathbf{e}_2, \ldots,$ $\alpha_{l-1} = \mathbf{e}_{l-1} - \mathbf{e}_l, \alpha_l = 2\mathbf{e}_l$. $W(C_l)$ is a semi-direct product of the symmetric group on $\mathbf{e}_i$ and the group $(\mathbb{Z}/2\mathbb{Z})^l$ operating by $\mathbf{e}_i \mapsto (\pm 1)_i \mathbf{e}_i$.

We shall write $w = \pi_w \sigma_w$ where π_w is the permutation and σ_w is the sign change.

Let w_l be the element sending α_l to $-\alpha_l$. The element w_l is the sign change $\mathbf{e}_i \mapsto \mathbf{e}_i$ for $i = 1, \ldots, l-1$ and $\mathbf{e}_l \mapsto -\mathbf{e}_l$. Let Φ_{k+1} be the sub-root system generated by $\{\, \alpha_{l-k}, \ldots, \alpha_l \,\}$ and $w_{\Phi_{k+1}}$ be the element sending Φ_{k+1}^+ to Φ_{k+1}^-. For $G = \mathrm{GSp}_{2l}$ we prove in Appendix B that for $k = 1, \ldots, l$,

$$P_G(X, Y) = (1 + X^{b_{k-1}/2} Y) \left(\sum_{w \in W(k)} X^{-\lambda(w)} \prod_{\alpha_j \in w(\Phi^-)} X^{b_j} Y^{c_j} \right)$$

$$= (1 + X^{b_{k-1}/2} Y) P_k(X, Y) ,$$

where

$$W(k) = \left\{ \begin{array}{c} w = \pi_w \sigma_w : w^{-1}(\alpha_{k-1}) \text{ and } (w_{\Phi_{l-k+1}} w w(k))^{-1}(\alpha_{k-1}) \\ \text{have the same sign and } (\sigma_{w^{-1}})_k = 1 \end{array} \right\}$$

$$\cup \left\{ \begin{array}{c} w = \pi_w \sigma_w : w^{-1}(\alpha_{k-1}) \text{ and } (w_{\Phi_{l-k+1}} w w(k))^{-1}(\alpha_{k-1}) \\ \text{have opposite signs and } (\sigma_{w^{-1}})_k = -1 \end{array} \right\}$$

$$= W(k)^+ \cup W(k)^- ,$$

where for each $w \in W$, $w(k)$ denotes the permutation of $\mathbf{e}_{\pi_{w^{-1}}(i)}$ for $i = k, \ldots, l$ which alters the order. Here it suffices to know the following: for $G = \mathrm{GSp}_{2l}$,

$$P_G(X, Y)$$

$$= (1 + X^{b_l-1/2} Y)(1 + X^{b_l-2/2} Y) P(X, Y)$$

$$= (1 + X^{b_l-1/2} Y)(1 + X^{b_l-2/2} Y)$$

$$\times \left(\sum_{w \in W(l) \cap W(l-1)} X^{-\lambda(w)} \prod_{\alpha_j \in w(\Phi^-)} X^{b_j} Y^{c_j} \right) .$$

The ghost of $P(X,Y)$ indicates that the first candidate natural boundary is at $\beta = b_{l-2}/2 + 1$. We see that Hypotheses 1 and 2 apply now to $P(X,Y)$. We make our change of variable $U = X^\beta Y$ and $V = X^{-1}$ so that $P(X,Y) = F(U,V)$. The corresponding polynomial $A(U) = 1 + U$ hence this satisfies Hypothesis 1 for the unique root $\omega = -1$.

To check Hypothesis 2 we need to determine

$$B_1(U) = \left.\frac{\partial}{\partial V}F(V,U)\right|_{V=0} = \sum_{\beta k - i = 1} a_{i,k}U^k .$$

We claim that for $i = \beta k - 1$, $a_{i,k} \neq 0$ if and only if $k = 1$. For $k = 1$, we are required to show that there is a monomial of the form $X^{b_{l-2}/2}Y$ in $P(X,Y)$. Now $X^{b_{l-2}/2}Y = X^{b_l-3}Y$. This can be realised by taking $w = w_l w_{l-1} w_{l-2}$, where w_i is the reflection defined by the root α_i. Now

$$w^{-1} : \mathbf{e}_l \mapsto -\mathbf{e}_{l-2}$$
$$: \mathbf{e}_{l-1} \mapsto \mathbf{e}_l$$
$$: \mathbf{e}_{l-2} \mapsto \mathbf{e}_{l-1}$$

We show that $w \in W(l)$. Now $(w_l w_{l-1} w_{l-2})^{-1}(\alpha_{l-1}) = \mathbf{e}_l + \mathbf{e}_{l-2} \in \Phi^+$ and $(w_l w_l w_{l-1} w_{l-2})^{-1}(\alpha_{l-1}) = \mathbf{e}_l - \mathbf{e}_{l-2} \in \Phi^-$. Since $(\sigma_{w^{-1}})_l = -1$ this implies that $w \in W(l)$.

Next we need that $w \in W(l-1)$. We have that $(w_l w_{l-1} w_{l-2})^{-1}(\alpha_{l-2}) = \alpha_{l-1} \in \Phi^+$. Now $w(l-1)$ is defined as the permutation which swaps $\mathbf{e}_{l-2} = \mathbf{e}_{\pi_{w^{-1}}(l)}$ and $\mathbf{e}_l = \mathbf{e}_{\pi_{w^{-1}}(l-1)}$ and w_{Φ_2} sends $\mathbf{e}_l$ to $-\mathbf{e}_l$ and $\mathbf{e}_{l-1}$ to $-\mathbf{e}_{l-1}$. Hence

$$w(l-1)^{-1}w^{-1}w_{\Phi_2}^{-1}(\alpha_{l-2}) = \mathbf{e}_{l-1} + \mathbf{e}_{l-2} .$$

Since $(\sigma_{w^{-1}})_{l-1} = 1$ this implies that $w \in W(l-1)$.

Finally $\{\alpha_j : \alpha_j \in w_l w_{l-1} w_{l-2}(\Phi^-)\} = \{\alpha_l\}$ and $\lambda(w_l w_{l-1} w_{l-2}) = 3$.

Consider any monomial term $X^r Y^{2j+\varepsilon}$ where $2j + \varepsilon > 1$ and $\varepsilon = 0$ or 1. Then

$$r = -\lambda(w) + \sum_{\alpha_i \in \Pi'} b_i ,$$

where w is an element of $W(l)$ such that w^{-1} sends Π' (a subset of the simple roots of size $j+\varepsilon$) to negative roots. Now $\beta = b_{l-2}/2+1 = b_{l-1}/2-1 = b_l - 2$ and b_i is strictly increasing for $i \leq l - 1$. Suppose first that $j \geq 2$ then since $\lambda(w) \geq j + \varepsilon$,

$$r = -\lambda(w) + \sum_{\alpha_i \in \Pi'} b_i$$
$$\leq (j-1)b_{l-2} + b_{l-1} + \varepsilon b_l - \lambda(w)$$
$$\leq 2j\beta + \varepsilon b_l - (\varepsilon + 1) - (j-1)$$
$$< (2j + \varepsilon)\beta - 1 .$$

Suppose that $j = 1$. Then (except if $\Pi' = \{\alpha_{l-1}\}$ or $\{\alpha_{l-1}, \alpha_l\}$)

$$r = -\lambda(w) + \sum_{\alpha_i \in \Pi'} b_i$$
$$\leq b_{l-2} + \varepsilon b_l - \lambda(w)$$
$$\leq (2j + \varepsilon)\beta - 2 .$$

This finishes the cases except for $\Pi' = \{\alpha_{l-1}\}$ or $\{\alpha_{l-1}, \alpha_l\}$.

If $w \in W(l) \cap W(l-1)$ and $\Pi' = \{\alpha_{l-1}\}$, then we are required to show that

$$1 < 2\beta - r$$
$$= (b_{l-1} - 2) - (b_{l-1} - \lambda(w)) ,$$

i.e. that $\lambda(w) \geq 4$. In this case $(\sigma_{w^{-1}})_l = 1$ and $w^{-1}(\alpha_{l-1}) \in \Phi^-$ hence $w^{-1}w_l^{-1}(\alpha_{l-1}) = w^{-1}(e_{l-1} + e_l) \in \Phi^-$. This in turn implies that $(\sigma_{w^{-1}})_{l-1} = -1$ and $\pi_{w^{-1}}(l-1) < \pi_{w^{-1}}(l)$ So we have already found three positive roots $(e_{l-1} + e_l, e_{l-1} - e_l$ and $2e_{l-1})$ that are sent to negative roots by w^{-1}. We just have to demonstrate a fourth such root to guarantee $\lambda(w) \geq 4$. Now since $(\sigma_{w^{-1}})_{l-1} = -1$ and $w \in W(l-1)$ we get that $w^{-1}(\alpha_{l-2})$ and $(w_{\Phi_2}ww(l-1))^{-1}(\alpha_{l-2}) = w^{-1}(ww(l-1)w^{-1}w_{\Phi_2})(\alpha_{l-2})$ have opposite signs. So we just need to know that $(ww(l-1)w^{-1}w_{\Phi_2})(\alpha_{l-2}) \neq e_{l-1} \pm e_l$ but is a positive root. Now $(ww(l-1)w^{-1}w_{\Phi_2})(e_{l-2}) = e_{l-2}$ whilst $(ww(l-1)w^{-1}w_{\Phi_2})(e_{l-1}) = -(\sigma_w)_{\pi_{w^{-1}}(l)}e_l$ which confirms both these facts. Hence we have a fourth positive root (either α_{l-2} or $(ww(l-1)w^{-1}w_{\Phi_2})(\alpha_{l-2})$) sent to a negative root by w^{-1}. This confirms that $\lambda(w) \geq 4$.

We show that if $w \in W(l) \cap W(l-1)$ then $\Pi' \neq \{\alpha_{l-1}, \alpha_l\}$. Suppose otherwise. In this case $(\sigma_{w^{-1}})_l = -1$ and $w^{-1}(\alpha_{l-1}) \in \Phi^-$ hence (1) $(\sigma_{w^{-1}})_{l-1} = -1$ and (2) $w^{-1}w_l^{-1}(\alpha_{l-1}) = w^{-1}(e_{l-1} + e_l) \in \Phi^+$ since $w \in W(l)$. But

$$w^{-1}(e_{l-1} + e_l) = -e_{\pi_{w^{-1}}(l-1)} - e_{\pi_{w^{-1}}(l)} \in \Phi^- .$$

Hence we have a contradiction.

This completes the analysis and confirms that $B_1(U) = a_{\beta-1,1}U$ where $a_{\beta-1,1} \geq 1$ (in fact it is possible to show that $a_{\beta-1,1} = 1$). Hence

$$-\frac{B_1(-1)}{(-1)A'(-1)} = -a_{\beta-1,1}$$

and so $\Re\left(-\frac{B_\gamma(\omega)}{\omega A'(\omega)}\right) < 0$, confirming Hypothesis 2. Therefore we can apply Theorem 5.13 to deduce that $P_G(s)$ has a natural boundary at $\beta_P = b_{l-2}/2 + 1 = l(l+1)/2 - 2$.

Corollary 6.11. *If $G = \mathrm{GSp}_{2l}$ of type C_l then $Z_G(s)$ has abscissa of convergence at $\alpha_G = b_l + 1$ and a natural boundary at $\beta_P = b_{l-2}/2 + 1 = l(l+1)/2 - 2$.*

Proof. We just have to add that $Q_G(s)^{-1} = \prod Q_G(p, p^{-s})^{-1}$ is a meromorphic function with abscissa of convergence at $\alpha_G = b_l + 1$. □

Note that had we not factored out $(1 + X^{b_{l-2}/2}Y)$ as well to define $P(X, Y)$ we would have got that $B_1(-1) = 0$. In the next example we only have to remove one factor.

6.3.2 $G = \mathrm{GO}_{2l}^+$ of Type D_l

We turn now to proving that Hypotheses 1 and 2 hold for D_l if $l \geq 4$. We recall the structure of the root system in this case. $D_l^+ = \{ \mathbf{e}_i \pm \mathbf{e}_j : 1 \leq i < j \leq l \}$ with simple roots $\alpha_1 = \mathbf{e}_1 - \mathbf{e}_2, \ldots, \alpha_{l-1} = \mathbf{e}_{l-1} - \mathbf{e}_l, \alpha_l = \mathbf{e}_{l-1} + \mathbf{e}_l$. $W(D_l)$ is a semi-direct product of the symmetric group on $\mathbf{e}_i$ and the group $(\mathbb{Z}/2\mathbb{Z})^{l-1}$ operating by $\mathbf{e}_i \mapsto (\pm 1)_i \mathbf{e}_i$ with $\prod_i (\pm 1)_i = 1$. Again we write an element of w as $\pi_w \sigma_w$.

In a similar fashion to the case of GSp_{2l} we prove for GO_{2l}^+ in Appendix B that $k = 1, \ldots, l - 1$

$$P_G(X, Y) = (1 + X^{b_k - 1/2}Y) \left(\sum_{w \in W(k)} X^{-\lambda(w)} \prod_{\alpha_j \in w(\Phi^-)} X^{b_j} Y^{c_j} \right)$$

$$= (1 + X^{b_k - 1/2}Y) P_k(X, Y) \,,$$

where

$$W(k) = \left\{ \begin{array}{c} w = \pi_w \sigma_w : w^{-1}(\alpha_{k-1}) \text{ and } (w_{\Phi_{l-k+1}} ww(k))^{-1}(\alpha_{k-1}) \\ \text{have the same sign and } (\sigma_{w^{-1}})_k = 1 \end{array} \right\}$$

$$\cup \left\{ \begin{array}{c} w = \pi_w \sigma_w : w^{-1}(\alpha_{k-1}) \text{ and } (w_{\Phi_{l-k+1}} ww(k))^{-1}(\alpha_{k-1}) \\ \text{have opposite signs and } (\sigma_{w^{-1}})_k = -1 \end{array} \right\}$$

$$= W(k)^+ \cup W(k)^- \,,$$

where for each $w \in W$, $w(k)$ denotes the permutation of $\mathbf{e}_{\pi_{w^{-1}}(i)}$ for $i = k, \ldots, l$ which alters the order. In this case we only need to know that

$$P_G(X, Y) = (1 + X^{b_{l-2}/2}Y) P(X, Y)$$

$$= (1 + X^{b_{l-2}/2}Y) \left(\sum_{w \in W(l-1)} X^{-\lambda(w)} \prod_{\alpha_j \in w(\Phi^-)} X^{b_j} Y^{c_j} \right) \,.$$

The ghost of $P(X, Y)$ indicates that the first candidate natural boundary is at $\beta = b_{l-2}/2$. We see that Hypotheses 1 and 2 apply now to $P(X, Y)$. We make our change of variable $U = X^\beta Y$ and $V = X^{-1}$ so that $P(X, Y) = F(U, V)$. The corresponding polynomial $A(U) = 1 + U$ hence this satisfies Hypothesis 1 for the unique root $\omega = -1$.

To check Hypothesis 2 we need to determine

$$B_1(U) = \left. \frac{\partial}{\partial V} F(V, U) \right|_{V=0} = \sum_{\beta k - i = 1} b_{i,k} U^k \, .$$

We claim that for $i = \beta k - 1$, $a_{i,k} \neq 0$ if and only if $k = 1$.

For $k = 1$, we are required to show that there is a monomial of the form $X^{b_l - 2/2 - 1} Y$ in $P(X, Y)$. If we rewrite $X^{b_l - 2/2 - 1} Y = X^{b_l - 2} Y = X^{b_l - 1 - 2} Y$ we see that we are looking for an element $w \in W(l-1)$ of length two with either $w^{-1}(\alpha_l)$ or $w^{-1}(\alpha_{l-1}) \in \Phi^-$. If we choose either $w = w_{l-1} w_{l-2}$ or $w = w_l w_{l-2}$ then we can satisfy these criterion.

Lemma 6.12.   1. *If $w = w_{l-1} w_{l-2}$ then $\{ \alpha_i \in \Pi : \alpha_i \in w(\Phi^-) \} = \{ \alpha_{l-1} \}$ and $w \in W(l-1)^+$.*
    2. *If $w = w_l w_{l-2}$ then $\{ \alpha_i \in \Pi : \alpha_i \in w(\Phi^-) \} = \{ \alpha_l \}$ and $w \in W(l-1)^-$.*

Proof.   1.

$$w_{l-2} w_{l-1} : \mathbf{e}_{l-2} - \mathbf{e}_{l-1} \mapsto \mathbf{e}_{l-1} - \mathbf{e}_l$$
$$: \mathbf{e}_{l-1} - \mathbf{e}_l \mapsto \mathbf{e}_l - \mathbf{e}_{l-2}$$
$$: \mathbf{e}_{l-1} + \mathbf{e}_l \mapsto \mathbf{e}_l + \mathbf{e}_{l-2} \, .$$

This is enough to check that $\{ \alpha_i \in \Pi : \alpha_i \in w(\Phi^-) \} = \{ \alpha_{l-1} \}$. The element $w(l-1)$ is the permutation of $\mathbf{e}_l = \mathbf{e}_{\pi_{w^{-1}}(l-1)}$ and $\mathbf{e}_{l-2} = \mathbf{e}_{\pi_{w^{-1}}(l)}$ whilst the element w_{Φ_2} maps $\mathbf{e}_{l-1}$ to $-\mathbf{e}_{l-1}$ and $\mathbf{e}_l$ to $-\mathbf{e}_l$. Hence

$$w(l-1) w^{-1} w_{\Phi_2} : \mathbf{e}_{l-2} - \mathbf{e}_{l-1} \mapsto \mathbf{e}_{l-1} + \mathbf{e}_{l-2} \in \Phi^+ \, .$$

Since $w_{l-2} w_{l-1} (\mathbf{e}_{l-2} - \mathbf{e}_{l-1}) \in \Phi^+$ and $(\sigma_{w^{-1}})_{l-1} = 1$ this confirms that $w \in W(l-1)^+$.
    2.

$$w_{l-2} w_l : \mathbf{e}_{l-2} - \mathbf{e}_{l-1} \mapsto \mathbf{e}_{l-1} + \mathbf{e}_l$$
$$: \mathbf{e}_{l-1} - \mathbf{e}_l \mapsto -\mathbf{e}_l + \mathbf{e}_{l-2}$$
$$: \mathbf{e}_{l-1} + \mathbf{e}_l \mapsto -\mathbf{e}_l - \mathbf{e}_{l-2} \, .$$

From this we can deduce $\{ \alpha_i \in \Pi : \alpha_i \in w(\Phi^-) \} = \{ \alpha_l \}$. The element $w(l-1)$ is again the permutation of $\mathbf{e}_l = \mathbf{e}_{\pi_{w^{-1}}(l-1)}$ and $\mathbf{e}_{l-2} = \mathbf{e}_{\pi_{w^{-1}}(l)}$. Hence

$$w(l-1) w^{-1} w_{\Phi_2} : \mathbf{e}_{l-2} - \mathbf{e}_{l-1} \mapsto \mathbf{e}_{l-1} - \mathbf{e}_{l-2} \in \Phi^- \, .$$

Since $w_{l-2} w_l (\mathbf{e}_{l-2} - \mathbf{e}_{l-1}) \in \Phi^+$ and $(\sigma_{w^{-1}})_{l-1} = -1$ this confirms that $w \in W(l-1)^-$.

So $a_{\beta-1,1} \geq 2$ (and in fact it is possible to show that $a_{\beta-1,1} = 2$). Now we need to show that we don't pick up any other terms.

Suppose we have a monomial term $X^r Y^{2j+\mathbf{e}_{l-1}+\mathbf{e}_l}$ corresponding to a $w \in W(l-1)$ where $\mathbf{e}_{l-1}$ (respectively, $\mathbf{e}_l$) $= 0$ or 1 according to whether α_{l-1} (respectively, α_l) $\in \{ \alpha_i \in \Pi : \alpha_i \in w(\Phi^-) \} = \Pi'$ and Π' is a set of size $j + \mathbf{e}_{l-1} + \mathbf{e}_l$.

Firstly assume $j > 1$. Then using the fact that b_i is a strictly increasing sequence for $i \le l-2$ and $b_l = b_{l-1} = b_{l-2}/2 + 1$ we can deduce

$$r = -\lambda(w) + \sum_{\alpha_i \in \Pi'} b_i$$
$$\le (j-1)b_{l-3} + b_{l-2} + \mathbf{e}_{l-1}b_{l-1} + \mathbf{e}_l b_l - (j + \mathbf{e}_{l-1} + \mathbf{e}_l)$$
$$< (2j + \mathbf{e}_{l-1} + \mathbf{e}_l)\beta - j$$
$$< (2j + \mathbf{e}_{l-1} + \mathbf{e}_l)\beta - 1 .$$

So we are left with the cases that $\Pi' = \{\alpha_{l-2}\}$, $\{\alpha_{l-2}, \alpha_{l-1}\}$, $\{\alpha_{l-2}, \alpha_l\}$ or $\{\alpha_{l-2}, \alpha_{l-1}, \alpha_l\}$. It suffices to show that $\lambda(w) > |\Pi'|$. Recall that $\lambda(w)$ is the number of positive roots sent to negative roots by w. Hence it suffices to show at least one positive root outside of Π' which gets sent to a negative root.

In the case that $\Pi' = \{\alpha_{l-2}\}$ we just have to demonstrate that $\lambda(w) > 1$. Now there is a unique element w of length one with $w^{-1}(\alpha_{l-2}) \in \Phi^-$, namely the reflection $w_{l-2} : \mathbf{e}_{l-2} - \mathbf{e}_{l-1} \mapsto \mathbf{e}_{l-1} - \mathbf{e}_{l-2}$. We need to show that this element is not in $W(l-1)$. Now $(\sigma_{w_{l-2}^{-1}})_{l-1} = 1$. So we just need to demonstrate that $w_{l-2}(l-1)w_{l-2}^{-1}w_{\Phi_2}(\alpha_{l-2}) \in \Phi^+$. The element $w_{l-2}(l-1)$ is again the element swapping $\mathbf{e}_{l-2}$ and $\mathbf{e}_l$. Then $w_{l-2}(l-1)w_{l-2}^{-1}w_{\Phi_2}(\alpha_{l-2}) = \mathbf{e}_{l-1} + \mathbf{e}_l \in \Phi^+$. Hence $w_{l-2} \notin W(l-1)$ and any element in $W(l-1)$ with $\Pi' = \{\alpha_{l-2}\}$ must have length greater than one. Recall that $\lambda(w)$ is the number of positive roots sent to negative roots by w. Hence it suffices to show at least one positive root outside of Π' which gets sent to a negative root. In the case that $\Pi' = \{\alpha_{l-2}, \alpha_{l-1}\}$, $\{\alpha_{l-2}, \alpha_l\}$ or $\{\alpha_{l-2}, \alpha_{l-1}, \alpha_l\}$ then since $\mathbf{e}_{l-2} - \mathbf{e}_{l-1}$ and $\mathbf{e}_{l-1} + \varepsilon\mathbf{e}_l$ are sent to negative roots (where $\varepsilon = \pm 1$ according to the choice of Π') then $\mathbf{e}_{l-2} + \varepsilon\mathbf{e}_l = (\mathbf{e}_{l-2} - \mathbf{e}_{l-1}) + (\mathbf{e}_{l-1} + \varepsilon\mathbf{e}_l)$ is also sent to a negative root. Hence $\lambda(w) > |\Pi'|$. $\qquad\square$

This completes the analysis and confirms that $B_1(U) = a_{\beta-1,1}U$ where $a_{\beta-1,1} \ge 1$ (in fact it is possible to show that $a_{\beta-1,1} = 2$). Hence

$$-\frac{B_1(-1)}{(-1)A'(-1)} = -a_{\beta-1,1}$$

and so $\Re\left(-\frac{B_\gamma(\omega)}{\omega A'(\omega)}\right) < 0$, confirming Hypothesis 2. Therefore we can apply Theorem 5.13 to deduce that $P_G(s)$ has a natural boundary at $\beta_P = b_{l-2}/2 = l(l-1)/2 - 1$.

Corollary 6.13. *If $G = \mathrm{GO}_{2l}^+$ of type D_l then $Z_G(s)$ has abscissa of convergence at $\alpha_G = b_l + 1$ and a natural boundary at $\beta_P = b_{l-2}/2 = l(l-1)/2 - 1$.*

7

Natural Boundaries III: Nilpotent Groups

7.1 Introduction

In the previous chapter, we found that Hypotheses 1 and 2 always held and the natural boundary for the zeta function of each algebraic group was an integer. Hence Corollary 5.9 ensured the existence natural boundary with no need to assume the Riemann Hypothesis. In this chapter, we consider the zeta functions of nilpotent groups and Lie rings listed in Chap. 2. We are not so lucky this time, since the candidate natural boundary is frequently non-integral, and in many cases the existence of the natural boundary requires us to assume the Riemann Hypothesis. We find that Hypotheses 1 and 2 continue to hold for all calculated examples, although there seems to be no good reason why these hypotheses should hold in general.

To simplify matters, we shall only consider those zeta functions for which the p-local factors are given by the same bivariate rational function for all primes p. To minimise repetition, we shall also only consider one of each pair of isospectral Lie rings.

A related topic to natural boundaries is that of the 'ghost' zeta functions. We additionally list whether the ghost of each zeta function is friendly or not.

7.2 Zeta Functions with Meromorphic Continuation

Below, we take the chance to list those zeta functions calculated in Chap. 2 which do not have a natural boundary, i.e. those with meromorphic continuation. Their 'ghosts' are equal to themselves, and hence automatically friendly. They will be of no further interest to us in this chapter.

Theorem 7.1. *For $r \in \mathbb{N}$, $m \in \mathbb{N}_{>0}$, the following zeta functions all have meromorphic continuation to $\mathbb{C}$:*

- *Counting ideals in the following Lie rings:*

$$\mathbb{Z}^r,\ \mathcal{H},\ \mathcal{H}^2,\ U_3(R_2),\ \mathcal{G}_3 \times \mathbb{Z}^r,\ G(m,r),\ \mathfrak{g}_{6,4},\ M_3 \times \mathbb{Z}^r,\ \mathfrak{g}_{5,3} \times \mathbb{Z}^r,$$
$$\mathcal{H} \times \mathfrak{g}_{5,3},\ \mathfrak{g}_{6,12},\ \mathfrak{g}_{6,14(\pm1)},\ \mathfrak{g}_{6,16},\ \mathfrak{g}_{17},\ \mathfrak{g}_{27A},\ \mathfrak{g}_{27B},\ \mathfrak{g}_{147A},\ \mathfrak{g}_{147B},\ \mathfrak{g}_{157},$$
$$\mathfrak{g}_{1357A}.$$

- *Counting all subrings in the Lie rings $\mathbb{Z}^r$, $\mathcal{H}$ and $G(1,r)$.*

7.3 Zeta Functions with Natural Boundaries

In this section we describe the local zeta functions from Chap. 2 for which we are able to prove the existence of the natural boundary, with, if necessary, an assumption of the Riemann Hypothesis.

The polynomials given in Chap. 2 are mostly irreducible. However, there are a small number which can be given more succinctly in reducible form, since an irreducible polynomial factor has more terms than the reducible polynomial given. In Table 7.1, we list all such reducible numerator polynomials, along with the cyclotomic factors that must be divided out. It is essential to ensure that these cyclotomic factors are removed prior to the calculation of the natural boundary.

The calculations required to determine the natural boundary and verify Hypothesis 1 and Hypothesis 2 are fairly similar for each case. We do not

Table 7.1. Factors of numerator polynomials of zeta functions

Ring	Counting	Page	Type	Factor(s)
$\mathcal{H}^2$	all subrings	35	III	$1 - X^2Y$
$G(2,0)$	all subrings	43	II	$1 - X^2Y$
$G(2,3)$	all subrings	43	III	$1 - X^2Y$
$G(2,5)$	all subrings	43	III	$1 - X^3Y$
$G(2,6)$	all subrings	43	III	$1 - X^3Y$
$\mathcal{G}_3 \times \mathbb{Z}^2$	all subrings	44	II	$1 - X^2Y$
M_3	all subrings	46	III	$1 - X^4Y^3$
$M_3 \times \mathbb{Z}$	all subrings	46	II	$1 - X^4Y^2$
$M_3 \times \mathbb{Z}^4$	all subrings	46	II	$1 - X^3Y$
$\mathcal{H} \times M_3$	ideals	47	I	$1 - XY$
$\mathcal{H}^2 \times M_3$	ideals	–	I	$1 - X^3Y^2$
$L_{(3,3)}$	ideals	49	III	$1 + XY,\ 1 + X^2Y^2$
$L_{(3,2)}$	ideals	49	III	$1 - XY$
$F_{3,2} \times \mathbb{Z}$	ideals	52	II	$1 - XY$
M_4	ideals	52	III	$1 - X^3Y^6$
$M_4 \times \mathbb{Z}$	ideals	53	III	$1 - X^4Y^6$
$\mathfrak{g}_{6,8}$	ideals	57	II	$1 + XY$
$\mathfrak{g}_{6,9}$	ideals	58	I	$1 + XY$
$\mathfrak{g}_{257K}$	ideals	66	II	$1 - XY$
$\mathfrak{g}_{1457A}$	ideals	186	III	$1 - X^2Y^3,\ 1 - X^5Y^8$

wish to keep repeating ourselves, so instead we shall tabulate relevant data for each example.

7.3.1 Type I

The numerator polynomials of the local zeta functions of the Lie rings in Tables 7.2 and 7.3, after dividing out cyclotomic factors if necessary, are all of Type I. For each zeta function we give the candidate natural boundary β, the squarefree polynomial $A(U)$ and a root ω of A with $|\omega| < 1$. Each root given is an approximation unless it is clear otherwise. We have chosen real roots whenever possible.

For clarity, the longer polynomials were omitted from Table 7.2. They are as follows:

$$A_{\mathcal{H} \times M_3}(U) = 1 + U + U^2 + U^3 + U^4 + 2U^5 + 2U^6 + 2U^7 + 2U^8,$$
$$A_{\mathfrak{g}_{6,7}}(U) = 1 + U^3 - U^9 - U^{10} + U^{16},$$
$$A_{\mathfrak{g}_{6,9}}(U) = 1 - U + U^2 + U^6 + U^9 + U^{11} + U^{13} - U^{14} + U^{15},$$
$$A_{\mathfrak{g}_{1357G}}(U) = A_{\mathfrak{g}_{6,7}}(U).$$

The ghost of each of these zeta functions is unfriendly by virtue of the fact that the polynomial $A(U)$ is the first factor of the ghost, and Type I implies that $A(U)$ is not cyclotomic.

7.3.2 Type II

The numerator polynomials of the local zeta functions of the Lie rings in Tables 7.4 and 7.5 are all of Type II. For each zeta function we give the

Table 7.2. Natural boundary data for polynomials of Type I, counting ideals

Ring	Page	β	$A(U)$	ω
$\mathcal{H}^3$	35	$13/8$	$1 - 2U^8$	$2^{-1/8}$
$\mathcal{H}^4$	179	$26/11$	$1 - 3U^{11}$	$3^{-1/11}$
$\mathcal{H} \times M_3$	47	1	$A_{\mathcal{H} \times M_3}(U)$	$0.67516 + 0.54041i$
$\mathcal{H}^2 \times M_3$	–	$17/10$	$1 + 2U^{20}$	$2^{-1/20}e^{\pi i/20}$
$\mathcal{G}_3 \times \mathfrak{g}_{5,3}$	182	2	$1 + U^3 - U^{10}$	-0.88712
$\mathfrak{g}_{6,7}$	57	1	$A_{\mathfrak{g}_{6,7}}(U)$	$0.26431 + 0.85097i$
$\mathfrak{g}_{6,9}$	58	1	$A_{\mathfrak{g}_{6,9}}(U)$	0.97827
$\mathfrak{g}_{1357G}$	184	1	$A_{\mathfrak{g}_{1357G}}(U)$	$0.26431 + 0.85097i$

Table 7.3. Natural boundary data for polynomials of Type I, counting all subrings

Ring	Page	β	$A(U)$	ω
$\mathcal{G}_4$	39	3	$1 + U^2 + U^3$	$0.23279 + 0.79255i$
$G(2,4)$	43	3	$1 + U^3 - U^4$	-0.81917

candidate natural boundary $\Re(s) = \beta$, the squarefree polynomial $A(U)$, a root ω of A, the constant γ and polynomial $B_\gamma(U)$ satisfying

$$\Re\left(-\frac{B_\gamma(\omega)}{\omega A'(\omega)}\right) < 0 \,.$$

Within Type II, ω will always be root of unity, so it can be given exactly. We have chosen roots ± 1 wherever possible.

As with Type I, there are some polynomials which are too long to fit in Tables 7.4 and 7.5. We list them below:

$$A_{F_{3,2}\times\mathbb{Z}}(U) = (1 - U^{11})/(1 - U) \,,$$
$$B_{F_{3,2}\times\mathbb{Z},1}(U) = U^4 + U^5 + U^6 - U^9 - U^{10} - U^{11} \,,$$
$$A_{\mathfrak{g}_{6,8}}(U) = (1 + U^7)/(1 + U) \,,$$
$$B_{\mathfrak{g}_{6,8},1}(U) = U^4 - 2U^7 - U^{11} \,,$$
$$A_{G(2,0)}(U) = 1 + U + U^2 + U^3 \,,$$

Table 7.4. Natural boundary data for polynomials of Type II, counting ideals

Ring	p.	β	$A(U)$	ω	γ	$B_\gamma(U)$
$F_{3,2} \times \mathbb{Z}$	52	1	$A_{F_{3,2}\times\mathbb{Z}}(U)$	$e^{8\pi i/11}$	1	$B_{F_{3,2}\times\mathbb{Z},1}(U)$
$\mathfrak{g}_{6,8}$	57	1	$A_{\mathfrak{g}_{6,8}}(U)$	$e^{5\pi i/7}$	1	$B_{\mathfrak{g}_{6,8},1}(U)$
$\mathfrak{g}_{6,13}$	60	1	$1 + U^3$	$e^{\pi i/3}$	3	$-U^{12}$
$\mathfrak{g}_{137C}$	64	9/8	$1 - U^8$	-1	19	U^{19}
$\mathfrak{g}_{257K}$	66	1	$A_{\mathfrak{g}_{257K}}(U)$	$e^{2\pi i/13}$	1	$B_{\mathfrak{g}_{257K},1}(U)$

Table 7.5. Natural boundary data for polynomials of Type II, counting all subrings

Ring	p.	β	$A(U)$	ω	γ	$B_\gamma(U)$
$\mathcal{G}_3$	38	2	$1 + U^2$	i	2	$U^2 - U^3$
$G(2,0)$	43	2	$A_{G(2,0)}(U)$	-1	1	U^3
$G(2,1)$	43	9/4	$1 - U^4$	-1	3	U^3
$G(2,2)$	43	5/2	$1 - U^4$	-1	2	U^3
$G(2,r)^{\text{a}}$	43	$(r+5)/3$	$1 + U^3$	-1	$r - 4$	$-U^4$
$\mathcal{G}_3 \times \mathbb{Z}$	44	5/2	$1 + U^2$	i	2	U^2
$\mathcal{G}_3 \times \mathbb{Z}^2$	44	3	$1 + U^2$	i	2	$U + U^2 + U^3$
$\mathcal{G}_3 \times \mathbb{Z}^{r\,\text{b}}$	44	$(r+4)/2$	$1 + U^2$	i	2	U^2
$M_3 \times \mathbb{Z}$	46	2	$1 - U + U^2$	$e^{\pi i/3}$	1	$U^2 - U^3 - U^5$
$M_3 \times \mathbb{Z}^4$	46	7/2	$1 + U^2 + U^4$	$e^{\pi i/3}$	2	$2U^2 + U^4$
$M_3 \times \mathbb{Z}^{r\,\text{c}}$	46	$(r+3)/2$	$1 + U^2 + U^4$	$e^{\pi i/3}$	2	$U^2 - U^6$
$\mathfrak{g}_{5,3}$	55	2	$1 + U^4$	$e^{\pi i/4}$	4	$B_{\mathfrak{g}_{5,3},4}(U)$

$^{\text{a}}\, r \geq 7$

$^{\text{b}}\, r \geq 3$

$^{\text{c}}\, r \geq 2, r \neq 4$

$$A_{\mathfrak{g}257\mathrm{K}}(U) = (1 - U^{13})/(1 - U) \, ,$$

$$B_{\mathfrak{g}257\mathrm{K},1}(U) = -U^5(1 + U^6)((1 + U)(1 + U^2)(1 + 2U^4) + U^8) \, ,$$

$$B_{\mathfrak{g}5,3,4}(U) = U^2 + U^3 + U^4 + U^5 - U^8 - U^{10} \, .$$

The value of j is the minimal nonzero exponent of U in the polynomial $A(U)$. In all but two cases it can be seen that $\gamma \geq j$, and hence Corollary 5.15 implies that the polynomial is Type II. The two exceptional cases are $G(2,1)$ and $G(2,2)$. We have that, for any $N \in \mathbb{N}$,

$$W^{\leq}_{G(2,1)}(X,Y) = (1 - X^9 Y^4) \prod_{k=0}^{N-1} (1 + X^{6+9k} Y^{3+4k}) + X^{6+9N} Y^{3+4N}$$

$$+ O(9/4) \, ,$$

$$W^{\leq}_{G(2,2)}(X,Y) = (1 - X^{10} Y^4) \prod_{k=0}^{N-1} (1 + X^{7+10k} Y^{3+4k}) + X^{7+10N} Y^{3+4N}$$

$$+ O(5/2) \, ,$$

where $X^n Y^m \in O(\beta)$ if $(n+1)/m \leq \beta$. Since $(12+1)/6 \leq 9/4$ and $(14+1)/6 \leq 5/2$, both expansions only have the one term with $c_{n,m} > 0$ and $(n+1)/m > \beta$. These are Type II by virtue of the fact that, although there exist infinitely many $c_{n,m} \neq 0$ with $(n+1)/m > \beta$, only one such $c_{n,m}$ is positive.

The ghosts of $\zeta^{\triangleleft}_{\mathfrak{g}6,8}(s)$, $\zeta^{\leq}_{G(2,7)}(s)$ and $\zeta^{\leq}_{M_3 \times \mathbb{Z}^r}(s)$ for $r = 2,3$ are unfriendly, with all other zeta functions listed in Tables 7.4 and 7.5 having friendly ghosts.

7.3.3 Type III

The numerator polynomials of the local zeta functions of the Lie rings in Tables 7.6 and 7.7 are all of Type III. For each zeta function we give the candidate natural boundary β, the squarefree polynomial $A(U)$, γ and $B_\gamma(U)$, and a root ω satisfying Hypothesis 2. Again, ω will always be a root of unity, so we give it exactly and choose ± 1 wherever possible.

In all but two cases, Lemma 5.12 applies, allowing us to easily confirm Hypothesis 2. The exceptions are $\zeta^{\leq}_{F_{2,3}}(s)$ and $\zeta^{\triangleleft}_{\mathfrak{g}37\mathrm{D}}(s)$. In both cases we have

$$\Re\left(-\frac{B_\gamma(\omega)}{\omega A'(\omega)}\right) = 0$$

for all roots ω of $A(U)$, so we must compute a further term of the power series expansion of U in terms of V near a root ω. For $\zeta^{\leq}_{F_{2,3}}(s)$, we have

$$U = \mathrm{i} + \tfrac{1}{2}V - \frac{17\mathrm{i}}{8}V^2 + \Omega_1(V)$$

Table 7.6. Natural boundary data for polynomials of Type III, counting ideals

Ring	p.	β	$A(U)$	ω	γ	$B_\gamma(U)$
$\mathcal{G}_4$	39	$9/5$	$1 + U^5$	-1	2	U^3
$\mathcal{G}_5$	39	$13/5$	$1 + U^5$	-1	4	U^3
$F_{2,3}$	41	$7/5$	$1 + U^5$	-1	1	U^3
T_4	45	$8/5$	$1 + U^5$	$e^{\pi i/5}$	4	$U^3 - U^8$
$M_3 \times M_3$	–	$10/9$	$1 + U^9$	$e^{7\pi i/9}$	4	$U^4 - 2U^{13} + U^{22}$
$M_3 \times_{\mathbb{Z}} M_3$	48	$14/17$	$1 + U^{17}$	$e^{\pi i/17}$	2	$-U^5$
$L_{(3,3)}$	49	$9/7$	$1 + U^7$	-1	3	U^5
$L_{(3,2)}$	49	$7/6$	$1 - U^{12}$	-1	2	$U + U^7 - U^{13}$
$\mathcal{H} \times L_{(3,2)}$	–	$11/6$	$1 - U^{12}$	1	4	$-U^{16}$
$L_{(3,2,2)}$	181	$13/7$	$1 - U^{14}$	-1	2	$U^5 - U^{12}$
$F_{3,2}$	51	$8/11$	$1 - U^{11}$	1	6	$-U^9$
M_4	52	$8/13$	$1 - U^{13}$	$e^{2\pi i/13}$	1	U^5
Fil_4	54	$8/13$	$1 - U^{13}$	$e^{2\pi i/13}$	1	U^5
$M_4 \times \mathbb{Z}$	53	$11/13$	$1 - U^{13}$	$e^{2\pi i/13}$	3	U^5
$\mathrm{Fil}_4 \times \mathbb{Z}$	54	$11/13$	$1 - U^{13}$	$e^{2\pi i/13}$	3	U^5
$\mathfrak{g}_{6,6}$	56	$7/6$	$1 - U^{12}$	-1	6	U^3
$\mathcal{H} \times \mathfrak{g}_{6,12}$	59	$5/4$	$1 + U^{16}$	$e^{\pi i/16}$	4	$-U^5$
$\mathfrak{g}_{6,15}$	61	$4/5$	$1 + U^5$	$e^{\pi i/5}$	1	$-U^9$
$\mathfrak{g}_{37D}$	63	$3/2$	$1 + U^6$	i	3	U^3
$\mathfrak{g}_{247B}$	–	$7/6$	$1 - U^{12}$	-1	2	$2U + U^7 - 2U^{13}$
$\mathfrak{g}_{257A}$	65	$4/3$	$1 + U^3$	-1	1	$-U^{10}$
$\mathfrak{g}_{257B}$	66	$13/10$	$1 - U^{10}$	1	3	$-U^{11}$
$\mathfrak{g}_{1357B}$	67	$5/6$	$1 - U^{12}$	1	2	$-U^{11}$
$\mathfrak{g}_{1457A}$	186	$14/15$	$1 - U^{15}$	1	10	$-U^5$
$\mathfrak{g}_{1457B}$	187	$14/15$	$1 - U^{15}$	1	10	$-U^5$

Table 7.7. Natural boundary data for polynomials of Type III, counting all subrings

Ring	p.	β	$A(U)$	ω	γ	$B_\gamma(U)$
$\mathcal{H}^2$	35	$7/3$	$1 - U^3 + U^6$	$e^{5\pi i/9}$	1	$U - 2U^4 + U^7$
$\mathcal{G}_5$	39	$13/3$	$1 + U^3$	$e^{\pi i/3}$	1	U^4
$F_{2,3}$	41	$5/2$	$1 + U^2$	i	1	U^5
$G(2,3)$	43	$11/4$	$1 - U^4$	-1	1	U^3
$G(2,5)$	43	$10/3$	$1 + U^3$	-1	1	U
$G(2,6)$	43	$11/3$	$1 + U^3$	-1	2	U^3
$\mathfrak{g}_{6,4}$	180	$13/5$	$1 - U^5$	$e^{4\pi i/5}$	2	U^4
T_4	180	$20/7$	$1 + U^7$	-1	1	$-2U^6$
M_3	46	$3/2$	$1 + U^2 + U^4$	$e^{\pi i/3}$	1	$U^3 - U^5$
$L_{(3,2)}$	50	$17/7$	$1 - U^7$	$e^{2\pi i/7}$	2	U^3
$F_{3,2}$	51	$15/8$	$1 - U^8$	1	1	$-U^7$
M_4	53	$13/7$	$1 - U^7$	1	1	$-U^6$
$\mathfrak{g}_{6,12}$	183	$7/3$	$1 - U^9$	$e^{4\pi i/3}$	3	$U + U^4 - U^{10}$

and for $\zeta_{\mathfrak{g}37D}^{\triangleleft}(s)$, we have

$$U = i + \tfrac{1}{6}V - \frac{13i}{72}V^2 + \Omega_2(V) ,$$

where $\Omega_1(V)$ and $\Omega_2(V)$ are power series in V^3 and higher. In both cases $|U| < 1$ for sufficiently small V, so Hypothesis 2 is satisfied.

For the majority of zeta functions in Tables 7.6 and 7.7, Corollaries 5.17 and 5.18 can be used to deduce that the numerator polynomials are of Type III. In Tables 7.8 and 7.9, we list the values of ε_1, ε_ℓ, n_ℓ and m_ℓ required by these corollaries. Corollary 5.18 applies if $\varepsilon_1 = -1$ and $\varepsilon_\ell = 1$, and Corollary 5.17 applies otherwise.

Two further cases can be dealt with by multiplying the numerator polynomial by a factor of the form $(1 \pm X^n Y^m)$.

Proposition 7.2. Let $P(X,Y) = W_{\widetilde{\mathcal{H}^2}}^{\leq}(X,Y)/(1 - X^2 Y)$. Then $P(X,Y)$ is irreducible and is of Type III.

Proof. We cannot apply Corollary 5.17 since $d = 3$. However, we can apply Corollary 5.17 to $Q(X,Y) := P(X,Y)(1 + X^7 Y^3)$ with $\varepsilon_1 = 1$, $\varepsilon_\ell = -1$, $n_\ell = 9$,

Table 7.8. Data for Corollaries 5.17 and 5.18, counting ideals

Ring	ε_1	ε_ℓ	n_ℓ	m_ℓ	Ring	ε_1	ε_ℓ	n_ℓ	m_ℓ
$\mathcal{G}_4$	1	1	5	3	$M_4 \times \mathbb{Z}$	-1	1	4	5
$\mathcal{G}_5$	1	1	7	3	$\mathrm{Fil}_4 \times \mathbb{Z}$	-1	1	4	5
$F_{2,3}$	1	1	4	3	$\mathfrak{g}_{6,6}$	-1	-1	11	10
$M_3 \times M_3$	1	1	5	5	$\mathcal{H} \times \mathfrak{g}_{6,12}$	1	-1	6	5
$M_3 \times_{\mathbb{Z}} M_3$	1	-1	4	5	$\mathfrak{g}_{6,15}$	1	-1	7	9
$L_{(3,3)}$	1	1	6	5	$\mathfrak{g}37D$	1	1	4	3
$L_{(3,2)}$	1	1	6	5	$\mathfrak{g}247B$	-1	1	8	7
$\mathcal{H} \times L_{(3,2)}$	-1	-1	29	16	$\mathfrak{g}257A$	-1	-1	13	10
$L_{(3,2,2)}$	-1	-1	22	12	$\mathfrak{g}257B$	-1	-1	14	11
$F_{3,2}$	-1	-1	6	9	$\mathfrak{g}1357B$	-1	-1	9	11
M_4	-1	1	3	5	$\mathfrak{g}1457A$	-1	-1	4	5
Fil_4	-1	1	3	5	$\mathfrak{g}1457B$	-1	-1	4	5

Table 7.9. Data for Corollaries 5.17 and 5.18, counting all subrings

Ring	ε_1	ε_ℓ	n_ℓ	m_ℓ	Ring	ε_1	ε_ℓ	n_ℓ	m_ℓ
$\mathcal{G}_5$	1	1	17	4	T_4	1	-1	17	6
$F_{2,3}$	1	-1	12	5	$L_{(3,2)}$	-1	1	8	7
$G(2,3)$	-1	1	8	3	$F_{3,2}$	-1	-1	13	7
$G(2,5)$	1	1	3	1	M_4	-1	-1	11	6
$G(2,6)$	1	1	3	1	$\mathfrak{g}_{6,12}$	-1	1	9	4
$\mathfrak{g}_{6,4}$	-1	-1	15	6					

$m_\ell = 4$. The cyclotomic expansion $Q(X,Y)$ has the factor $(1 + X^{21}Y^9)$, so it is legitimate to divide by $(1 + X^7Y^3)$. Thus $P(X,Y) = (1 + X^7Y^3)^{-1}Q(X,Y)$ is of Type III. $\square$

Remark 7.3. If one tries to compute the cyclotomic expansions of $P(X,Y)$ and $Q(X,Y)$ directly using the method outlined in Lemma 5.5, the only difference is that $P(X,Y)$ will feature factors $(1 - X^7Y^3)$, $(1 + X^{14}Y^6)$, $(1 + X^{28}Y^{12}), \dots,$ $(1 + X^{7 \times 2^k}Y^{3 \times 2^k}), \dots$. These factors comprise the geometric expansion of $(1 + X^7Y^3)^{-1}$, and so by premultiplying by $(1 + X^7Y^3)^{-1}$, we can avoid its expansion cluttering up the calculation.

Proposition 7.4. *Let* $P(X,Y) = W^{\leq}_{M_3}(X,Y)/(1 - X^4Y^3)$. *Then* $P(X,Y)$ *is of Type III.*

Proof. Corollary 5.17 applies to $P(X,Y)(1 - X^3Y^2)$ with $\varepsilon_1 = \varepsilon_\ell = -1$, $n_\ell = 7$, $m_\ell = 5$. $\square$

The last case is one where we must compute the cyclotomic expansion explicitly.

Proposition 7.5. $W^{\triangleleft}_{T_4}(X,Y)$ *is of Type III.*

Proof. The congruence (5.16) has no unique solutions, so we cannot apply Corollary 5.17. However, it can easily be shown that, for any $N \in \mathbb{N}$,

$$W^{\triangleleft}_{T_4}(X,Y)$$

$$= (1 + X^4Y^3)(1 + X^8Y^5)(1 - X^9Y^6) \prod_{k=0}^{N-1} (1 - (-1)^k X^{12+8k}Y^{8+5k})^2$$

$$- (-1)^N 2 X^{12+8N}Y^{8+5N} + O(8/5) ,$$

where $X^nY^m \in O(8/5)$ if $(n+1)/m \leq 8/5$. $\square$

Remark 7.6. Note that

$$W^{\triangleleft}_{T_4}(X,Y) = 1 + X^4Y^3 + X^8Y^5 - X^9Y^6 - X^{12}Y^8 - X^{17}Y^{11} + O(8/5) .$$
$$(7.1)$$

Consider instead a polynomial of the form

$$W(X,Y) = 1 + X^4Y^3 + X^8Y^5 - X^9Y^6 + X^{12}Y^8 - X^{17}Y^{11} + O(8/5) ,$$

which has been obtained from (7.1) by doing nothing more than changing the sign of the term $-X^{12}Y^8$. This polynomial also fails the congruence condition, and hence Corollary 5.17 cannot be applied to this polynomial for exactly the same reason as with $W^{\triangleleft}_{T_4}(X,Y)$. However, in this case,

$$W(X,Y) = (1 + X^8Y^5)(1 + X^4Y^3 - X^9Y^6) + O(8/5) ,$$

and, provided $1 + X^8Y^5$ is not a factor of $W(X,Y)$, $W(X,Y)$ is of Type II.

Finally we note that $\zeta_L^{\triangleleft}(s)$ has an unfriendly ghost if L is one of $M_3 \times M_3$, $L_{(3,3)}$, $\mathfrak{g}_8$, $F_{3,2}$, $\mathfrak{g}_{6,6}$ or $\mathfrak{g}_{247B}$, and $\zeta_{\overline{L}}^{\leq}(s)$ has an unfriendly ghost if L is M_3, $F_{3,2}$, M_4 or $\mathfrak{g}_{6,12}$. All other zeta functions listed in Tables 7.6 and 7.7 have friendly ghosts.

7.4 Other Types

7.4.1 Types IIIa and IIIb

In Sect. 5.3 we introduced two subcases of Type III where it is possible to remove the dependence on the Riemann Hypothesis. The following zeta functions are of Type IIIa:

- Counting ideals in $F_{2,3}$, $M_3 \times_{\mathbb{Z}} M_3$, $L_{(3,2)}$, $L_{(3,2,2)}$, M_4, Fil$_4$, $M_4 \times \mathbb{Z}$, Fil$_4 \times \mathbb{Z}$, $\mathcal{H} \times \mathfrak{g}_{6,12}$, $\mathfrak{g}_{6,15}$, $\mathfrak{g}_{1357B}$, $\mathfrak{g}_{1357C}$ and $\mathfrak{g}_{257B}$
- Counting all subrings in T_4, $G(2,3)$, $L_{(3,2)}$, $F_{3,2}$, M_4 and $\mathfrak{g}_{5,3} \times \mathbb{Z}$

However, none of the examples calculated in Chap. 2 is of Type IIIb.

7.4.2 Types IV, V and VI

Types I, II and III account for all the calculated examples of zeta functions in Chap. 2. There are no examples of Types IV, V nor VI arising from zeta functions of Lie rings.

The zeta functions in the following lists are of Type III-IV but not Type IIIa nor Type IIIb. Hence their natural boundaries are as prescribed if one assumes rational independence of Riemann zeros instead of the Riemann Hypothesis:

- Counting ideals in $\mathcal{G}_4$, $M_3 \times M_3$, $L_{(3,3)}$, $\mathcal{H} \times L_{(3,2)}$, $L_{(3,2,2)}$ and $\mathfrak{g}_{257A}$
- Counting all subrings in $\mathcal{G}_5$, $G(2,5)$, $\mathfrak{g}_{6,4}$ and $\mathfrak{g}_{6,12}$

A

Large Polynomials

In this appendix, we quarantine off some of the larger polynomials which would otherwise disrupt the flow of the text of Chaps. 2 and 3.

A.1 $\mathcal{H}^4$, Counting Ideals

The following polynomial is $W^{\triangleleft}_{\mathcal{H}^4}(X, Y)$, mentioned on p. 36:

$$
\begin{aligned}
1 &- 6X^8Y^5 + 5X^9Y^5 + 4X^8Y^7 - 8X^9Y^7 + 3X^{10}Y^7 + 4X^{16}Y^8 - 8X^{17}Y^8 \\
&+ 3X^{18}Y^8 - X^8Y^9 + 3X^9Y^9 - 3X^{10}Y^9 - X^{16}Y^{10} + 5X^{17}Y^{10} - 6X^{18}Y^{10} \\
&+ X^{19}Y^{10} - X^{24}Y^{11} + 3X^{25}Y^{11} - 3X^{26}Y^{11} + 8X^{17}Y^{12} - 10X^{18}Y^{12} \\
&+ 3X^{19}Y^{12} + 8X^{25}Y^{13} - 10X^{26}Y^{13} + 3X^{27}Y^{13} - 5X^{17}Y^{14} + 15X^{18}Y^{14} \\
&- 9X^{19}Y^{14} - 19X^{25}Y^{15} + 43X^{26}Y^{15} - 24X^{27}Y^{15} + 2X^{28}Y^{15} - 3X^{18}Y^{16} \\
&+ 5X^{19}Y^{16} - X^{20}Y^{16} - 5X^{33}Y^{16} + 15X^{34}Y^{16} - 9X^{35}Y^{16} + 8X^{25}Y^{17} \\
&- 30X^{26}Y^{17} + 32X^{27}Y^{17} - 7X^{28}Y^{17} - X^{29}Y^{17} + 8X^{33}Y^{18} - 30X^{34}Y^{18} \\
&+ 32X^{35}Y^{18} - 7X^{36}Y^{18} - X^{37}Y^{18} + 3X^{26}Y^{19} - 9X^{27}Y^{19} + 7X^{28}Y^{19} \\
&- 3X^{42}Y^{19} + 5X^{43}Y^{19} - X^{44}Y^{19} - 3X^{33}Y^{20} + 8X^{34}Y^{20} - 17X^{35}Y^{20} \\
&+ 15X^{36}Y^{20} - 3X^{37}Y^{20} - 3X^{27}Y^{21} + X^{28}Y^{21} + X^{29}Y^{21} + 3X^{42}Y^{21} \\
&- 9X^{43}Y^{21} + 7X^{44}Y^{21} + 4X^{34}Y^{22} - 12X^{35}Y^{22} - 2X^{36}Y^{22} + 13X^{37}Y^{22} \\
&- 5X^{38}Y^{22} + 4X^{42}Y^{23} - 12X^{43}Y^{23} - 2X^{44}Y^{23} + 13X^{45}Y^{23} - 5X^{46}Y^{23} \\
&+ 9X^{35}Y^{24} - 10X^{36}Y^{24} - 10X^{37}Y^{24} + 9X^{38}Y^{24} - 3X^{51}Y^{24} + X^{52}Y^{24} \\
&+ X^{53}Y^{24} - 3X^{42}Y^{25} + 18X^{43}Y^{25} - 16X^{44}Y^{25} - 16X^{45}Y^{25} + 18X^{46}Y^{25} \\
&- 3X^{47}Y^{25} + X^{36}Y^{26} + X^{37}Y^{26} - 3X^{38}Y^{26} + 9X^{51}Y^{26} - 10X^{52}Y^{26} \\
&- 10X^{53}Y^{26} + 9X^{54}Y^{26} - 5X^{43}Y^{27} + 13X^{44}Y^{27} - 2X^{45}Y^{27} - 12X^{46}Y^{27} \\
&+ 4X^{47}Y^{27} - 5X^{51}Y^{28} + 13X^{52}Y^{28} - 2X^{53}Y^{28} - 12X^{54}Y^{28} + 4X^{55}Y^{28}
\end{aligned}
$$

$$+ 7X^{45}Y^{29} - 9X^{46}Y^{29} + 3X^{47}Y^{29} + X^{60}Y^{29} + X^{61}Y^{29} - 3X^{62}Y^{29}$$
$$- 3X^{52}Y^{30} + 15X^{53}Y^{30} - 17X^{54}Y^{30} + 8X^{55}Y^{30} - 3X^{56}Y^{30} - X^{45}Y^{31}$$
$$+ 5X^{46}Y^{31} - 3X^{47}Y^{31} + 7X^{61}Y^{31} - 9X^{62}Y^{31} + 3X^{63}Y^{31} - X^{52}Y^{32}$$
$$- 7X^{53}Y^{32} + 32X^{54}Y^{32} - 30X^{55}Y^{32} + 8X^{56}Y^{32} - X^{60}Y^{33} - 7X^{61}Y^{33}$$
$$+ 32X^{62}Y^{33} - 30X^{63}Y^{33} + 8X^{64}Y^{33} - 9X^{54}Y^{34} + 15X^{55}Y^{34} - 5X^{56}Y^{34}$$
$$- X^{69}Y^{34} + 5X^{70}Y^{34} - 3X^{71}Y^{34} + 2X^{61}Y^{35} - 24X^{62}Y^{35} + 43X^{63}Y^{35}$$
$$- 19X^{64}Y^{35} - 9X^{70}Y^{36} + 15X^{71}Y^{36} - 5X^{72}Y^{36} + 3X^{62}Y^{37} - 10X^{63}Y^{37}$$
$$+ 8X^{64}Y^{37} + 3X^{70}Y^{38} - 10X^{71}Y^{38} + 8X^{72}Y^{38} - 3X^{63}Y^{39} + 3X^{64}Y^{39}$$
$$- X^{65}Y^{39} + X^{70}Y^{40} - 6X^{71}Y^{40} + 5X^{72}Y^{40} - X^{73}Y^{40} - 3X^{79}Y^{41}$$
$$+ 3X^{80}Y^{41} - X^{81}Y^{41} + 3X^{71}Y^{42} - 8X^{72}Y^{42} + 4X^{73}Y^{42} + 3X^{79}Y^{43}$$
$$- 8X^{80}Y^{43} + 4X^{81}Y^{43} + 5X^{80}Y^{45} - 6X^{81}Y^{45} + X^{89}Y^{50}.$$

A.2 $\mathfrak{g}_{6,4}$, Counting All Subrings

The following polynomial is $W^{\leq}_{\mathfrak{g}_{6,4}}(X,Y)$, mentioned on p. 44:

$$1 + X^4Y^2 - X^5Y^3 + X^6Y^3 - X^6Y^4 - X^7Y^4 - X^9Y^4 + X^{10}Y^4 - X^9Y^5$$
$$- 2X^{10}Y^5 - 3X^{11}Y^5 - 2X^{12}Y^5 - X^{13}Y^5 + X^{10}Y^6 + X^{11}Y^6 + 2X^{12}Y^6$$
$$+ X^{13}Y^6 + X^{14}Y^6 - X^{15}Y^6 - X^{13}Y^7 - X^{14}Y^7 - 2X^{15}Y^7 - X^{16}Y^7$$
$$- X^{17}Y^7 + X^{14}Y^8 + 2X^{15}Y^8 + 3X^{16}Y^8 + 3X^{17}Y^8 + X^{18}Y^8 - X^{19}Y^8$$
$$+ X^{20}Y^8 - X^{17}Y^9 + X^{18}Y^9 + 2X^{19}Y^9 + 2X^{20}Y^9 + 2X^{21}Y^9 + X^{22}Y^9$$
$$+ X^{22}Y^{10} + X^{23}Y^{10} + X^{24}Y^{10} - X^{21}Y^{11} - X^{22}Y^{11} + X^{26}Y^{11}$$
$$+ X^{27}Y^{11} - X^{24}Y^{12} - X^{25}Y^{12} - X^{26}Y^{12} - X^{26}Y^{13} - 2X^{27}Y^{13}$$
$$- 2X^{28}Y^{13} - 2X^{29}Y^{13} - X^{30}Y^{13} + X^{31}Y^{13} - X^{28}Y^{14} + X^{29}Y^{14}$$
$$- X^{30}Y^{14} - 3X^{31}Y^{14} - 3X^{32}Y^{14} - 2X^{33}Y^{14} - X^{34}Y^{14} + X^{31}Y^{15}$$
$$+ X^{32}Y^{15} + 2X^{33}Y^{15} + X^{34}Y^{15} + X^{35}Y^{15} + X^{33}Y^{16} - X^{34}Y^{16}$$
$$- X^{35}Y^{16} - 2X^{36}Y^{16} - X^{37}Y^{16} - X^{38}Y^{16} + X^{35}Y^{17} + 2X^{36}Y^{17}$$
$$+ 3X^{37}Y^{17} + 2X^{38}Y^{17} + X^{39}Y^{17} - X^{38}Y^{18} + X^{39}Y^{18} + X^{41}Y^{18}$$
$$+ X^{42}Y^{18} - X^{42}Y^{19} + X^{43}Y^{19} - X^{44}Y^{20} - X^{48}Y^{22}.$$

A.3 T_4, Counting All Subrings

The following polynomial is $W^{\leq}_{T_4}(X,Y)$, mentioned on p. 45:

$$1 + X^4Y^2 + X^5Y^2 - 2X^5Y^3 - 3X^6Y^3 + X^6Y^4 - X^8Y^4 - 2X^9Y^4 - 2X^{10}Y^4$$

$$+ X^{11}Y^5 - 2X^{12}Y^5 - 4X^{13}Y^5 - X^{14}Y^5 + X^{11}Y^6 + X^{12}Y^6 + 4X^{13}Y^6$$
$$+ 2X^{14}Y^6 + 2X^{15}Y^6 - 2X^{16}Y^6 - 2X^{17}Y^6 + X^{14}Y^7 + 3X^{15}Y^7 + 2X^{16}Y^7$$
$$+ 4X^{17}Y^7 + 2X^{18}Y^7 - X^{19}Y^7 + X^{20}Y^7 - X^{15}Y^8 + 3X^{18}Y^8 + 6X^{19}Y^8$$
$$+ 4X^{20}Y^8 + 3X^{21}Y^8 - X^{18}Y^9 - 4X^{19}Y^9 - 4X^{20}Y^9 - 2X^{21}Y^9 + 5X^{22}Y^9$$
$$+ 2X^{23}Y^9 + 4X^{24}Y^9 + 2X^{25}Y^9 - X^{20}Y^{10} - 3X^{22}Y^{10} - 8X^{23}Y^{10}$$
$$- 3X^{24}Y^{10} - X^{25}Y^{10} + 2X^{26}Y^{10} + 2X^{27}Y^{10} + 2X^{23}Y^{11} - 2X^{24}Y^{11}$$
$$- 4X^{25}Y^{11} - 8X^{26}Y^{11} - 11X^{27}Y^{11} - 4X^{28}Y^{11} + X^{29}Y^{11} + 3X^{30}Y^{11}$$
$$+ X^{25}Y^{12} + 4X^{26}Y^{12} + 3X^{27}Y^{12} + 2X^{28}Y^{12} - 6X^{29}Y^{12} - 11X^{30}Y^{12}$$
$$- 6X^{31}Y^{12} - 4X^{32}Y^{12} + X^{33}Y^{12} + 2X^{29}Y^{13} + 6X^{30}Y^{13} + 5X^{31}Y^{13}$$
$$- 5X^{33}Y^{13} - 6X^{34}Y^{13} - 2X^{35}Y^{13} - X^{31}Y^{14} + 4X^{32}Y^{14} + 6X^{33}Y^{14}$$
$$+ 11X^{34}Y^{14} + 6X^{35}Y^{14} - 2X^{36}Y^{14} - 3X^{37}Y^{14} - 4X^{38}Y^{14} - X^{39}Y^{14}$$
$$- 3X^{34}Y^{15} - X^{35}Y^{15} + 4X^{36}Y^{15} + 11X^{37}Y^{15} + 8X^{38}Y^{15} + 4X^{39}Y^{15}$$
$$+ 2X^{40}Y^{15} - 2X^{41}Y^{15} - 2X^{37}Y^{16} - 2X^{38}Y^{16} + X^{39}Y^{16} + 3X^{40}Y^{16}$$
$$+ 8X^{41}Y^{16} + 3X^{42}Y^{16} + X^{44}Y^{16} - 2X^{39}Y^{17} - 4X^{40}Y^{17} - 2X^{41}Y^{17}$$
$$- 5X^{42}Y^{17} + 2X^{43}Y^{17} + 4X^{44}Y^{17} + 4X^{45}Y^{17} + X^{46}Y^{17} - 3X^{43}Y^{18}$$
$$- 4X^{44}Y^{18} - 6X^{45}Y^{18} - 3X^{46}Y^{18} + X^{49}Y^{18} - X^{44}Y^{19} + X^{45}Y^{19}$$
$$- 2X^{46}Y^{19} - 4X^{47}Y^{19} - 2X^{48}Y^{19} - 3X^{49}Y^{19} - X^{50}Y^{19} + 2X^{47}Y^{20}$$
$$+ 2X^{48}Y^{20} - 2X^{49}Y^{20} - 2X^{50}Y^{20} - 4X^{51}Y^{20} - X^{52}Y^{20} - X^{53}Y^{20}$$
$$+ X^{50}Y^{21} + 4X^{51}Y^{21} + 2X^{52}Y^{21} - X^{53}Y^{21} + 2X^{54}Y^{22} + 2X^{55}Y^{22}$$
$$+ X^{56}Y^{22} - X^{58}Y^{22} + 3X^{58}Y^{23} + 2X^{59}Y^{23} - X^{59}Y^{24} - X^{60}Y^{24} - X^{64}Y^{26}.$$

A.4 $L_{(3,2,2)}$, Counting Ideals

The following polynomial is $W^{\triangleleft}_{L_{(3,2,2)}}(X, Y)$, mentioned on p. 50:

$$1 - X^3Y^2 + X^4Y^3 + X^5Y^3 + X^6Y^4 - X^6Y^5 - X^7Y^5 + X^9Y^5 - X^{10}Y^7$$
$$- X^{11}Y^8 - X^{12}Y^8 + X^{13}Y^8 - X^{12}Y^9 + X^{13}Y^9 - 2X^{14}Y^9 - X^{15}Y^9$$
$$+ X^{14}Y^{10} - X^{16}Y^{10} - X^{17}Y^{10} + X^{15}Y^{11} - 2X^{16}Y^{11} - X^{18}Y^{11} + X^{20}Y^{11}$$
$$+ X^{16}Y^{12} + X^{18}Y^{12} - X^{19}Y^{12} + X^{20}Y^{12} - X^{21}Y^{12} - X^{22}Y^{12} + X^{19}Y^{13}$$
$$- X^{20}Y^{13} - 2X^{22}Y^{13} - X^{23}Y^{13} + 3X^{22}Y^{14} - 2X^{23}Y^{14} + X^{24}Y^{14}$$
$$- X^{26}Y^{14} + X^{22}Y^{15} + X^{23}Y^{15} + X^{25}Y^{15} + X^{23}Y^{16} + X^{24}Y^{16} - 2X^{25}Y^{16}$$
$$+ 2X^{26}Y^{16} - X^{27}Y^{16} - X^{25}Y^{17} + 2X^{26}Y^{17} + X^{27}Y^{17} - X^{28}Y^{17} - X^{30}Y^{17}$$
$$- X^{26}Y^{18} + X^{27}Y^{18} + 2X^{28}Y^{18} + 2X^{29}Y^{18} - X^{30}Y^{18} + X^{31}Y^{18} - X^{29}Y^{19}$$
$$+ 2X^{30}Y^{19} + X^{33}Y^{19} - X^{30}Y^{20} + X^{31}Y^{20} - X^{32}Y^{20} + 3X^{33}Y^{20} + X^{35}Y^{20}$$

$$- 2X^{33}Y^{21} + 2X^{34}Y^{21} - X^{37}Y^{21} - X^{34}Y^{22} + 3X^{35}Y^{22} - 2X^{36}Y^{22}$$
$$+ 3X^{37}Y^{22} + X^{39}Y^{22} - X^{34}Y^{23} - X^{35}Y^{23} - X^{37}Y^{23} + X^{38}Y^{23} + 3X^{40}Y^{23}$$
$$- X^{41}Y^{23} - X^{38}Y^{24} + X^{39}Y^{24} - X^{40}Y^{24} + X^{41}Y^{24} - X^{42}Y^{24} - X^{38}Y^{25}$$
$$- 2X^{39}Y^{25} - X^{40}Y^{25} - X^{41}Y^{25} + 2X^{42}Y^{25} - 2X^{43}Y^{25} + 2X^{44}Y^{25}$$
$$+ X^{42}Y^{26} - X^{43}Y^{26} + 2X^{45}Y^{26} - 3X^{44}Y^{27} - 2X^{45}Y^{27} + 2X^{46}Y^{27}$$
$$+ X^{48}Y^{27} - X^{45}Y^{28} - 2X^{47}Y^{28} - X^{48}Y^{28} + X^{49}Y^{28} + X^{44}Y^{29} - X^{45}Y^{29}$$
$$- X^{46}Y^{29} - X^{49}Y^{29} - X^{50}Y^{29} - X^{49}Y^{30} + X^{50}Y^{30} - X^{51}Y^{30} - X^{52}Y^{30}$$
$$+ X^{53}Y^{30} + X^{48}Y^{31} - X^{50}Y^{31} - 2X^{51}Y^{31} - 2X^{52}Y^{31} + 2X^{53}Y^{31}$$
$$- X^{54}Y^{31} + 2X^{51}Y^{32} - X^{53}Y^{32} - X^{55}Y^{32} - X^{56}Y^{32} + X^{52}Y^{33} - 3X^{56}Y^{33}$$
$$- X^{54}Y^{34} + 2X^{55}Y^{34} + 2X^{56}Y^{34} + X^{56}Y^{35} - X^{58}Y^{35} - X^{60}Y^{35} + X^{57}Y^{36}$$
$$+ X^{60}Y^{36} - X^{63}Y^{36} + X^{61}Y^{37} + X^{62}Y^{37} + X^{62}Y^{38} + X^{63}Y^{38} + X^{64}Y^{38}$$
$$- X^{65}Y^{38} + X^{66}Y^{38} - X^{65}Y^{40} + X^{66}Y^{40} + X^{67}Y^{40} + X^{68}Y^{40} - X^{69}Y^{40}$$
$$+ X^{69}Y^{41} - X^{69}Y^{42} + X^{72}Y^{42} + X^{71}Y^{43} - X^{72}Y^{43} - X^{73}Y^{45} - X^{74}Y^{45}$$
$$+ X^{75}Y^{45} - X^{78}Y^{47}.$$

A.5 $\mathcal{G}_3 \times \mathfrak{g}_{5,3}$, Counting Ideals

The following polynomial is $W^{\triangleleft}_{\mathcal{G}_3 \times \mathfrak{g}_{5,3}}(X, Y)$, mentioned on p. 56:

$$1 + X^6Y^3 - X^6Y^5 - X^7Y^7 - X^{12}Y^7 - X^{14}Y^8 - X^{13}Y^9 - X^{15}Y^{10}$$
$$- X^{20}Y^{10} + X^{13}Y^{11} - X^{14}Y^{11} - X^{15}Y^{11} + X^{14}Y^{12} + X^{15}Y^{12} - X^{16}Y^{12}$$
$$+ X^{20}Y^{12} - X^{21}Y^{12} + X^{19}Y^{14} + 2X^{21}Y^{14} - X^{22}Y^{14} - X^{23}Y^{14} + X^{23}Y^{15}$$
$$+ X^{26}Y^{15} + 2X^{22}Y^{16} + X^{26}Y^{16} + X^{27}Y^{16} - X^{28}Y^{16} - X^{26}Y^{17} + X^{27}Y^{17}$$
$$+ X^{28}Y^{17} + X^{29}Y^{17} + X^{23}Y^{18} + X^{28}Y^{18} - X^{27}Y^{19} + X^{28}Y^{19} + X^{30}Y^{19}$$
$$+ X^{35}Y^{19} + X^{29}Y^{20} - X^{28}Y^{21} + X^{31}Y^{21} - X^{33}Y^{21} + X^{36}Y^{21} - X^{35}Y^{22}$$
$$- X^{29}Y^{23} - X^{34}Y^{23} - X^{36}Y^{23} + X^{37}Y^{23} - X^{36}Y^{24} - X^{41}Y^{24} - X^{35}Y^{25}$$
$$- X^{36}Y^{25} - X^{37}Y^{25} + X^{38}Y^{25} + X^{36}Y^{26} - X^{37}Y^{26} - X^{38}Y^{26} - 2X^{42}Y^{26}$$
$$- X^{38}Y^{27} - X^{41}Y^{27} + X^{41}Y^{28} + X^{42}Y^{28} - 2X^{43}Y^{28} - X^{45}Y^{28} + X^{43}Y^{30}$$
$$- X^{44}Y^{30} + X^{48}Y^{30} - X^{49}Y^{30} - X^{50}Y^{30} + X^{49}Y^{31} + X^{50}Y^{31} - X^{51}Y^{31}$$
$$+ X^{44}Y^{32} + X^{49}Y^{32} + X^{51}Y^{33} + X^{50}Y^{34} + X^{52}Y^{35} + X^{57}Y^{35} + X^{58}Y^{37}$$
$$- X^{58}Y^{39} - X^{64}Y^{42}.$$

A.6 $\mathfrak{g}_{6,12}$, Counting All Subrings

The following polynomial is $W^{\leq}_{\mathfrak{g}_{6,12}}(X,Y)$, mentioned on p. 59:

$$
\begin{aligned}
&1 + X^2Y + 2X^4Y^2 - X^5Y^3 + 2X^6Y^3 - X^6Y^4 - X^7Y^4 + 4X^8Y^4 + X^9Y^4 \\
&- 3X^8Y^5 - 5X^9Y^5 + X^{10}Y^5 + X^9Y^6 - X^{10}Y^6 - 3X^{11}Y^6 + 3X^{12}Y^6 \\
&- X^{13}Y^6 + X^{11}Y^7 - 5X^{12}Y^7 - 8X^{13}Y^7 - 2X^{14}Y^7 - 3X^{15}Y^7 + 5X^{13}Y^8 \\
&+ X^{14}Y^8 - X^{15}Y^8 - 2X^{16}Y^8 - 6X^{17}Y^8 + 2X^{15}Y^9 - 2X^{16}Y^9 + X^{17}Y^9 \\
&- 2X^{18}Y^9 - 7X^{19}Y^9 - 2X^{20}Y^9 - X^{21}Y^9 + 4X^{17}Y^{10} + 3X^{18}Y^{10} + 8X^{19}Y^{10} \\
&- X^{20}Y^{10} - 4X^{21}Y^{10} - X^{23}Y^{10} - X^{18}Y^{11} + X^{20}Y^{11} + 11X^{21}Y^{11} \\
&+ X^{22}Y^{11} - 4X^{23}Y^{11} - 4X^{24}Y^{11} - 2X^{25}Y^{11} + 2X^{22}Y^{12} + 13X^{23}Y^{12} \\
&+ 8X^{24}Y^{12} + 8X^{25}Y^{12} + X^{26}Y^{12} - X^{27}Y^{12} - 3X^{23}Y^{13} - 2X^{24}Y^{13} \\
&+ 8X^{25}Y^{13} + 3X^{26}Y^{13} + 5X^{27}Y^{13} + X^{29}Y^{13} - 3X^{25}Y^{14} - 2X^{26}Y^{14} \\
&+ 6X^{27}Y^{14} + 8X^{28}Y^{14} + 13X^{29}Y^{14} + 3X^{30}Y^{14} + 2X^{31}Y^{14} - 5X^{27}Y^{15} \\
&- 5X^{28}Y^{15} - 4X^{29}Y^{15} - 3X^{30}Y^{15} + 9X^{31}Y^{15} + 6X^{32}Y^{15} + 4X^{33}Y^{15} \\
&- 2X^{29}Y^{16} - 3X^{30}Y^{16} - 8X^{31}Y^{16} - 5X^{32}Y^{16} + 6X^{33}Y^{16} + 2X^{34}Y^{16} \\
&+ 6X^{35}Y^{16} + 2X^{36}Y^{16} - 2X^{31}Y^{17} - X^{32}Y^{17} - 11X^{33}Y^{17} - 11X^{34}Y^{17} \\
&- X^{36}Y^{17} + 4X^{37}Y^{17} + X^{38}Y^{17} - 12X^{35}Y^{18} - 11X^{36}Y^{18} - 8X^{37}Y^{18} \\
&- 6X^{38}Y^{18} + 6X^{39}Y^{18} + 2X^{40}Y^{18} + 2X^{35}Y^{19} + 6X^{36}Y^{19} - 6X^{37}Y^{19} \\
&- 8X^{38}Y^{19} - 11X^{39}Y^{19} - 12X^{40}Y^{19} + X^{37}Y^{20} + 4X^{38}Y^{20} - X^{39}Y^{20} \\
&- 11X^{41}Y^{20} - 11X^{42}Y^{20} - X^{43}Y^{20} - 2X^{44}Y^{20} + 2X^{39}Y^{21} + 6X^{40}Y^{21} \\
&+ 2X^{41}Y^{21} + 6X^{42}Y^{21} - 5X^{43}Y^{21} - 8X^{44}Y^{21} - 3X^{45}Y^{21} - 2X^{46}Y^{21} \\
&+ 4X^{42}Y^{22} + 6X^{43}Y^{22} + 9X^{44}Y^{22} - 3X^{45}Y^{22} - 4X^{46}Y^{22} - 5X^{47}Y^{22} \\
&- 5X^{48}Y^{22} + 2X^{44}Y^{23} + 3X^{45}Y^{23} + 13X^{46}Y^{23} + 8X^{47}Y^{23} + 6X^{48}Y^{23} \\
&- 2X^{49}Y^{23} - 3X^{50}Y^{23} + X^{46}Y^{24} + 5X^{48}Y^{24} + 3X^{49}Y^{24} + 8X^{50}Y^{24} \\
&- 2X^{51}Y^{24} - 3X^{52}Y^{24} - X^{48}Y^{25} + X^{49}Y^{25} + 8X^{50}Y^{25} + 8X^{51}Y^{25} \\
&+ 13X^{52}Y^{25} + 2X^{53}Y^{25} - 2X^{50}Y^{26} - 4X^{51}Y^{26} - 4X^{52}Y^{26} + X^{53}Y^{26} \\
&+ 11X^{54}Y^{26} + X^{55}Y^{26} - X^{57}Y^{26} - X^{52}Y^{27} - 4X^{54}Y^{27} - X^{55}Y^{27} \\
&+ 8X^{56}Y^{27} + 3X^{57}Y^{27} + 4X^{58}Y^{27} - X^{54}Y^{28} - 2X^{55}Y^{28} - 7X^{56}Y^{28} \\
&- 2X^{57}Y^{28} + X^{58}Y^{28} - 2X^{59}Y^{28} + 2X^{60}Y^{28} - 6X^{58}Y^{29} - 2X^{59}Y^{29} \\
&- X^{60}Y^{29} + X^{61}Y^{29} + 5X^{62}Y^{29} - 3X^{60}Y^{30} - 2X^{61}Y^{30} - 8X^{62}Y^{30} \\
&- 5X^{63}Y^{30} + X^{64}Y^{30} - X^{62}Y^{31} + 3X^{63}Y^{31} - 3X^{64}Y^{31} - X^{65}Y^{31} \\
&+ X^{66}Y^{31} + X^{65}Y^{32} - 5X^{66}Y^{32} - 3X^{67}Y^{32} + X^{66}Y^{33} + 4X^{67}Y^{33} \\
&- X^{68}Y^{33} - X^{69}Y^{33} + 2X^{69}Y^{34} - X^{70}Y^{34} + 2X^{71}Y^{35} + X^{73}Y^{36} + X^{75}Y^{37}.
\end{aligned}
$$

A.7 $\mathfrak{g}_{1357G}$, Counting Ideals

The following polynomial is $W^{\lhd}_{\mathfrak{g}_{1357B}}(X,Y)$, mentioned on p. 67:

$$1 + X^3Y^3 - X^3Y^5 - 2X^6Y^7 - 2X^4Y^8 + X^5Y^8 - X^7Y^8 + X^4Y^9 - 2X^5Y^9$$
$$- X^9Y^9 - X^7Y^{10} - X^{10}Y^{10} - X^7Y^{11} - X^8Y^{11} + X^9Y^{11} - X^{10}Y^{11}$$
$$+ 3X^7Y^{12} - 3X^8Y^{12} - 2X^9Y^{12} + X^{10}Y^{12} - X^7Y^{13} + 3X^8Y^{13} + X^{10}Y^{13}$$
$$- 3X^{11}Y^{13} + X^9Y^{14} + X^{10}Y^{14} - X^{11}Y^{14} + X^{13}Y^{14} + 5X^{11}Y^{15} - 2X^{12}Y^{15}$$
$$- X^{14}Y^{15} + X^{11}Y^{16} + 5X^{12}Y^{16} - 2X^{14}Y^{16} + X^{16}Y^{16} + X^9Y^{17} - X^{12}Y^{17}$$
$$+ X^{13}Y^{17} + 7X^{14}Y^{17} - X^{16}Y^{17} + X^{12}Y^{18} + 2X^{14}Y^{18} + 2X^{15}Y^{18}$$
$$+ X^{16}Y^{18} + 2X^{12}Y^{19} - 4X^{14}Y^{19} + 6X^{15}Y^{19} + X^{16}Y^{19} + 3X^{17}Y^{19}$$
$$- X^{12}Y^{20} + 4X^{13}Y^{20} - 3X^{15}Y^{20} + 2X^{16}Y^{20} + X^{17}Y^{20} + 3X^{18}Y^{20}$$
$$- X^{12}Y^{21} - 2X^{13}Y^{21} + 4X^{14}Y^{21} + X^{15}Y^{21} - 2X^{17}Y^{21} + X^{18}Y^{21}$$
$$+ 2X^{19}Y^{21} + X^{20}Y^{21} - 2X^{14}Y^{22} + X^{16}Y^{22} + X^{20}Y^{22} + X^{21}Y^{22}$$
$$- 3X^{15}Y^{23} + X^{16}Y^{23} + 3X^{17}Y^{23} - 2X^{18}Y^{23} - X^{19}Y^{23} - X^{20}Y^{23}$$
$$+ 2X^{21}Y^{23} - 5X^{16}Y^{24} - 3X^{18}Y^{24} + 6X^{19}Y^{24} - X^{20}Y^{24} - 6X^{21}Y^{24}$$
$$+ X^{22}Y^{24} + X^{16}Y^{25} - 5X^{17}Y^{25} - X^{18}Y^{25} - 8X^{19}Y^{25} + 6X^{20}Y^{25}$$
$$+ X^{21}Y^{25} - 2X^{22}Y^{25} + X^{17}Y^{26} - 2X^{18}Y^{26} - 4X^{19}Y^{26} - 6X^{20}Y^{26}$$
$$+ X^{21}Y^{26} - X^{23}Y^{26} - 2X^{24}Y^{26} + X^{18}Y^{27} - X^{19}Y^{27} - 9X^{20}Y^{27}$$
$$- 3X^{21}Y^{27} - 2X^{22}Y^{27} + X^{23}Y^{27} - X^{25}Y^{27} - X^{17}Y^{28} + 2X^{19}Y^{28}$$
$$+ X^{20}Y^{28} - 7X^{21}Y^{28} - 8X^{22}Y^{28} - 3X^{23}Y^{28} + 2X^{24}Y^{28} - X^{26}Y^{28}$$
$$- X^{27}Y^{28} - X^{19}Y^{29} + 4X^{21}Y^{29} + X^{22}Y^{29} - 11X^{23}Y^{29} - 4X^{24}Y^{29}$$
$$- 2X^{25}Y^{29} + X^{26}Y^{29} - 2X^{21}Y^{30} + 4X^{22}Y^{30} - X^{23}Y^{30} - 5X^{24}Y^{30}$$
$$- 5X^{25}Y^{30} - 4X^{26}Y^{30} + X^{27}Y^{30} - X^{21}Y^{31} - X^{22}Y^{31} + 8X^{23}Y^{31}$$
$$- X^{24}Y^{31} - 9X^{26}Y^{31} - X^{27}Y^{31} + X^{20}Y^{32} + X^{21}Y^{32} - 2X^{22}Y^{32}$$
$$- 2X^{23}Y^{32} + 3X^{24}Y^{32} + 5X^{25}Y^{32} + 7X^{26}Y^{32} - 10X^{27}Y^{32} - X^{28}Y^{32}$$
$$- 2X^{29}Y^{32} + X^{21}Y^{33} + 2X^{22}Y^{33} - 4X^{25}Y^{33} + 6X^{26}Y^{33} + 8X^{27}Y^{33}$$
$$- 5X^{28}Y^{33} - X^{29}Y^{33} - X^{30}Y^{33} + 3X^{23}Y^{34} + X^{24}Y^{34} + 3X^{25}Y^{34}$$
$$- X^{26}Y^{34} + X^{27}Y^{34} + 6X^{28}Y^{34} + 2X^{29}Y^{34} - 2X^{30}Y^{34} - X^{31}Y^{34}$$
$$+ 5X^{24}Y^{35} + X^{26}Y^{35} + X^{27}Y^{35} + 3X^{28}Y^{35} + 9X^{29}Y^{35} - X^{30}Y^{35}$$
$$- X^{32}Y^{35} - 2X^{24}Y^{36} + 5X^{25}Y^{36} + 4X^{26}Y^{36} + 10X^{27}Y^{36} - 8X^{28}Y^{36}$$
$$- 5X^{29}Y^{36} + 12X^{30}Y^{36} + X^{31}Y^{36} + 4X^{32}Y^{36} - X^{33}Y^{36} - X^{25}Y^{37}$$
$$+ 2X^{26}Y^{37} + 15X^{28}Y^{37} - X^{29}Y^{37} - 2X^{30}Y^{37} + 4X^{31}Y^{37} + X^{32}Y^{37}$$
$$+ 3X^{33}Y^{37} - X^{26}Y^{38} - X^{27}Y^{38} + 2X^{28}Y^{38} + 13X^{29}Y^{38} + 2X^{30}Y^{38}$$
$$+ X^{31}Y^{38} - X^{32}Y^{38} + 4X^{33}Y^{38} + X^{34}Y^{38} + X^{35}Y^{38} - 2X^{27}Y^{39} - X^{28}Y^{39}$$

$$+ 2X^{29}Y^{39} + 9X^{30}Y^{39} + 5X^{31}Y^{39} - X^{33}Y^{39} + 3X^{34}Y^{39} + X^{36}Y^{39}$$
$$+ X^{27}Y^{40} - X^{28}Y^{40} - 6X^{29}Y^{40} - 5X^{30}Y^{40} + 13X^{31}Y^{40} + 12X^{32}Y^{40}$$
$$- 3X^{33}Y^{40} - X^{34}Y^{40} + X^{35}Y^{40} + X^{36}Y^{40} + X^{37}Y^{40} + X^{29}Y^{41} - 2X^{30}Y^{41}$$
$$- 10X^{31}Y^{41} + 8X^{33}Y^{41} + 9X^{34}Y^{41} - 2X^{35}Y^{41} - X^{36}Y^{41} + X^{30}Y^{42}$$
$$- 5X^{31}Y^{42} - 4X^{32}Y^{42} - X^{33}Y^{42} + 3X^{34}Y^{42} + 7X^{35}Y^{42} - 2X^{36}Y^{42}$$
$$+ X^{37}Y^{42} - X^{29}Y^{43} + 2X^{31}Y^{43} - 5X^{32}Y^{43} - 6X^{33}Y^{43} - 12X^{34}Y^{43}$$
$$+ 6X^{35}Y^{43} + 6X^{36}Y^{43} - X^{39}Y^{43} - X^{30}Y^{44} - X^{31}Y^{44} + X^{32}Y^{44}$$
$$+ 2X^{33}Y^{44} - 6X^{34}Y^{44} - 18X^{35}Y^{44} + 4X^{36}Y^{44} + 2X^{37}Y^{44} + 3X^{38}Y^{44}$$
$$- X^{31}Y^{45} - 3X^{32}Y^{45} - 3X^{33}Y^{45} + 6X^{34}Y^{45} + X^{35}Y^{45} - 15X^{36}Y^{45}$$
$$- 8X^{37}Y^{45} - X^{38}Y^{45} + 5X^{39}Y^{45} - X^{40}Y^{45} - 2X^{33}Y^{46} - 4X^{34}Y^{46}$$
$$- X^{36}Y^{46} - 6X^{37}Y^{46} - 6X^{38}Y^{46} - 2X^{39}Y^{46} + X^{40}Y^{46} - 3X^{34}Y^{47}$$
$$- 7X^{35}Y^{47} + 4X^{36}Y^{47} + 3X^{37}Y^{47} - 12X^{38}Y^{47} - 5X^{39}Y^{47} - 5X^{40}Y^{47}$$
$$+ 2X^{41}Y^{47} + 2X^{35}Y^{48} - 12X^{36}Y^{48} + 9X^{38}Y^{48} - 6X^{39}Y^{48} - 10X^{41}Y^{48}$$
$$+ X^{42}Y^{48} + X^{35}Y^{49} + 4X^{36}Y^{49} - 10X^{37}Y^{49} - 5X^{38}Y^{49} + X^{39}Y^{49}$$
$$+ X^{41}Y^{49} - 7X^{42}Y^{49} - X^{43}Y^{49} + X^{36}Y^{50} + 4X^{37}Y^{50} - 4X^{38}Y^{50}$$
$$- 5X^{39}Y^{50} - 5X^{40}Y^{50} + 6X^{41}Y^{50} - X^{42}Y^{50} - 2X^{43}Y^{50} - 2X^{44}Y^{50}$$
$$+ X^{37}Y^{51} + 6X^{38}Y^{51} - 2X^{39}Y^{51} - 6X^{40}Y^{51} - 3X^{41}Y^{51} + 5X^{42}Y^{51}$$
$$- X^{44}Y^{51} - 2X^{45}Y^{51} + X^{37}Y^{52} + 9X^{39}Y^{52} + 4X^{40}Y^{52} - 7X^{41}Y^{52}$$
$$- 7X^{42}Y^{52} + X^{43}Y^{52} + 5X^{44}Y^{52} - X^{45}Y^{52} - X^{46}Y^{52} - 2X^{39}Y^{53}$$
$$+ 5X^{40}Y^{53} + 8X^{41}Y^{53} + 4X^{42}Y^{53} - 8X^{43}Y^{53} - 2X^{44}Y^{53} + 3X^{45}Y^{53}$$
$$+ X^{40}Y^{54} + 3X^{41}Y^{54} + 6X^{42}Y^{54} + 7X^{43}Y^{54} - 5X^{44}Y^{54} - X^{45}Y^{54}$$
$$+ X^{46}Y^{54} + 2X^{47}Y^{54} - X^{48}Y^{54} + X^{39}Y^{55} + X^{40}Y^{55} - X^{41}Y^{55} + X^{42}Y^{55}$$
$$+ 10X^{43}Y^{55} + 5X^{44}Y^{55} - X^{45}Y^{55} - 2X^{46}Y^{55} - 2X^{47}Y^{55} + 3X^{48}Y^{55}$$
$$+ X^{41}Y^{56} - 2X^{42}Y^{56} - X^{43}Y^{56} + 8X^{44}Y^{56} + 7X^{45}Y^{56} + 5X^{46}Y^{56}$$
$$- 4X^{47}Y^{56} - X^{48}Y^{56} + X^{49}Y^{56} + 3X^{42}Y^{57} - 4X^{44}Y^{57} + 2X^{45}Y^{57}$$
$$+ 9X^{46}Y^{57} + 7X^{47}Y^{57} - 2X^{48}Y^{57} + 3X^{44}Y^{58} - 2X^{46}Y^{58} + 4X^{47}Y^{58}$$
$$+ 5X^{48}Y^{58} + X^{44}Y^{59} + X^{45}Y^{59} - 3X^{46}Y^{59} + 2X^{48}Y^{59} + 7X^{49}Y^{59}$$
$$- X^{44}Y^{60} + X^{45}Y^{60} + 2X^{46}Y^{60} - X^{48}Y^{60} - 6X^{49}Y^{60} + 7X^{50}Y^{60}$$
$$+ X^{51}Y^{60} - X^{45}Y^{61} + 3X^{48}Y^{61} - X^{49}Y^{61} - 5X^{50}Y^{61} + 4X^{51}Y^{61}$$
$$+ X^{52}Y^{61} - 2X^{46}Y^{62} - X^{47}Y^{62} - 3X^{48}Y^{62} + 4X^{49}Y^{62} - 4X^{51}Y^{62}$$
$$- X^{52}Y^{62} + 2X^{53}Y^{62} - 2X^{47}Y^{63} + X^{48}Y^{63} - 2X^{50}Y^{63} + 2X^{51}Y^{63}$$
$$- 3X^{52}Y^{63} - 2X^{53}Y^{63} + 2X^{54}Y^{63} + X^{47}Y^{64} - 3X^{48}Y^{64} - 5X^{49}Y^{64}$$
$$- X^{50}Y^{64} + X^{51}Y^{64} + 3X^{52}Y^{64} - 2X^{53}Y^{64} - 2X^{54}Y^{64} - 6X^{51}Y^{65}$$

$$+ X^{53}Y^{65} - X^{54}Y^{65} - X^{51}Y^{66} - 6X^{52}Y^{66} - X^{56}Y^{66} - X^{50}Y^{67} - 2X^{53}Y^{67}$$
$$- X^{54}Y^{67} - X^{55}Y^{67} + X^{56}Y^{67} - X^{57}Y^{67} + 2X^{52}Y^{68} - 4X^{54}Y^{68}$$
$$- 3X^{55}Y^{68} + X^{56}Y^{68} - X^{53}Y^{69} + 3X^{54}Y^{69} - 3X^{56}Y^{69} - X^{57}Y^{69}$$
$$- X^{58}Y^{70} + 2X^{55}Y^{71} + X^{57}Y^{71} - X^{59}Y^{71} + 2X^{58}Y^{72} - X^{59}Y^{72} + X^{56}Y^{73}$$
$$- X^{57}Y^{73} + 2X^{59}Y^{73} + 2X^{57}Y^{74} + X^{60}Y^{74} - X^{60}Y^{75} + 2X^{61}Y^{75}$$
$$+ X^{59}Y^{76} + X^{60}Y^{76} - X^{61}Y^{76} + X^{62}Y^{76} + X^{60}Y^{77} + X^{63}Y^{78} - X^{63}Y^{81}$$
$$- X^{66}Y^{83}.$$

A.8 $\mathfrak{g}_{1457A}$, Counting Ideals

The following polynomial is $W^{\lhd}_{\mathfrak{g}_{1457A}}(X, Y)$, mentioned on p. 68:

$$1 - X^4Y^5 - X^4Y^8 + X^5Y^8 - X^5Y^9 - X^8Y^{10} + X^8Y^{11} - 2X^9Y^{11} + X^8Y^{12}$$
$$- X^9Y^{12} - X^{10}Y^{12} + 2X^9Y^{13} - 2X^{10}Y^{13} + X^{10}Y^{14} - X^9Y^{15} + 2X^{13}Y^{15}$$
$$- X^{14}Y^{15} + X^9Y^{16} - 2X^{10}Y^{16} - X^{13}Y^{16} + 2X^{14}Y^{16} + X^{10}Y^{17} - X^{11}Y^{17}$$
$$+ X^{14}Y^{17} + 2X^{13}Y^{18} - 2X^{14}Y^{18} + 3X^{14}Y^{19} - 2X^{15}Y^{19} - X^{13}Y^{20}$$
$$+ 3X^{14}Y^{20} - X^{14}Y^{21} + 4X^{15}Y^{21} - X^{16}Y^{21} + X^{18}Y^{21} - X^{15}Y^{22} + X^{16}Y^{22}$$
$$- X^{17}Y^{22} + X^{18}Y^{22} + X^{19}Y^{22} + X^{14}Y^{23} - X^{15}Y^{23} - 3X^{18}Y^{23} + 4X^{19}Y^{23}$$
$$+ 2X^{15}Y^{24} - X^{16}Y^{24} - X^{18}Y^{24} - 2X^{19}Y^{24} + 2X^{20}Y^{24} + X^{16}Y^{25}$$
$$+ X^{18}Y^{25} - X^{19}Y^{25} - X^{18}Y^{26} + 4X^{19}Y^{26} - 2X^{20}Y^{26} - X^{23}Y^{26} - X^{18}Y^{27}$$
$$- X^{19}Y^{27} + 4X^{20}Y^{27} - X^{23}Y^{27} - 3X^{19}Y^{28} + 3X^{20}Y^{28} + X^{21}Y^{28}$$
$$- X^{24}Y^{28} - 3X^{20}Y^{29} + 2X^{21}Y^{29} - X^{23}Y^{29} + X^{24}Y^{29} - X^{21}Y^{30}$$
$$- 3X^{23}Y^{30} + X^{24}Y^{30} + X^{20}Y^{31} - 5X^{24}Y^{31} + 2X^{25}Y^{31} - X^{20}Y^{32}$$
$$+ X^{21}Y^{32} + X^{23}Y^{32} - X^{24}Y^{32} - 3X^{25}Y^{32} - X^{23}Y^{33} + X^{24}Y^{33} - X^{25}Y^{33}$$
$$- X^{28}Y^{33} - 3X^{24}Y^{34} + 2X^{25}Y^{34} + X^{27}Y^{34} - X^{29}Y^{34} - X^{24}Y^{35}$$
$$- 2X^{25}Y^{35} + X^{26}Y^{35} + X^{27}Y^{35} + X^{28}Y^{35} - X^{29}Y^{35} - 3X^{25}Y^{36} - X^{26}Y^{37}$$
$$- X^{28}Y^{37} + X^{27}Y^{38} + X^{28}Y^{38} - 3X^{29}Y^{38} - X^{30}Y^{38} + X^{33}Y^{38} + 3X^{28}Y^{39}$$
$$- 3X^{30}Y^{39} - X^{25}Y^{40} + X^{28}Y^{40} + 3X^{29}Y^{40} - X^{30}Y^{40} - X^{31}Y^{40} + X^{30}Y^{41}$$
$$+ X^{32}Y^{41} + 3X^{33}Y^{42} + X^{29}Y^{43} - X^{30}Y^{43} - X^{31}Y^{43} - X^{32}Y^{43} + 2X^{33}Y^{43}$$
$$+ X^{34}Y^{43} + X^{29}Y^{44} - X^{31}Y^{44} - 2X^{33}Y^{44} + 3X^{34}Y^{44} + X^{30}Y^{45} + X^{33}Y^{45}$$
$$- X^{34}Y^{45} + X^{35}Y^{45} + 3X^{33}Y^{46} + X^{34}Y^{46} - X^{35}Y^{46} - X^{37}Y^{46} + X^{38}Y^{46}$$
$$- 2X^{33}Y^{47} + 5X^{34}Y^{47} - X^{38}Y^{47} - X^{34}Y^{48} + 3X^{35}Y^{48} + X^{37}Y^{48}$$
$$- X^{34}Y^{49} + X^{35}Y^{49} - 2X^{37}Y^{49} + 3X^{38}Y^{49} + X^{34}Y^{50} - X^{37}Y^{50}$$
$$- 3X^{38}Y^{50} + 3X^{39}Y^{50} + X^{35}Y^{51} - 4X^{38}Y^{51} + X^{39}Y^{51} + X^{40}Y^{51}$$

$+ X^{35}Y^{52} + 2X^{38}Y^{52} - 4X^{39}Y^{52} + X^{40}Y^{52} + X^{39}Y^{53} - X^{40}Y^{53} - X^{42}Y^{53}$

$- 2X^{38}Y^{54} + 2X^{39}Y^{54} + X^{40}Y^{54} + X^{42}Y^{54} - 2X^{43}Y^{54} - 4X^{39}Y^{55}$

$+ 3X^{40}Y^{55} + X^{43}Y^{55} - X^{44}Y^{55} - X^{39}Y^{56} - X^{40}Y^{56} + X^{41}Y^{56} - X^{42}Y^{56}$

$+ X^{43}Y^{56} - X^{40}Y^{57} + X^{42}Y^{57} - 4X^{43}Y^{57} + X^{44}Y^{57} - 3X^{44}Y^{58} + X^{45}Y^{58}$

$+ 2X^{43}Y^{59} - 3X^{44}Y^{59} + 2X^{44}Y^{60} - 2X^{45}Y^{60} - X^{44}Y^{61} + X^{47}Y^{61}$

$- X^{48}Y^{61} - 2X^{44}Y^{62} + X^{45}Y^{62} + 2X^{48}Y^{62} - X^{49}Y^{62} + X^{44}Y^{63}$

$- 2X^{45}Y^{63} + X^{49}Y^{63} - X^{48}Y^{64} + 2X^{48}Y^{65} - 2X^{49}Y^{65} + X^{48}Y^{66}$

$+ X^{49}Y^{66} - X^{50}Y^{66} + 2X^{49}Y^{67} - X^{50}Y^{67} + X^{50}Y^{68} + X^{53}Y^{69} - X^{53}Y^{70}$

$+ X^{54}Y^{70} + X^{54}Y^{73} - X^{58}Y^{78}.$

A.9 $\mathfrak{g}_{1457B}$, Counting Ideals

The following polynomial is $W^{\triangleleft}_{\mathfrak{g}_{1457B}}(X, Y)$, mentioned on p. 68:

$1 - X^4Y^5 - X^4Y^8 + X^5Y^8 - X^5Y^9 - X^8Y^{10} + X^8Y^{11} - 2X^9Y^{11} + X^8Y^{12}$

$- X^9Y^{12} - X^{10}Y^{12} + 2X^9Y^{13} - 2X^{10}Y^{13} + X^{10}Y^{14} + 2X^{13}Y^{15} - X^{14}Y^{15}$

$- X^{10}Y^{16} - X^{13}Y^{16} + 2X^{14}Y^{16} + X^{10}Y^{17} - X^{11}Y^{17} + X^{14}Y^{17} + X^{13}Y^{18}$

$- X^{14}Y^{18} + 3X^{14}Y^{19} - 2X^{15}Y^{19} + X^{15}Y^{20} + 3X^{15}Y^{21} - X^{16}Y^{21}$

$+ X^{18}Y^{21} - X^{15}Y^{22} + X^{16}Y^{22} + X^{19}Y^{22} - X^{17}Y^{23} - X^{18}Y^{23} + 3X^{19}Y^{23}$

$- X^{18}Y^{24} - 2X^{19}Y^{24} + 2X^{20}Y^{24} + X^{15}Y^{25} - X^{18}Y^{25} + X^{19}Y^{26} - X^{20}Y^{26}$

$- X^{23}Y^{26} + X^{20}Y^{27} - X^{22}Y^{27} - X^{19}Y^{28} + X^{20}Y^{28} + X^{21}Y^{28} + X^{22}Y^{28}$

$- X^{23}Y^{28} - X^{24}Y^{28} - X^{19}Y^{29} + X^{21}Y^{29} - X^{20}Y^{30} - X^{23}Y^{30} - 3X^{23}Y^{31}$

$+ X^{25}Y^{31} + 2X^{23}Y^{32} - 3X^{24}Y^{32} - X^{25}Y^{32} - X^{25}Y^{33} - X^{27}Y^{33} + X^{24}Y^{34}$

$- X^{25}Y^{34} + 2X^{27}Y^{34} - 2X^{28}Y^{34} - X^{24}Y^{35} + X^{27}Y^{35} + X^{28}Y^{35} - X^{29}Y^{35}$

$- X^{25}Y^{36} + 3X^{28}Y^{36} - 2X^{29}Y^{36} - X^{25}Y^{37} - 2X^{28}Y^{37} + 2X^{29}Y^{37}$

$- X^{29}Y^{38} + X^{32}Y^{38} + 2X^{28}Y^{39} - 2X^{29}Y^{39} - X^{30}Y^{39} - X^{32}Y^{39} + X^{33}Y^{39}$

$+ 4X^{29}Y^{40} - 3X^{30}Y^{40} + X^{29}Y^{41} + X^{30}Y^{41} - X^{31}Y^{41} + X^{32}Y^{41} - X^{33}Y^{41}$

$+ X^{30}Y^{42} - X^{32}Y^{42} + 4X^{33}Y^{42} - X^{34}Y^{42} + 2X^{34}Y^{43} - X^{35}Y^{43}$

$- 2X^{33}Y^{44} + 3X^{34}Y^{44} - 2X^{34}Y^{45} + 2X^{35}Y^{45} + X^{34}Y^{46} - X^{37}Y^{46}$

$+ X^{38}Y^{46} + 2X^{34}Y^{47} - X^{35}Y^{47} - 2X^{38}Y^{47} + X^{39}Y^{47} - X^{34}Y^{48}$

$+ 2X^{35}Y^{48} + X^{38}Y^{48} - X^{39}Y^{48} + X^{38}Y^{49} - 2X^{38}Y^{50} + 2X^{39}Y^{50}$

$- X^{38}Y^{51} - X^{39}Y^{51} + X^{40}Y^{51} - 2X^{39}Y^{52} + X^{40}Y^{52} - X^{40}Y^{53} - X^{43}Y^{54}$

$+ X^{43}Y^{55} - X^{44}Y^{55} - X^{44}Y^{58} + X^{48}Y^{63}.$

A.10 $\mathfrak{tr}_6(\mathbb{Z})$, Counting Ideals

The following polynomial is $W^{\lhd}_{\mathfrak{tr}_6(\mathbb{Z})}(Y)$ mentioned on p. 78:

$$1 + 2Y^2 + 3Y^4 + 2Y^5 + 4Y^6 + 4Y^7 + 7Y^8 + 8Y^9 + 10Y^{10} + 13Y^{11} + 16Y^{12}$$
$$+ 19Y^{13} + 24Y^{14} + 27Y^{15} + 34Y^{16} + 37Y^{17} + 44Y^{18} + 48Y^{19} + 56Y^{20}$$
$$+ 59Y^{21} + 70Y^{22} + 72Y^{23} + 81Y^{24} + 83Y^{25} + 90Y^{26} + 91Y^{27} + 95Y^{28}$$
$$+ 93Y^{29} + 99Y^{30} + 91Y^{31} + 92Y^{32} + 82Y^{33} + 80Y^{34} + 63Y^{35} + 62Y^{36}$$
$$+ 38Y^{37} + 34Y^{38} + 9Y^{39} - 27Y^{41} - 38Y^{42} - 68Y^{43} - 75Y^{44} - 105Y^{45}$$
$$- 115Y^{46} - 139Y^{47} - 146Y^{48} - 173Y^{49} - 171Y^{50} - 195Y^{51} - 188Y^{52}$$
$$- 206Y^{53} - 194Y^{54} - 206Y^{55} - 188Y^{56} - 195Y^{57} - 171Y^{58} - 173Y^{59}$$
$$- 146Y^{60} - 139Y^{61} - 115Y^{62} - 105Y^{63} - 75Y^{64} - 68Y^{65} - 38Y^{66} - 27Y^{67}$$
$$+ 9Y^{69} + 34Y^{70} + 38Y^{71} + 62Y^{72} + 63Y^{73} + 80Y^{74} + 82Y^{75} + 92Y^{76}$$
$$+ 91Y^{77} + 99Y^{78} + 93Y^{79} + 95Y^{80} + 91Y^{81} + 90Y^{82} + 83Y^{83} + 81Y^{84}$$
$$+ 72Y^{85} + 70Y^{86} + 59Y^{87} + 56Y^{88} + 48Y^{89} + 44Y^{90} + 37Y^{91} + 34Y^{92}$$
$$+ 27Y^{93} + 24Y^{94} + 19Y^{95} + 16Y^{96} + 13Y^{97} + 10Y^{98} + 8Y^{99} + 7Y^{100}$$
$$+ 4Y^{101} + 4Y^{102} + 2Y^{103} + 3Y^{104} + 2Y^{106} + Y^{108}.$$

A.11 $\mathfrak{tr}_7(\mathbb{Z})$, Counting Ideals

The following polynomial is $W^{\lhd}_{\mathfrak{tr}_7(\mathbb{Z})}(Y)$ mentioned on p. 78:

$$1 + 3Y^2 + 5Y^4 + 3Y^5 + 7Y^6 + 9Y^7 + 13Y^8 + 18Y^9 + 25Y^{10} + 32Y^{11}$$
$$+ 44Y^{12} + 56Y^{13} + 75Y^{14} + 94Y^{15} + 125Y^{16} + 153Y^{17} + 199Y^{18} + 242Y^{19}$$
$$+ 305Y^{20} + 367Y^{21} + 459Y^{22} + 545Y^{23} + 673Y^{24} + 793Y^{25} + 958Y^{26}$$
$$+ 1124Y^{27} + 1337Y^{28} + 1553Y^{29} + 1834Y^{30} + 2106Y^{31} + 2458Y^{32}$$
$$+ 2806Y^{33} + 3228Y^{34} + 3656Y^{35} + 4172Y^{36} + 4668Y^{37} + 5290Y^{38}$$
$$+ 5867Y^{39} + 6573Y^{40} + 7245Y^{41} + 8028Y^{42} + 8767Y^{43} + 9642Y^{44}$$
$$+ 10421Y^{45} + 11360Y^{46} + 12183Y^{47} + 13136Y^{48} + 13963Y^{49} + 14921Y^{50}$$
$$+ 15683Y^{51} + 16609Y^{52} + 17279Y^{53} + 18089Y^{54} + 18627Y^{55} + 19271Y^{56}$$
$$+ 19582Y^{57} + 20023Y^{58} + 20038Y^{59} + 20192Y^{60} + 19882Y^{61} + 19663Y^{62}$$
$$+ 18961Y^{63} + 18352Y^{64} + 17163Y^{65} + 16125Y^{66} + 14444Y^{67} + 12905Y^{68}$$
$$+ 10732Y^{69} + 8700Y^{70} + 5995Y^{71} + 3517Y^{72} + 305Y^{73} - 2612Y^{74}$$
$$- 6241Y^{75} - 9546Y^{76} - 13535Y^{77} - 17095Y^{78} - 21361Y^{79} - 25071Y^{80}$$

$$- 29441Y^{81} - 33196Y^{82} - 37522Y^{83} - 41121Y^{84} - 45290Y^{85} - 48557Y^{86}$$
$$- 52361Y^{87} - 55180Y^{88} - 58427Y^{89} - 60607Y^{90} - 63191Y^{91} - 64544Y^{92}$$
$$- 66322Y^{93} - 66778Y^{94} - 67583Y^{95} - 67068Y^{96} - 66871Y^{97} - 65267Y^{98}$$
$$- 64071Y^{99} - 61396Y^{100} - 59142Y^{101} - 55484Y^{102} - 52239Y^{103}$$
$$- 47622Y^{104} - 43560Y^{105} - 38095Y^{106} - 33306Y^{107} - 27241Y^{108}$$
$$- 21857Y^{109} - 15362Y^{110} - 9666Y^{111} - 2883Y^{112} + 2883Y^{113} + 9666Y^{114}$$
$$+ 15362Y^{115} + 21857Y^{116} + 27241Y^{117} + 33306Y^{118} + 38095Y^{119}$$
$$+ 43560Y^{120} + 47622Y^{121} + 52239Y^{122} + 55484Y^{123} + 59142Y^{124}$$
$$+ 61396Y^{125} + 64071Y^{126} + 65267Y^{127} + 66871Y^{128} + 67068Y^{129}$$
$$+ 67583Y^{130} + 66778Y^{131} + 66322Y^{132} + 64544Y^{133} + 63191Y^{134}$$
$$+ 60607Y^{135} + 58427Y^{136} + 55180Y^{137} + 52361Y^{138} + 48557Y^{139}$$
$$+ 45290Y^{140} + 41121Y^{141} + 37522Y^{142} + 33196Y^{143} + 29441Y^{144}$$
$$+ 25071Y^{145} + 21361Y^{146} + 17095Y^{147} + 13535Y^{148} + 9546Y^{149}$$
$$+ 6241Y^{150} + 2612Y^{151} - 305Y^{152} - 3517Y^{153} - 5995Y^{154} - 8700Y^{155}$$
$$- 10732Y^{156} - 12905Y^{157} - 14444Y^{158} - 16125Y^{159} - 17163Y^{160}$$
$$- 18352Y^{161} - 18961Y^{162} - 19663Y^{163} - 19882Y^{164} - 20192Y^{165}$$
$$- 20038Y^{166} - 20023Y^{167} - 19582Y^{168} - 19271Y^{169} - 18627Y^{170}$$
$$- 18089Y^{171} - 17279Y^{172} - 16609Y^{173} - 15683Y^{174} - 14921Y^{175}$$
$$- 13963Y^{176} - 13136Y^{177} - 12183Y^{178} - 11360Y^{179} - 10421Y^{180}$$
$$- 9642Y^{181} - 8767Y^{182} - 8028Y^{183} - 7245Y^{184} - 6573Y^{185} - 5867Y^{186}$$
$$- 5290Y^{187} - 4668Y^{188} - 4172Y^{189} - 3656Y^{190} - 3228Y^{191} - 2806Y^{192}$$
$$- 2458Y^{193} - 2106Y^{194} - 1834Y^{195} - 1553Y^{196} - 1337Y^{197} - 1124Y^{198}$$
$$- 958Y^{199} - 793Y^{200} - 673Y^{201} - 545Y^{202} - 459Y^{203} - 367Y^{204}$$
$$- 305Y^{205} - 242Y^{206} - 199Y^{207} - 153Y^{208} - 125Y^{209} - 94Y^{210} - 75Y^{211}$$
$$- 56Y^{212} - 44Y^{213} - 32Y^{214} - 25Y^{215} - 18Y^{216} - 13Y^{217} - 9Y^{218}$$
$$- 7Y^{219} - 3Y^{220} - 5Y^{221} - 3Y^{223} - Y^{225}.$$

B

Factorisation of Polynomials Associated to Classical Groups

In this appendix we are concerned with the proof of Theorem 6.9. The proof depends on extending the following classical identity on root systems: let w_i be the reflection in the root defined by α_i, then

$$\lambda(w_i w) = \begin{cases} \lambda(w) + 1 & \text{if } w^{-1}(\alpha_i) \in \Phi^+ , \\ \lambda(w) - 1 & \text{if } w^{-1}(\alpha_i) \in \Phi^- . \end{cases}$$

To explain our generalisation to the root systems $X_l = C_l$ or D_l, we set up some notation. Let Φ_{k+1} be the sub-root system generated by $\{\alpha_{l-k}, \ldots, \alpha_l\}$ of type X_{k+1}. Let $w_{\Phi_{k+1}}$ be the element sending Φ_{k+1}^+ to Φ_{k+1}^-.

Let us recall the structure of the root systems C_l and D_l and their corresponding Weyl groups. Let $\mathbf{e}_i$ be the standard basis for the l-dimensional vector space $\mathbb{R}^l$.

$C_l^+ = \{\, 2\mathbf{e}_i, \mathbf{e}_i \pm \mathbf{e}_j : 1 \le i < j \le l \,\}$ with simple roots $\alpha_1 = \mathbf{e}_1 - \mathbf{e}_2$, ..., $\alpha_{l-1} = \mathbf{e}_{l-1} - \mathbf{e}_l$, $\alpha_l = 2\mathbf{e}_l$. $W(C_l)$ is the semi-direct product of the symmetric group on $\mathbf{e}_i$ and the group $(\mathbb{Z}/2\mathbb{Z})^l$ operating by $\mathbf{e}_i \mapsto (\pm 1)_i \mathbf{e}_i$.

$D_l^+ = \{\, \mathbf{e}_i \pm \mathbf{e}_j : 1 \le i < j \le l \,\}$ with simple roots $\alpha_1 = \mathbf{e}_1 - \mathbf{e}_2$, ..., $\alpha_{l-1} = \mathbf{e}_{l-1} - \mathbf{e}_l$, $\alpha_l = \mathbf{e}_{l-1} + \mathbf{e}_l$. $W(D_l)$ is the semi-direct product of the symmetric group on $\mathbf{e}_i$ and the group $(\mathbb{Z}/2\mathbb{Z})^{l-1}$ operating by $\mathbf{e}_i \mapsto (\pm 1)_i \mathbf{e}_i$ with $\prod_i (\pm 1)_i = 1$.

We shall write $w = \pi_w \sigma_w$ where π_w is the permutation and σ_w is the sign change (where we employ the convention that we implement the sign change followed by the permutation). For each $w \in W$, let $w(k)$ be the permutation of $\mathbf{e}_{\pi_{w^{-1}}(i)}$ for $i = k, \ldots, l$ which alters the order. For $k = 1, \ldots, r+1$ let

$$W(k) = \left\{ \begin{array}{c} w = \pi_w \sigma_w : w^{-1}(\alpha_{k-1}) \text{ and } (w_{\Phi_{l-k+1}} w w(k))^{-1}(\alpha_{k-1}) \\ \text{have the same sign and } (\sigma_{w^{-1}})_k = 1 \end{array} \right\}$$

$$\cup \left\{ \begin{array}{c} w = \pi_w \sigma_w : w^{-1}(\alpha_{k-1}) \text{ and } (w_{\Phi_{l-k+1}} w w(k))^{-1}(\alpha_{k-1}) \\ \text{have opposite signs and } (\sigma_{w^{-1}})_k = -1 \end{array} \right\}$$

$$= W(k)^+ \cup W(k)^-$$

and

$$J_k^-(w) = \{\, j : k \leq j \leq r, (\sigma_{w^{-1}})_j = 1, (\sigma_{w^{-1}})_{j+1} = -1 \,\} \ ,$$
$$J_k^+(w) = \{\, j : k \leq j \leq r, (\sigma_{w^{-1}})_j = -1, (\sigma_{w^{-1}})_{j+1} = 1 \,\} \ .$$

Note that we put $W(1) = \{\, w = \pi_w \sigma_w : (\sigma_{w^{-1}})_1 = 1 \,\} = W(1)^+$ since there is no α_0.

Theorem B.1. *For* $k = 1, \ldots, r+1,$

1. *The map* $w \mapsto w_{\Phi_{l-k+1}} ww(k)$ *is a bijection from* $W(k)$ *to* $W \setminus W(k)$;
2. *If* $w \in W(k)^+$ *then*

$$\lambda(w_{\Phi_{l-k+1}} ww(k))$$
$$= \lambda(w) - b_{k-1}/2 - \sum_{j \in J_k^+(w)} b_j + \sum_{j \in J_k^-(w)} b_j + (\sigma_{w^{-1}})_{r+1} b_{r+1} \ ;$$

3. *If* $w \in W(k)^-$ *then*

$$\lambda(w_{\Phi_{l-k+1}} ww(k))$$
$$= \lambda(w) - b_{k-1}/2 - \sum_{j \in J_k^+(w)} b_j + b_{k-1} + \sum_{j \in J_k^-(w)} b_j + (\sigma_{w^{-1}})_{r+1} b_{r+1} \ .$$

Note that part 1 implies that parts 2 and 3 can be used to provide an identity valid on the whole of W. Although complicated, taking $X_l = C_l$ and $k = l$ reduces to the classical identity for $i = l$. To see this note that $J_l^+(w) = J_l^-(w) = \varnothing$, $b_l - b_{l-1}/2 = 1$, and $w_{\Phi_{l-k+1}} ww(k) = w_l w$.

Having set up this notation, we can extend Theorem 6.9 to describe more precisely the factorisation:

Theorem B.2. *If* $G = \mathrm{GSp}_{2l}$ *of type* C_l *or* $G = \mathrm{GO}_{2l}^+$ *of type* D_l *then for* $k = 1, \ldots, r+1$

$$P_{G,\rho}(X, Y) = (1 + X^{b_{k-1}/2}Y) \left(\sum_{w \in W(k)} X^{-\lambda(w)} \prod_{\alpha_j \in w(\Phi^-)} X^{b_j} Y^{c_j} \right)$$

$$= (1 + Y) \left(\prod_{i=1}^{r} (1 + X^{b_i/2}Y) \right) R_G(X, Y) \ ,$$

where

$$R_G(X, Y) = \left(\sum_{w \in \tilde{W}} X^{-\lambda(w)} \prod_{\alpha_j \in w(\Phi^-)} X^{b_j} Y^{c_j} \right)$$

and

$$\tilde{W} = \bigcap_{k=1}^{r+1} W(k) \ .$$

It is important therefore in establishing natural boundaries to remove the cyclotomic factors and provide a description of the resulting polynomial. This is precisely the goal of Theorem B.2 in the case of $P_{\mathrm{GSp}_{2l}}(X,Y)$ and $P_{\mathrm{GO}_{2l}^+}(X,Y)$. In this appendix we establish the following:

Theorem B.3. *If $G = \mathrm{GSp}_{2l}$ of type C_l or $G = \mathrm{GO}_{2l}^+$ of type D_l then $P_G(X,Y)$ has a factor of the form*

$$(1+Y)\prod_{i=1}^{r}(1+X^{b_i/2}Y),$$

where $r = l-1$ for $G = \mathrm{GSp}_{2l}$ and $r = l-2$ for $G = \mathrm{GO}_{2l}^+$.

Proof. For convenience, let us use the notation that $b_0 = 0$. Let X_l denote either the Dynkin diagram C_l or D_l. We shall use the following identities: for C_l we have

$$b_l - \mathrm{card}(C_{k+1}^+) + \mathrm{card}(A_k^+) = b_{l-(k+1)}/2 \text{ for } k = 0,\dots,l-1\,,$$

for D_l we have

$$b_l - \mathrm{card}(D_{k+1}^+) + \mathrm{card}(A_k^+) = b_{l-(k+1)}/2 \text{ for } k = 2,\dots,l-1\,,$$
$$b_l - \mathrm{card}(D_1^+) = b_{l-2}/2\,.$$

The element $w_{\varPhi_{k+1}}$ is the sign change $\mathbf{e}_i \mapsto \mathbf{e}_i$ for $i = 1,\dots,l-k-1$ and $\mathbf{e}_i \mapsto -\mathbf{e}_i$ for $i = l-k,\dots,r+1$ (note that in the case of D_l this then determines the sign change $\mathbf{e}_l$, namely $\mathbf{e}_l \mapsto (-1)^k \mathbf{e}_l$.) For each $w \in W$, let $w(k)$ be the permutation of $\mathbf{e}_{\pi_{w^{-1}}(i)}$ for $i = k,\dots,l$ which alters the order.

For $k = 1,\dots,r+1$ let

$$W(k) = \left\{ \begin{array}{c} w = \pi_w \sigma_w : w^{-1}(\alpha_{k-1}) \text{ and } (w_{\varPhi_{l-k+1}}ww(k))^{-1}(\alpha_{k-1}) \\ \text{have the same sign and } (\sigma_{w^{-1}})_k = 1 \end{array} \right\}$$
$$\cup \left\{ \begin{array}{c} w = \pi_w \sigma_w : w^{-1}(\alpha_{k-1}) \text{ and } (w_{\varPhi_{l-k+1}}ww(k))^{-1}(\alpha_{k-1}) \\ \text{have opposite signs and } (\sigma_{w^{-1}})_k = -1 \end{array} \right\}$$
$$= W(k)^+ \cup W(k)^-\,.$$

Note that we shall put $W(1) = \{\, w = \pi_w \sigma_w : (\sigma_{w^{-1}})_1 = 1 \,\} = W(1)^+$ since there is no α_0. The point is that things are going to work out because this means that in the second case actually it forces $\alpha_{k-1} \in w(\varPhi_l^+)$. We're trying to divide W up into two pieces so that $w \mapsto w_{\varPhi_{l-k+1}}ww(k)$ is a bijection and the difference in the polynomial is effected by multiplication by $X^{b_{k-1}/2}Y$.

Then the claim is that $w \mapsto w_{\varPhi_{l-k+1}}ww(k)$ is a bijection between $W(k)$ and $W \setminus W(k)$.

Note first of all that $w_{\varPhi_{l-k+1}}(w_{\varPhi_{l-k+1}}ww(k))(w_{\varPhi_{l-k+1}}ww(k))(k) = w$, since $(w_{\varPhi_{l-k+1}}ww(k))(k) = w(k)$. Secondly, since $w(k)$ is just a permutation and

$w_{\Phi_{l-k+1}}$ changes the sign of $\mathbf{e}_k$, $(\sigma_{w^{-1}})_k = -(\sigma_{(w_{\Phi_{l-k+1}}ww(k))^{-1}})_k$. Hence $w \mapsto w_{\Phi_{l-k+1}}ww(k)$ maps $W(k)$ into $W \setminus W(k)$ and also maps $W \setminus W(k)$ into $W(k)$. It is straightforward to see, using this second map, that $w \mapsto w_{\Phi_{l-k+1}}ww(k)$ is then a bijection between $W(k)$ and $W \setminus W(k)$.

We claim now that the correspondence $w \mapsto w_{\Phi_{l-k+1}}ww(k)$ behaves in the following manner:

$$X^{b_{k-1}/2}Y\left(X^{-\lambda(w)}\prod_{\alpha_j\in w(\Phi^-)}X^{b_j}Y^{c_j}\right)$$
$$= X^{-\lambda(w_{\Phi_{l-k+1}}ww(k))}\prod_{\alpha_j\in w_{\Phi_{l-k+1}}ww(k)(\Phi^-)}X^{b_j}Y^{c_j}\,.$$

Let

$$J_k^-(w) = \{\,j: k\le j\le r, (\sigma_{w^{-1}})_j = 1, (\sigma_{w^{-1}})_{j+1} = -1\,\}\,,$$
$$J_k^+(w) = \{\,j: k\le j\le r, (\sigma_{w^{-1}})_j = -1, (\sigma_{w^{-1}})_{j+1} = 1\,\}\,.$$

Then divide $J(w) = \{\,j\le r: w^{-1}\alpha_j\in\Phi^-\,\}$ into $J_k^+(w)$ and its complement $J(w)\setminus J_k^+(w)$. The first claim is then that for $w\in W(k)^+$

$$J(w_{\Phi_{l-k+1}}ww(k)) = (J(w)\setminus J_k^+(w))\cup J_k^-(w)$$

and for $w\in W(k)^-$,

$$J(w_{\Phi_{l-k+1}}ww(k)) = (J(w)\setminus J_k^+(w))\cup J_k^-(w)\cup\{k-1\}\,.$$

For $1\le j\le r$,

$$w^{-1}\alpha_j = w^{-1}(\mathbf{e}_j - \mathbf{e}_{j+1}) = (\sigma_{w^{-1}})_j\,\mathbf{e}_{\pi_{w^{-1}}(j)} - (\sigma_{w^{-1}})_{j+1}\mathbf{e}_{\pi_{w^{-1}}(j+1)}\,.$$

So firstly $J(w)\supset J_k^+(w)$ and $J(w)\cap J_k^-(w) = \varnothing$.

For $k\le j\le r$,

$$\left(w_{\Phi_{l-k+1}}ww(k)\right)^{-1}\alpha_j$$
$$= -(\sigma_{w^{-1}})_j\,w(k)\left(\mathbf{e}_{\pi_{w^{-1}}(j)}\right) + (\sigma_{w^{-1}})_{j+1}\,w(k)\left(\mathbf{e}_{\pi_{w^{-1}}(j+1)}\right)\,.$$

If $(\sigma_{w^{-1}})_j = -(\sigma_{w^{-1}})_{j+1}$, (i.e. $j\in J_k^+(w)\cup J_k^-(w)$) then $w^{-1}\alpha_j\in\Phi^-$ if and only if $\left(w_{\Phi_{l-k+1}}ww(k)\right)^{-1}\alpha_j\notin\Phi^-$. If $(\sigma_{w^{-1}})_j = (\sigma_{w^{-1}})_{j+1}$, then

$$(\sigma_{w^{-1}})_j\,\mathbf{e}_{\pi_{w^{-1}}(j)} - (\sigma_{w^{-1}})_{j+1}\,\mathbf{e}_{\pi_{w^{-1}}(j+1)} = (\sigma_{w^{-1}})_j\left(\mathbf{e}_{\pi_{w^{-1}}(j)} - \mathbf{e}_{\pi_{w^{-1}}(j+1)}\right)\,.$$

The point of using $w(k)$ now comes into effect because

$$\left(w_{\Phi_{l-k+1}}ww(k)\right)^{-1}\alpha_j = (\sigma_{w^{-1}})_j\left(-w(k)\mathbf{e}_{\pi_{w^{-1}}(j)} + w(k)\mathbf{e}_{\pi_{w^{-1}}(j+1)}\right)$$

will have the same sign as $w^{-1}\alpha_j$. This is because $\mathbf{e}_{i_1} - \mathbf{e}_{i_2} \in \Phi^-$ if and only if $i_2 < i_1$ and $w(k)$ has the effect of altering the order of $\pi_{w^{-1}}(i)$ for $i = k, \ldots, l$.

$w \mapsto w_{\Phi_{l-k+1}} ww(k)$ has no effect on those $j < k - 1$ since $w^{-1}\alpha_j = \left(w_{\Phi_{l-k+1}} ww(k)\right)^{-1} \alpha_j$.

The only root we haven't taken account of is $\alpha_{k-1} = \mathbf{e}_{k-1} - \mathbf{e}_k$. If $w \in W(k)^+$ then we are assuming that $k - 1 \in J(w)$ if and only if $k - 1 \in J(w_{\Phi_{l-k+1}} ww(k))$. So the only issue here is that if $w \in W(k)^-$, then $w^{-1}\alpha_{k-1} \notin \Phi^-$, i.e. $k - 1 \notin J(w)$. Then by definition of $W(k)^-$, $k - 1 \in J(w_{\Phi_{l-k+1}} ww(k))$. Now

$$w^{-1}\alpha_{k-1} = (\sigma_{w^{-1}})_{k-1} \, \mathbf{e}_{\pi_{w^{-1}}(k-1)} - (\sigma_{w^{-1}})_k \, \mathbf{e}_{\pi_{w^{-1}}k}$$

$$\left(w_{\Phi_{l-k+1}} ww(k)\right)^{-1} \alpha_{k-1} = (\sigma_{w^{-1}})_{k-1} \, \mathbf{e}_{\pi_{w^{-1}}(k-1)} + (\sigma_{w^{-1}})_k \, w(k) \left(\mathbf{e}_{\pi_{w^{-1}}k}\right)$$

Then $w \in W(k)^-$ (i.e. that these two elements have different signs) implies that the sign of $w^{-1}\alpha_{k-1}$ is $-(\sigma_{w^{-1}})_k = 1$, by definition of $W(k)^-$.

We start with the case C_l. For ease of notation, set $\Phi^1 = \Phi^+$, $\Phi^{-1} = \Phi^-$. Let us suppose first that $w \in W(k)^+$. We have to prove that:

$$\lambda(w_{\Phi_{l-k+1}} ww(k)) = \lambda(w) - b_{k-1}/2 - \sum_{j \in J_k^+(w)} b_j + \sum_{j \in J_k^-(w)} b_j + \varepsilon_w b_l , \quad \text{(B.1)}$$

where $w^{-1}\alpha_l \in \Phi^{\varepsilon_w}$ and $\varepsilon_w \in \{\pm 1\}$. Notice that the powers of Y are correct since if $\varepsilon_w = 1$, then card $J_k^+(w) = $ card $J_k^-(w)$ (look at the string of signs in $\sigma_{w^{-1}}$ from k to l which by hypothesis begins and ends with $+$, then card $J_k^+(w)$ is the number of sign changes $-$ to $+$, and card $J_k^-(w)$ is the number of sign changes $+$ to $-$). Then the degree of Y in the monomial corresponding to w is $2\,\text{card}\, J(w)$ and to $w_{\Phi_{l-k+1}} ww(k)$ is

$$2\,\text{card}\, J(w_{\Phi_{l-k+1}} ww(k)) + 1 = 2\,\text{card}\left((J(w) \setminus J_k^+(w)) \cup J_k^-(w)\right) + 1$$
$$= 2\,\text{card}\, J(w) + 1 .$$

If $\varepsilon_w = -1$, then card $J_k^-(w) = $ card $J_k^+(w) - 1$, and the degree of Y in the monomial corresponding to w is $2\,\text{card}\, J(w) + 1$ and to $w_{\Phi_{l-k+1}} ww(k)$ is

$$2\,\text{card}\, J(w_{\Phi_{l-k+1}} ww(k)) = 2\,\text{card}\left((J(w) \setminus J_k^+(w)) \cup J_k^-(w)\right)$$
$$= 2\,(\text{card}\, J(w) + 1)$$
$$= 2\,\text{card}\, J(w) + 2 .$$

Recall that the length of a word is the number of positive roots sent to negative roots by that word. It is the same as the length of its inverse. We look first at the effect of w^{-1} and $(w_{\Phi_{l-k+1}} ww(k))^{-1}$ on $\mathbf{e}_i \pm \mathbf{e}_j$ for $k \le i \le j \le l$. Define

$$K_i(w) = \left\{ \alpha = \mathbf{e}_i \pm \mathbf{e}_j : i \le j \le l, w^{-1}(\alpha) \in \Phi^- \right\} .$$

Lemma B.4. *If $i \geq k$ and $(\sigma_{w^{-1}})_i = \varepsilon_i$ then*

$$\operatorname{card} K_i(w) = \operatorname{card} K_i(w_{\Phi_{l-k+1}} ww(k)) + \varepsilon_i(l - i + 1) .$$

Proof. The point here is that in $w^{-1}(\mathbf{e}_i \pm \mathbf{e}_j) = \varepsilon_i \mathbf{e}_{i'} \pm \mathbf{e}_{j'}$ there is always one root with both signs of the basis elements equal to ε_i, and one with alternate signs. As we have explained the first root then changes sign under $(w_{\Phi_{l-k+1}} ww(k))^{-1}$ whilst the second retains its sign. So if $(\sigma_{w^{-1}})_i = \varepsilon_i$ then there are $l - i + 1$ roots (including $2\mathbf{e}_i$) which get mapped by w^{-1} into Φ^{ε_i} but get mapped by $(w_{\Phi_{l-k+1}} ww(k))^{-1}$ into $-\Phi^{\varepsilon_i}$; the other $l - i$ roots will keep the same sign.   □

So in the roots $\mathbf{e}_i \pm \mathbf{e}_j$ for $k \leq i \leq j \leq l$ we get a change of length

$$\sum_{i=k}^{l} \varepsilon_i(l - i + 1) = \sum_{i=k}^{l}(l - i + 1) - 2 \sum_{k \leq i, \varepsilon_i = -1}(l - i + 1) .$$

Now, $b_l - b_{k-1}/2 = \sum_{i=1}^{l}(l - i + 1) - \sum_{i=1}^{k-1}(l - i + 1) = \sum_{i=k}^{l}(l - i + 1)$. Also we have

$$2 \sum_{k \leq i, \varepsilon_i = -1}(l - i + 1)$$

$$= \sum_{j \in J_k^+(w)} 2\sum_{i=1}^{j}(l - i + 1) - \sum_{j \in J_k^-(w)} 2\sum_{i=1}^{j}(l - i + 1) + 2\delta\sum_{i=1}^{l}(l - i + 1) ,$$

where $\delta = 0$ if $\varepsilon_w = 1$ and $\delta = 1$ if $\varepsilon_w = -1$. One can see this by looking at the string of $+$s and $-$s. A string of $-$s starts at a $j_1 + 1$ where $j_1 \in J_k^-(w)$ and ends at a j_2 where $j_2 \in J_k^+(w)$. If the last term in the string is a $-$ then since $l \notin J(w)$ we need to add the last term as appropriate. But

$$\sum_{j \in J_k^+(w)} 2\sum_{i=1}^{j}(l - i + 1) - \sum_{j \in J_k^-(w)} 2\sum_{i=1}^{j}(l - i + 1) + 2\delta\sum_{i=1}^{l}(l - i + 1)$$

$$= \sum_{j \in J_k^+(w)} b_j - \sum_{j \in J_k^-(w)} b_j + 2\delta b_l .$$

Hence we have got a contribution to the change in length between w and $w_{\Phi_{l-k+1}} ww(k)$ by looking at the roots $\mathbf{e}_i \pm \mathbf{e}_j$ for $k \leq i \leq j \leq l$ of

$$b_l - b_{k-1}/2 - \sum_{j \in J_k^+(w)} b_j + \sum_{j \in J_k^-(w)} b_j - 2\delta b_l .$$

So our claim is that the other roots don't contribute any change in length. That is certainly true of $\mathbf{e}_i \pm \mathbf{e}_j$ for $i \leq j \leq k - 1$ since the elements w^{-1} and $(w_{\Phi_{l-k+1}} ww(k))^{-1}$ act in the same way on these roots.

The last case where $i \leq k - 1 < j$, if $\mathbf{e}_{\pi_{w-1}(j)}$ and $w(k)\mathbf{e}_{\pi_{w-1}(j)}$ are both on the same side of i then there is no change in the number of roots being sent to negative roots. If however they are on different sides then to see that there is no change in the number of positive roots changing sign we have to consider the four positive roots $\mathbf{e}_i \pm \mathbf{e}_j$ and $\mathbf{e}_i \pm \mathbf{e}_{\pi_{w-1}(j)}$ if $\pi_{w-1}(j) > i$ (and otherwise $\mathbf{e}_i \pm \mathbf{e}_j$ and $\mathbf{e}_i \pm w(k)\mathbf{e}_{\pi_{w-1}(j)}$).

Let us suppose now that $w \in W(k)^-$. We have to prove that:

$$\lambda(w_{\Phi_{l-k+1}} ww(k)) = \lambda(w) - b_{k-1}/2 - \sum_{j \in J_k^+(w)} b_j + b_{k-1} + \sum_{j \in J_k^-(w)} b_j + \varepsilon_w b_l .$$

$$\text{(B.2)}$$

Check first that the powers of Y match up again. If $\varepsilon_w = 1$, then card $J_k^+(w) - 1 = $ card $J_k^-(w)$ (look at the string of signs in σ_{w-1} from k to l which by hypothesis begins with $-$ and ends with $+$, then card $J_k^+(w)$ is the number of sign changes $-$ to $+$, and card $J_k^-(w)$ is the number of sign changes $+$ to $-$). Then the degree of Y in the monomial corresponding to w is $2 \operatorname{card} J(w)$ and to $w_{\Phi_{l-k+1}} ww(k)$ is

$$2 \operatorname{card} J(w_{\Phi_{l-k+1}} ww(k)) + 1$$
$$= 2 \operatorname{card} \left((J(w) \setminus J_k^+(w)) \cup J_k^-(w) \cup \{k-1\} \right) + 1$$
$$= 2 \operatorname{card} J(w) + 1 .$$

If $\varepsilon_w = -1$, then card $J_k^-(w) = $ card $J_k^+(w)$, and the degree of Y in the monomial corresponding to w is $2 \operatorname{card} J(w) + 1$ and to $w_{\Phi_{l-k+1}} ww(k)$ is

$$2 \operatorname{card} J(w_{\Phi_{l-k+1}} ww(k)) = 2 \operatorname{card} \left((J(w) \setminus J_k^+(w)) \cup J_k^-(w) \cup \{k-1\} \right)$$
$$= 2 \left(\operatorname{card} J(w) + 1 \right)$$
$$= 2 \operatorname{card} J(w) + 2 .$$

Again we look at the effect of w^{-1} and $(w_{\Phi_{l-k+1}} ww(k))^{-1}$ on $\mathbf{e}_i \pm \mathbf{e}_j$ for $k \leq i \leq j \leq l$ and with the same argument we get a change of length

$$\sum_{i=k}^{l} \varepsilon_i(l - i + 1) = \sum_{i=k}^{l}(l - i + 1) - 2 \sum_{k \leq i, \varepsilon_i = -1} (l - i + 1) .$$

Now, this time since the string of $+$'s and $-$'s starts with a $-$ we need to add an extra term to get

$$2 \sum_{k \leq i, \varepsilon_i = -1} (l - i + 1) = \sum_{j \in J_k^+(w)} 2 \sum_{i=1}^{j}(l - i + 1) - \sum_{j \in J_k^-(w)} 2 \sum_{i=1}^{j}(l - i + 1)$$

$$- 2 \sum_{i=1}^{k-1}(l - i + 1) + 2\delta \sum_{i=1}^{l}(l - i + 1)$$

$$= \sum_{j \in J_k^+(w)} b_j - \sum_{j \in J_k^-(w)} b_j - b_{k-1} + 2\delta b_l ,$$

where $\delta = 0$ if $\varepsilon_w = 1$ and $\delta = 1$ if $\varepsilon_w = -1$.

Hence we have got a contribution to the change in length between w and $w_{\Phi_{l-k+1}} ww(k)$ by looking at the roots $\mathbf{e}_i \pm \mathbf{e}_j$ for $k \le i \le j \le l$ of

$$b_l - b_{k-1}/2 - \sum_{j \in J_k^+(w)} b_j + \sum_{j \in J_k^-(w)} b_j - 2\delta b_l .$$

The same argument as above shows that the other roots don't contribute to a change in length.

Note that these identities B.1 and B.2 are generalisations of the classical identities:

$$\lambda(w_l w) = \lambda(w) + 1 \text{ if } w^{-1}(\alpha_l) \in \Phi^+ ,$$
$$\lambda(w_l w) = \lambda(w) - 1 \text{ if } w^{-1}(\alpha_l) \in \Phi^- .$$

This establishes the proof of Theorem B.1 detailed in the Introduction.

These identities therefore suffice in the case of C_l to show that our claim that the correspondence $w \mapsto w_{\Phi_{l-k+1}} ww(k)$ behaves in the following manner:

$$X^{b_{k-1}/2} Y \left(X^{-\lambda(w)} \prod_{\alpha_j \in w(\Phi^-)} X^{b_j} Y^{c_j} \right)$$
$$= X^{-\lambda(w_{\Phi_{l-k+1}} ww(k))} \prod_{\alpha_j \in w_{\Phi_{l-k+1}} ww(k)(\Phi^-)} X^{b_j} Y^{c_j} .$$

Hence

$$P_G(X,Y) = (1 + X^{b_{k-1}/2} Y) \left(\sum_{w \in W(k)} X^{-\lambda(w)} \prod_{\alpha_j \in w(\Phi^-)} X^{b_j} Y^{c_j} \right) ,$$

where

$$W(k) = \left\{ \begin{array}{c} w = \pi_w \sigma_w : w^{-1}(\alpha_{k-1}) \text{ and } \left(w_{\Phi_{l-k+1}} ww(k) \right)^{-1} (\alpha_{k-1}) \\ \text{have the same sign and } (\sigma_{w^{-1}})_k = 1 \end{array} \right\}$$
$$\cup \left\{ \begin{array}{c} w = \pi_w \sigma_w : w^{-1}(\alpha_{k-1}) \text{ and } \left(w_{\Phi_{l-k+1}} ww(k) \right)^{-1} (\alpha_{k-1}) \\ \text{have opposite signs and } (\sigma_{w^{-1}})_k = -1 \end{array} \right\}$$

and

$$P_G(X,Y) = (1 + Y) \prod_{i=1}^{r} (1 + X^{b_i/2} Y) R_G(X,Y) ,$$

where

$$R_G(X,Y) = \left(\sum_{w \in \tilde{W}} X^{-\lambda(w)} \prod_{\alpha_j \in w(\Phi^-)} X^{b_j} Y^{c_j} \right)$$

and

$$\tilde{W} = \bigcap_{k=1}^{r+1} W(k) .$$

This concludes the proof of Theorem B.3 for the case of C_l and establishes the description of the resulting factor detailed in B.2.

Next consider the case D_l. We start with looking at the effect of w and $w_{\Phi_{l-k+1}} ww(k)$ on the roots $\mathbf{e}_i \pm \mathbf{e}_j$ for $k \leq i < j \leq l$. Define again

$$K_i(w) = \{ \alpha = \mathbf{e}_i \pm \mathbf{e}_j : i < j \leq l, w^{-1}(\alpha) \in \Phi^- \} .$$

Lemma B.5. *If $i \geq k$ and $(\sigma_{w^{-1}})_i = \varepsilon_i$ then*

$$\operatorname{card} K_i(w) = \operatorname{card} K_i(w_{\Phi_{l-k+1}} ww(k)) + \varepsilon_i(l - i) .$$

If $i < k$ then $\operatorname{card} K_i(w) = \operatorname{card} K_i(w_{\Phi_{l-k+1}} ww(k))$.

The same proof works here with the observation that in D_l we don't have roots $2\mathbf{e}_i$ so our counting arguments for C_l here and elsewhere will generally be effected by a drop of one everywhere.

Note taking $i = l - 1$, that this lemma implies in particular for the roots simple α_{l-1} and α_l that we get one more or one less of these roots in the monomial corresponding to $w_{\Phi_{l-k+1}} ww(k)$ according to whether ε_{l-1} is respectively 1 or -1. Note that in the combinatorial data for D_l, $c_{l-1} = c_l = 1$. Therefore the proof that the degree of Y in the monomial corresponding to $w_{\Phi_{l-k+1}} ww(k)$ is one more than that for $w \in W(k)$ is the same as for C_l except that we look just at the string of $+$'s and $-$'s in $\sigma_{w^{-1}}$ from k to $l - 1$.

Let us suppose first that $w \in W(k)^+$. We have to prove that:

$$\lambda(w_{\Phi_{l-k+1}} ww(k)) = \lambda(w) - b_{k-1}/2 - \sum_{j \in J_k^+(w)} b_j + \sum_{j \in J_k^-(w)} b_j + \varepsilon_{l-1} b_{l-1} .$$

$$(\mathrm{B.3})$$

Note that since $b_{l-1} = b_l$, the last term takes account of the change of degree in X corresponding to the action of w and $w_{\Phi_{l-k+1}} ww(k)$ on the roots α_{l-1} and α_l.

By Lemma B.5,

$$\lambda(w_{\Phi_{l-k+1}} ww(k)) - \lambda(w) = \sum_{i=k}^{l-1} \varepsilon_i (l - i)$$

$$= \sum_{i=k}^{l-1} (l - i) - 2 \sum_{k \leq i < l, \varepsilon_i = -1} (l - i)$$

$$= b_{l-1} - b_{k-1}/2 - \sum_{j \in J_k^+(w)}{}' b_j + \sum_{j \in J_k^-(w)} b_j - 2\delta b_{l-1} ,$$

where $\delta = 0$ if $\varepsilon_{l-1} = 1$ and $\delta = 1$ if $\varepsilon_{l-1} = -1$. The last equality just follows the same argument as for C_l with the observation that for D_l

$$b_{l-1} = \sum_{i=1}^{l-1}(l-i) ,$$

$$b_j = 2\sum_{i=1}^{j}(l-i) .$$

If $w \in W(k)^-$ then by a similar adaptation of the argument for C_l one can prove that

$$\lambda(w_{\Phi_{l-k+1}} ww(k))$$
$$= \lambda(w) - b_{k-1}/2 - \sum_{j\in J_k^+(w)} b_j + b_{k-1} + \sum_{j\in J_k^-(w)} b_j + \varepsilon_{l-1}b_{l-1} . \qquad (B.4)$$

Again the identities B.3 and B.4 prove that in the case of D_l,

$$P_G(X,Y) = (1+Y)\prod_{i=1}^{r}(1+X^{b_i/2}Y)R_G(X,Y) ,$$

where

$$R_G(X,Y) = \left(\sum_{w\in \tilde{W}} X^{-\lambda(w)} \prod_{\alpha_j \in w(\Phi^-)} X^{b_j}Y^{c_j} \right)$$

and

$$\tilde{W} = \bigcap_{k=1}^{r+1} W(k) .$$

This concludes the proof of Theorem B.3 for the case of D_l and establishes the description of the resulting factor detailed in B.2.   □

References

1. C. Breuil, B. Conrad, F. Diamond, and R. Taylor. On the modularity of elliptic curves over $\mathbb{Q}$: wild 3-adic exercises. *J. Amer. Math. Soc.*, 14(4):843–939, 2001.
2. J. Coates and A. Wiles. On the conjecture of Birch and Swinnerton-Dyer. *Invent. Math.*, 39:223–251, 1977.
3. J.B. Conrey. More than two-fifths of the zeros of the Riemann zeta function are on the critical line. *J. Reine. Angew. Math.*, 399:1–26, 1989.
4. G. Dahlquist. On the analytic continuation of Eulerian products. *Arkiv für Matematik, Band 1*, 36:533–554, 1952.
5. H. Darmon. Wiles' theorem and the arithmetic of elliptic curves. In *Modular forms and Fermat's Last Theorem*, pages 549–569. Springer, Berlin, 1997.
6. J. Denef. The rationality of the Poincaré series associated to the p-adic points on a variety. *Invent. Math.*, 77:1–23, 1984.
7. J. Denef. On the degree of Igusa's local zeta function. *Amer. J. Math.*, 109:991–1008, 1987.
8. J. Denef and D. Meuser. A functional equation of Igusa's local zeta function. *Amer. J. Math.*, 113:1135–1152, 1991.
9. M.P.F. du Sautoy. Functional equations for local zeta functions of algebraic groups. In preparation.
10. M.P.F. du Sautoy. Finitely generated groups, p-adic analytic groups and Poincaré series. *Ann. Math.*, 137:639–670, 1993.
11. M.P.F. du Sautoy. Counting p-groups and nilpotent groups. *Publ. Math. IHES*, 92:63–112, 2000.
12. M.P.F. du Sautoy. The zeta function of $\mathfrak{sl}_2(\mathbb{Z})$. *Forum Math.*, 12:197–221, 2000.
13. M.P.F. du Sautoy. A nilpotent group and its elliptic curve: non-uniformity of local zeta functions of groups. *Isr. J. Math.*, 126:269–288, 2001.
14. M.P.F. du Sautoy. Counting subgroups in nilpotent groups and points on elliptic curves. *J. Reine. Angew. Math.*, 549:1–21, 2002.
15. M.P.F. du Sautoy. Zeta functions of groups: the quest for order versus the flight from ennui. In *Groups St. Andrews 2001 in Oxford, Vol. I*, number 304 in London Math. Soc. Lecture Note Series, pages 150–189, 2002.
16. M.P.F. du Sautoy and F.J. Grunewald. Zeta functions of classical groups and their friendly ghosts. *C. R. Acad. Sci. Paris*, 327(Série I):1–6, 1998.
17. M.P.F. du Sautoy and F.J. Grunewald. Analytic properties of zeta functions and subgroup growth. *Ann. Math.*, 152:793–833, 2000.

18. M.P.F. du Sautoy and F.J. Grunewald. Zeta functions of groups: zeros and friendly ghosts. *Amer. J. Math.*, 124:1–48, 2002.

19. M.P.F. du Sautoy and F.J. Grunewald. Zeta functions of groups and rings. In Proceedings of the ICM Madrid 2006 volume II, 131–149, 2006.

20. M.P.F. du Sautoy and F. Loeser. Motivic zeta functions of infinite dimensional Lie algebras. *Selecta Math. (N.S.)*, 10(2):253–303, 2004.

21. M.P.F. du Sautoy and A. Lubotzky. Functional equations and uniformity for local zeta functions of nilpotent groups. *Amer. J. Math.*, 118:39–90, 1996.

22. M.P.F. du Sautoy, J.J. McDermott, and G.C. Smith. Zeta functions of crystallographic groups. *Proc. London Math. Soc.*, 79:511–534, 1999.

23. M.P.F. du Sautoy and D. Segal. Zeta functions of groups. In *New horizons in pro-p groups*, volume 184 of *Progress In Mathematics*, pages 249–286. Birkhaüser, Boston, MA, 2000.

24. M.P.F. du Sautoy and G. Taylor. The zeta function of $\mathfrak{sl}_2(\mathbb{Z})$ and resolution of singularities. *Math. Proc. Camb. Phil. Soc.*, 132:57–73, 2002.

25. T. Estermann. On certain functions represented by Dirichlet series. *Proc. London Math. Soc.*, 27:435–448, 1928.

26. M.-P. Gong. *Classification of Nilpotent Lie Algebras Of Dimension 7 (Over Algebraically Closed Fields and $\mathbb{R}$)*. PhD thesis, University of Waterloo, 1998.

27. X. Gourdon. The 10^{13} first zeros of the Riemann zeta function and zeros computation at very large height. Electronic resource, available at http://numbers.computation.free.fr/Constants/Miscellaneous/.

28. D. Grenham. *Some topics in nilpotent group theory*. PhD thesis, University of Oxford, 1998.

29. C. Griffin. *Subgroups of infinite groups: Interactions between group theory and number theory*. PhD thesis, University of Nottingham, 2002.

30. B. Grünbaum. *Convex polytopes*, volume XVI of *Pure And Applied Mathematics*. InterScience, London, 1967.

31. F.J. Grunewald and D. Segal. Reflections on the classification of torsion-free nilpotent groups. In *Group theory: Essays for Philip Hall*, pages 121–158. Academic Press, London, 1984.

32. F.J. Grunewald, D. Segal, and G.C. Smith. Subgroups of finite index in nilpotent groups. *Invent. Math.*, 93:185–223, 1988.

33. G.H. Hardy and J.E. Littlewood. The zeros of Riemann's zeta-function on the critical line. *Math. Zeitschr.*, 10:283–317, 1921.

34. H. Hauser. The Hironaka Theorem on Resolution of Singularities (Or: A proof we always wanted to understand). *Bull. AMS*, 40(3):323–403, 2003.

35. K. Hey. *Analytische Zahlentheorie in Systemen hyperkomplexer Zahlen*. PhD thesis, Hamburg, 1929.

36. J.-I. Igusa. Universal p-adic zeta functions and their functional equations. *Amer. J. Math.*, 111:671–716, 1989.

37. S. Iyanaga and Y. Kawada, editors. *Encyclopedic Dictionary of Mathematics*. M.I.T. Press, Cambridge, MA, 1977.

38. N. Kurokawa. On the meromorphy of Euler products, I. *Proc. London Math. Soc.*, 53:1–47, 1986.

39. N. Kurokawa. On the meromorphy of Euler products, II. *Proc. London Math. Soc.*, 53:209–236, 1986.

40. A. Lubotzky and A. Mann. On groups of polynomial subgroup growth. *Invent. Math.*, 104:521–533, 1991.

41. A. Lubotzky, A. Mann, and D. Segal. Finitely generated groups of polynomial subgroup growth. *Isr. J. Math.*, 82:363–371, 1993.

42. A. Lubotzky and D. Segal. *Subgroup Growth*, volume 212 of *Progress In Mathematics*. Birkhaüser, Basel, 2003.

43. I.G. Macdonald. *Symmetric functions and Hall polynomials*. Clarendon, Oxford, 1979.

44. L. Magnin. Sur les algébres de Lie nilpotentes de dimension ≤7. *J. Geom. Phys.*, 3:119–144, 1986.

45. W. Magnus, A. Karrass, and S. Solitar. *Combinatorial Group Theory*. Wiley, Chichester, UK, 1966.

46. V.V. Morozov. Classification of nilpotent Lie algebras of sixth order (in Russian). *Isv. Vysš. Učebn. Zaved. Matematika*, 5:161–171, 1958.

47. W. Narkiewicz. *Number Theory*. World Scientific, Singapore, 1983.

48. M.F. Newman and E.A. O'Brien. Classifying 2-groups by coclass. *Trans. Amer. Math. Soc.*, 351:131–169, 1999.

49. P. Paajanen. The geometry of the Segre surface and zeta functions of groups. To appear in *Math. Proc. Camb. Phil. Soc.*

50. I. Satake. Theory of spherical functions on reductive algebraic groups over p-adic fields. *Inst. Hautes Études Sci. Publ. Math.*, 18:5–70, 1963.

51. D. Segal. *Polycyclic groups*, volume 82 of *Cambridge Tracts in Mathematics*. CUP, Cambridge, 1983.

52. D. Segal and A. Shalev. Profinite groups with polynomial subgroup growth. *J. London Math. Soc.*, 55:320–334, 1997.

53. J.G. Semple and G.T. Kneebone. *Algebraic Curves*. OUP, London, 1959.

54. R.P. Stanley. Linear homogeneous diophantine equations and magic labelings of graphs. *Duke Math. J.*, 40:607–632, 1973.

55. R.P. Stanley. Combinatorial reciprocity theorems. *Adv. Math.*, 14:194–253, 1974.

56. T. Tamagawa. On the ζ-function of a division algebra. *Ann. Math.*, 77:387–405, 1963.

57. G. Taylor. *Zeta Functions of Algebras and Resolution of Singularities*. PhD thesis, University of Cambridge, 2001.

58. C. Voll. Counting subgroups in a family of nilpotent semi-direct products. *Bull. London Math. Soc.*, 38:743–752, 2006.

59. C. Voll. Functional equations for zeta functions of groups and rings. Preprint available at http://www.arxiv.org/abs/math/0612511.

60. C. Voll. Zeta functions of groups and enumeration in Bruhat-Tits buildings. *Amer. J. Math.*, 126(5):1005–1032, 2004.

61. C. Voll. Functional equations for local normal zeta functions of nilpotent groups. *Geom. Funct. Anal.*, 15(1):274–295, 2005.

62. C. Voll. Normal subgroup growth in free class-2-nilpotent groups. *Math. Ann.*, 332(1):67–79, 2005.

63. H. von Mangoldt. Zur Verteilung der Nullstellen der Riemannschen Funktion $\zeta(s)$. *Math. Ann.*, 60:1–19, 1905.

64. L. Woodward. *Zeta functions of groups: computer calculations and functional equations*. PhD thesis, University of Oxford, 2005.

Index

Index of Notation

Lecture Notes in Mathematics

For information about earlier volumes
please contact your bookseller or Springer
LNM Online archive: springerlink.com

Vol. 1887: K. Habermann, L. Habermann, Introduction to Symplectic Dirac Operators (2006)

Vol. 1888: J. van der Hoeven, Transseries and Real Differential Algebra (2006)

Vol. 1889: G. Osipenko, Dynamical Systems, Graphs, and Algorithms (2006)

Vol. 1890: M. Bunge, J. Funk, Singular Coverings of Toposes (2006)

Vol. 1891: J.B. Friedlander, D.R. Heath-Brown, H. Iwaniec, J. Kaczorowski, Analytic Number Theory, Cetraro, Italy, 2002. Editors: A. Perelli, C. Viola (2006)

Vol. 1892: A. Baddeley, I. Bárány, R. Schneider, W. Weil, Stochastic Geometry, Martina Franca, Italy, 2004. Editor: W. Weil (2007)

Vol. 1893: H. Hanßmann, Local and Semi-Local Bifurcations in Hamiltonian Dynamical Systems, Results and Examples (2007)

Vol. 1894: C.W. Groetsch, Stable Approximate Evaluation of Unbounded Operators (2007)

Vol. 1895: L. Molnár, Selected Preserver Problems on Algebraic Structures of Linear Operators and on Function Spaces (2007)

Vol. 1896: P. Massart, Concentration Inequalities and Model Selection, Ecole d'Été de Probabilités de Saint-Flour XXXIII-2003. Editor: J. Picard (2007)

Vol. 1897: R. Doney, Fluctuation Theory for Lévy Processes, Ecole d'Été de Probabilités de Saint-Flour XXXV-2005. Editor: J. Picard (2007)

Vol. 1898: H.R. Beyer, Beyond Partial Differential Equations, On linear and Quasi-Linear Abstract Hyperbolic Evolution Equations (2007)

Vol. 1899: Séminaire de Probabilités XL. Editors: C. Donati-Martin, M. Émery, A. Rouault, C. Stricker (2007)

Vol. 1900: E. Bolthausen, A. Bovier (Eds.), Spin Glasses (2007)

Vol. 1901: O. Wittenberg, Intersections de deux quadriques et pinceaux de courbes de genre 1, Intersections of Two Quadrics and Pencils of Curves of Genus 1 (2007)

Vol. 1902: A. Isaev, Lectures on the Automorphism Groups of Kobayashi-Hyperbolic Manifolds (2007)

Vol. 1903: G. Kresin, V. Maz'ya, Sharp Real-Part Theorems (2007)

Vol. 1904: P. Giesl, Construction of Global Lyapunov Functions Using Radial Basis Functions (2007)

Vol. 1905: C. Prévôt, M. Röckner, A Concise Course on Stochastic Partial Differential Equations (2007)

Vol. 1906: T. Schuster, The Method of Approximate Inverse: Theory and Applications (2007)

Vol. 1907: M. Rasmussen, Attractivity and Bifurcation for Nonautonomous Dynamical Systems (2007)

Vol. 1908: T.J. Lyons, M. Caruana, T. Lévy, Differential Equations Driven by Rough Paths, Ecole d'Été de Probabilités de Saint-Flour XXXIV-2004 (2007)

Vol. 1909: H. Akiyoshi, M. Sakuma, M. Wada, Y. Yamashita, Punctured Torus Groups and 2-Bridge Knot Groups (I) (2007)

Vol. 1910: V.D. Milman, G. Schechtman (Eds.), Geometric Aspects of Functional Analysis. Israel Seminar 2004-2005 (2007)

Vol. 1911: A. Bressan, D. Serre, M. Williams, K. Zumbrun, Hyperbolic Systems of Balance Laws. Lectures given at the C.I.M.E. Summer School held in Cetraro, Italy, July 14–21, 2003. Editor: P. Marcati (2007)

Vol. 1912: V. Berinde, Iterative Approximation of Fixed Points (2007)

Vol. 1913: J.E. Marsden, G. Misiołek, J.-P. Ortega, M. Perlmutter, T.S. Ratiu, Hamiltonian Reduction by Stages (2007)

Vol. 1914: G. Kutyniok, Affine Density in Wavelet Analysis (2007)

Vol. 1915: T. Bıyıkoğlu, J. Leydold, P.F. Stadler, Laplacian Eigenvectors of Graphs. Perron-Frobenius and Faber-Krahn Type Theorems (2007)

Vol. 1916: C. Villani, F. Rezakhanlou, Entropy Methods for the Boltzmann Equation. Editors: F. Golse, S. Olla (2008)

Vol. 1917: I. Veselić, Existence and Regularity Properties of the Integrated Density of States of Random Schrödinger (2008)

Vol. 1918: B. Roberts, R. Schmidt, Local Newforms for GSp(4) (2007)

Vol. 1919: R.A. Carmona, I. Ekeland, A. Kohatsu-Higa, J.-M. Lasry, P.-L. Lions, H. Pham, E. Taflin, Paris-Princeton Lectures on Mathematical Finance 2004. Editors: R.A. Carmona, E. Çinlar, I. Ekeland, E. Jouini, J.A. Scheinkman, N. Touzi (2007)

Vol. 1920: S.N. Evans, Probability and Real Trees. Ecole d'Été de Probabilités de Saint-Flour XXXV-2005 (2008)

Vol. 1921: J.P. Tian, Evolution Algebras and their Applications (2008)

Vol. 1922: A. Friedman (Ed.), Tutorials in Mathematical BioSciences IV. Evolution and Ecology (2008)

Vol. 1923: J.P.N. Bishwal, Parameter Estimation in Stochastic Differential Equations (2008)

Vol. 1924: M. Wilson, Littlewood-Paley Theory and Exponential-Square Integrability (2008)

Vol. 1925: M. du Sautoy, L. Woodward, Zeta Functions of Groups and Rings (2008)

Vol. 1926: L. Barreira, V. Claudia, Stability of Nonautonomous Differential Equations (2008)

Vol. 1927: L. Ambrosio, L. Caffarelli, M.G. Crandall, L.C. Evans, N. Fusco, Calculus of Variations and Non-Linear Partial Differential Equations. Lectures given at the C.I.M.E. Summer School held in Cetraro, Italy, June 27–July 2, 2005. Editors: B. Dacorogna, P. Marcellini (2008)

Vol. 1928: J. Jonsson, Simplicial Complexes of Graphs (2008)

Recent Reprints and New Editions

Vol. 1618: G. Pisier, Similarity Problems and Completely Bounded Maps. 1995 – 2nd exp. edition (2001)

Vol. 1629: J.D. Moore, Lectures on Seiberg-Witten Invariants. 1997 – 2nd edition (2001)

Vol. 1638: P. Vanhaecke, Integrable Systems in the realm of Algebraic Geometry. 1996 – 2nd edition (2001)

Vol. 1702: J. Ma, J. Yong, Forward-Backward Stochastic Differential Equations and their Applications. 1999 – Corr. 3rd printing (2007)

Vol. 830: J.A. Green, Polynomial Representations of GL_n, with an Appendix on Schensted Correspondence and Littelmann Paths by K. Erdmann, J.A. Green and M. Schocker 1980 – 2nd corr. and augmented edition (2007)

4. Careful preparation of the manuscripts will help keep production time short besides ensuring satisfactory appearance of the finished book in print and online. After acceptance of the manuscript authors will be asked to prepare the final LaTeX source files (and also the corresponding dvi-, pdf- or zipped ps-file) together with the final printout made from these files. The LaTeX source files are essential for producing the full-text online version of the book (see http://www.springerlink.com/openurl.asp?genre=journal&issn=0075-8434 for the existing online volumes of LNM).

    The actual production of a Lecture Notes volume takes approximately 8 weeks.

5. Authors receive a total of 50 free copies of their volume, but no royalties. They are entitled to a discount of 33.3 % on the price of Springer books purchased for their personal use, if ordering directly from Springer.

6. Commitment to publish is made by letter of intent rather than by signing a formal contract. Springer-Verlag secures the copyright for each volume. Authors are free to reuse material contained in their LNM volumes in later publications: a brief written (or e-mail) request for formal permission is sufficient.

Addresses:

Professor J.-M. Morel, CMLA,
École Normale Supérieure de Cachan,
61 Avenue du Président Wilson, 94235 Cachan Cedex, France
E-mail: Jean-Michel.Morel@cmla.ens-cachan.fr

Professor F. Takens, Mathematisch Instituut,
Rijksuniversiteit Groningen, Postbus 800,
9700 AV Groningen, The Netherlands
E-mail: F.Takens@math.rug.nl

Professor B. Teissier, Institut Mathématique de Jussieu,
UMR 7586 du CNRS, Équipe "Géométrie et Dynamique",
175 rue du Chevaleret
75013 Paris, France
E-mail: teissier@math.jussieu.fr

For the "Mathematical Biosciences Subseries" of LNM:

Professor P. K. Maini, Center for Mathematical Biology,
Mathematical Institute, 24-29 St Giles,
Oxford OX1 3LP, UK
E-mail : maini@maths.ox.ac.uk

Springer, Mathematics Editorial, Tiergartenstr. 17,
69121 Heidelberg, Germany,
Tel.: +49 (6221) 487-8410
Fax: +49 (6221) 487-8355
E-mail: lnm@springer.com